Rare Diseases of the Immune System

Series Editors: Lorenzo Emmi, Domenico Prisco

According to the European Parliament, a disease is considered rare when it affects no more than 5 out of 10,000 people. People with rare diseases are not that scarce, however, since we are talking about a phenomenon that affects hundreds of thousands of people in Italy and several millions all over Europe. Many associations, both national and international, have been created in recent years to inform and safeguard patients affected by these syndromes. At the same time, across various disciplines the medical and scientific awareness of the challenges posed by these diseases has risen inexorably. Rare diseases due to alterations in the immune system represent a field of medical science of great interest that is undergoing continuous expansion; these syndromes have a great social and economic impact because they often affect young or very young people at the height of their social activities and relationships. In recent years, much information has been acquired on pathogenesis, diagnosis, and therapy, allowing us to confront these syndromes in a more holistic manner. The series editors are acknowledged experts on immune diseases who have published extensively on various facets of this branch of medicine. This series will be an invaluable resource for immunologists, rheumatologists, hematologists and medical practitioners looking to update their knowledge and a useful educational tool for residents and students.

Indexed in Scopus

Interested authors should contact the corresponding series editor, Lorenzo Emmi at: lorenzoemmi@yahoo.it

Lorenzo Emmi • Vassilios Fanos
Editors

Epigenetics and Rare Diseases

Springer

Editors
Lorenzo Emmi
Clinical Immunology
Consultant at the University of Florence
Firenze, Italy

Vassilios Fanos
Department of Medical Sciences
AOU Cagliari & University of Cagliari
Monserrato, Cagliari, Italy

ISSN 2282-6505 ISSN 2283-6403 (electronic)
Rare Diseases of the Immune System
ISBN 978-3-032-02901-0 ISBN 978-3-032-02902-7 (eBook)
https://doi.org/10.1007/978-3-032-02902-7

This Springer imprint is published by the registered company Springer Nature Switzerland AG
The registered company address is: Gewerbestrasse 11, 6330 Cham, Switzerland

Sentiero rosa. Laura Maddii Emmi. (Private collection)

Preface

Why a book on "epigenetics and rare diseases?" The answer is that epigenetics, as well as all -omics fields of study, such as transcriptomics, proteomics, and metabolomics, have allowed a real revolution in all branches of medicine in recent years. On the other hand, the molecular mechanisms responsible for the epigenetic modifications of the genome due to environmental stress factors—and the part that lifestyle plays for each of us—are well known today. We gradually switched from a linear medicine model having the genetic background and the altered immune-inflammatory response as key actors (possibly activated by exogenous infectious agents) to a much more articulated model, which involves the participation of more players, including the microbiome, the role of the environment and epigenetic modulations, all within a continuous cross-talk between complex biological systems. A new interpretation of pathological events has been emerging, related not so much to organ damage, which also exists, but rather to alterations of the information circuits within the network. Hence, disease can be read as a "rupture" of a balance of information that runs within complex systems. If this is to be the medicine of the future, I would then argue that medicine should right away aim to collect the greatest number of genetic/epigenetic/metabolic information possible, interpretable by means of machine learning and deep learning techniques.

The new interpretation of neurodevelopment also appears to be of enormous biological, clinical, and epistemological importance, in light of the possible epigenetic modifications already occurring in the early stages of fetal development itself. On the basis of this interpretation, we can begin to believe that many pathological events taking place in the course of adult life are, so to speak, already "written" in the first moments of development. However, this should not lead to a deterministic view, but to the ability to perceive a window of opportunity of great importance in this all.

In regard to immune-mediated diseases, which represent the subject of the present series, of which Prof. Domenico Prisco and I are the Series Editors, many works have recently demonstrated the role that epigenetic modifications can play in the responses given by immune systems and the inflammatory cascade. Therefore, an epigenetic reading of these pathologies could lead to a revision of the pathogenesis of the latter, and could lead to innovative and hitherto unthinkable therapeutic strategies. Finally, the "reasoned" use of the -omics in science could help us build a personalized and precise medicine around the patient.

This, notwithstanding, allows us to conclude by saying that all this should never make us lose sight of the need, indeed I daresay the ethical imperative, to establish a relationship that is as empathetic as possible with the patient.

Firenze, Italy Lorenzo Emmi
Monserrato, Cagliari, Italy Vassilios Fanos

Contents

1 **From Genetics to Epigenetics: The Steps of a Path** 1
Cristina Dallabona and Maria Carla Gerra

2 **Molecular Basis of Epigenetic Mechanisms** 15
Lucia Migliore and Marianna Giangreco

3 **Epigenetics and Endocrine Disruptors** 29
Mauro Bologna and Ilaria Bologna

4 **New Insights into the Fetal Programming Hypothesis of the Linkage Between Neurodevelopmental Disorders and Impaired Neurogenesis** 47
G. Faa, P. Paribello, M. Fraschini, M. Castagnola, and M. Manchia

5 **The Placenta and the Development of Human Health** 57
Antonio Ragusa, Antonino Lo Re, Maria Segata, Denise Rinaldo, Giovanni la Motta, Marika De Vito, Caterina De Luca, and Alessandro Svelato

6 **Epigenetics and Cancer** 77
Laura Giunti, Marco Tellini, and Iacopo Sardi

7 **Epigenetics in Autoimmune Diseases** 97
Giacomo Bagni, Edoardo Biancalana, Emanuele Chiara, Iole Costanzo, Danilo Malandrino, Giacomo Emmi, Elena Silvestri, and Lorenzo Emmi

8 **Epigenetics and Systemic Vasculitis** 111
Michelangelo Tesi and Augusto Vaglio

9 **Epigenetics and Allergic Diseases** 145
Beatrice Andrenacci, Amelia Licari, and Gian Luigi Marseglia

10 **The Role of Epigenetics in Autoinflammatory Diseases** 163
Maria Vincenza Mastrolia, Sarah Abu-Rumeileh, Valerio Maniscalco, Ilaria Maccora, Edoardo Marrani, Ilaria Pagnini, and Gabriele Simonini

11 Epigenetics and Immune-Mediated Diseases 179
A. A. Senatore, P. Currao, E. Arnaudo, R. Castagnoli, and S. Savasta

12 Epigenetics and Inherited Metabolic Disorders 193
Catia Cavicchi, Jean-Louis Guéant, and Amelia Morrone

13 Epigenetics and Psychiatric Diseases 219
M. Manchia, P. Paribello, and B. Chaumette

14 The Metabolomics Revolution Toward Personalized Medicine: The Case of Gestational Diabetes 233
Roberta Pintus, Angelica Dessì, Alice Bosco, and Vassilios Fanos

15 Epigenetics and Microbiome 251
Niccolò Meriggi, Mariela Mejia Monroy, Sofia Chioccioli, and Carlotta De Filippo

Index 269

From Genetics to Epigenetics: The Steps of a Path

1

Cristina Dallabona and Maria Carla Gerra

Contents

1.1 Historical Overview and Milestones of Genetics 1
1.2 What Is Epigenetics? 3
1.3 Epigenetics: How Does It Work? 6
1.4 From Epigenetics to Epigenomics 10
1.5 Epigenetics: What's Next? 11
References 13

1.1 Historical Overview and Milestones of Genetics

Since ancient times, the observation of the presence of particular traits in a specific plant or animal had led breeders to selectively reproduce specific organisms to obtain offspring with the desired characteristics. However, there was a lack of understanding regarding the laws governing this phenomenon. Thus, the official beginning of genetics dates back to the discovery made by Gregor Mendel in 1865, elucidating how hereditary traits are transmitted from one generation to the next. Mendel had the ability to interpret the results and transform his observations into numerical and analyzable data. Only in the early twentieth century was Mendel recognized for his work on the transmission of traits, which profoundly influenced subsequent research.

C. Dallabona (✉) · M. C. Gerra
Department of Chemistry, Life Sciences and Environmental Sustainability, University of Parma, Parma, Italy
e-mail: cristina.dallabona@unipr.it

L. Emmi, V. Fanos (eds.), *Epigenetics and Rare Diseases*, Rare Diseases of the Immune System, https://doi.org/10.1007/978-3-032-02902-7_1

In the early 1900s, the question of which structures within the cell were responsible for transmitting hereditary traits was addressed. In 1902, Walter Sutton proposed that hereditary traits were located on chromosomes, and this hypothesis was demonstrated by Thomas H. Morgan's research group, who worked on *Drosophila melanogaster*. Important genetic principles were established, and many discoveries rapidly followed: the chromosomal theory of inheritance, which localized genes on chromosomes; the inheritance of sex-linked genes; the presence of linkage between genes located on the same chromosome; the possibility of recombination between genes on the same chromosome through crossing-over events; the creation of the first genetic map, representing the linear order of genes on the chromosome.

Moreover, advances in cytological observations allowed for correlating anomalies in chromosome structure with phenotypic changes. However, only in the 1940s and 1950s the DNA was identified as the carrier of genetic information and, from that point onward, the study of DNA, its structure, replication, and transcription, became a central focus of research, paving the way for Molecular Genetics and Molecular Biology.

In 1941, George Beadle and Edward Tatum demonstrated that genes perform specific biochemical functions by using auxotrophic mutants of the orange bread mold *Neurospora crassa*. Specifically, they assigned specific enzymatic functions to many genes. From these studies, the famous hypothesis "one gene-one enzyme" emerged [1]. Subsequently, there were a series of fundamental discoveries, such as the fact that some genes, like hemoglobin, encode proteins without enzymatic functions, and how genes can "regulate" their activity in response to intra- or extracellular signals.

Genetics has caused a revolution by converting descriptive observations in biology, particularly from the study of microorganisms, into an understanding of the function and regulation of genetic information. Additionally, genetics has facilitated the development of other disciplines, notably biochemistry, biotechnology, and bioinformatics. This transformation was accelerated by the advent of DNA sequencing technologies. The "complete" sequence of the human genome was published in 2004 [2]. The reduction in sequencing costs and time now allows for the sequencing of an entire human genome in a few hours. But the most significant changes began in 2011 with the rise of massive DNA sequencing, which has expanded into every area of human genetics, involving the widespread use of parallel sequencing technologies and, more recently, studies in single-cell genomics.

Bioinformatic analysis has enabled the development of comparative genomics, which compares our genome with that of an increasing number of other organisms. This understanding sheds light on how our genome has been modified during evolution. In this context, the studies in the field of human genetics have greatly benefited from over a century of research on model organisms; since studying the human being at the genetic level is challenging, comparisons with other species have allowed significant progress in the study of human genetics.

The post-genomic era has brought about a fundamental shift in functional analysis. Forward genetics involved screening of populations containing random genomic modifications that alter gene function, identifying the gene with the mutation and uncovering the encoded protein. This approach has played a crucial role in advancing our understanding of human genetics. However, efficient mutagenesis methods, comprehensive sequence data, and the availability of valid model organisms enable moving away from this traditional genetic method to embrace reverse genetics. This new approach deciphers gene function by unraveling the DNA sequence and studying the effects of altered gene expression on the phenotype [3]. From the conventional path of moving from mutant to gene, the exploration has shifted from the traditional path of mutant-to-gene to a gene-to-mutant approach (Fig. 1.1).

Thanks to these fundamental discoveries and technological advancements, the study of the genetic bases of human diseases has been significantly enhanced. Many complex human diseases were revealed to have a strong environmental component in addition to causative genetic factors. But how can we explain the effect of the environment from a molecular point of view? To answer this question, we need to move into another discipline: epigenetics.

1.2 What Is Epigenetics?

Starting from Mendel's laws (1865), through the isolation of the DNA molecule (1869) up to the resolution of its double helix structure (1953), and more recently to the decoding of the human genome, the general principles of genetics and heredity have been established.

The twentieth century was extremely DNA-centric, focused on the genotype and on how DNA could explain who we are. But the picture was not yet complete; there was certainly something missing that could not be explained merely by the genotype.

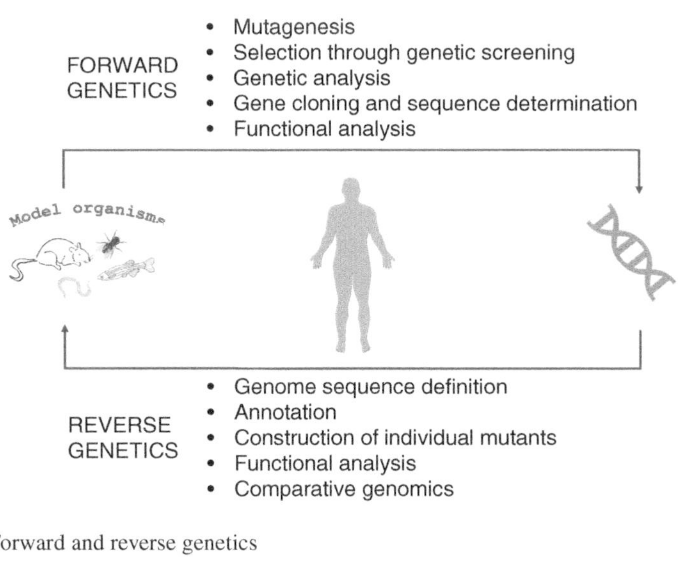

Fig. 1.1 Forward and reverse genetics

In nature, there are many examples where genotype and phenotype, the DNA and its characteristics, do not perfectly align. Consider the different types of cells in our body: they share the same genome but are phenotypically and functionally very different. Similarly, monozygotic twins are genetically identical, yet phenotypic differences can arise and these differences tend to increase with the subjects' age and tend to become more pronounced as individuals' lifestyles diverge.

The term "epigenetics" was coined in the 1940s by Conrad Hal Waddington, who defined it as "the branch of biology which studies the causal interactions between genes and their products which bring the phenotype into being" [4]. Waddington was a developmental biologist, and he was interested in how a zygote can give rise to hundreds of different cell types present in the body. Over time, the concept of epigenetics has changed and evolved as molecular knowledge has expanded. Figure 1.2 shows a timeline of some of the major discoveries in this research area. Over the years, different definitions have been proposed, leading to extensive debates. For a comprehensive overview of the various ways in which epigenetics has been conceived and defined historically, see the review by Nicoglou and Merlin [5]. Currently, one of the most quoted and commonly used definitions of epigenetics in molecular and cellular biology is the one proposed by Arthur D. Riggs, which closely resembles those independently proposed by Robin Holliday: "the study of mitotically and/or meiotically heritable changes in gene function that cannot be explained by changes in DNA sequence" [6]. This definition emphasizes three key aspects: (i) the cellular stability or the "memory" of epigenetic changes, (ii) the impact of epigenetic changes on gene function (i.e., gene expression), and (iii) the fact that these changes do not involve modifications of the DNA sequence. However, the term epigenetics is now often used also for modifications that are not necessarily mitotically heritable (such as histone modifications like acetylation). A more inclusive meaning of what epigenetics is, at the molecular level, has thus emerged: it applies to the alterations in gene activity that are not due to changes in the DNA sequence but rather to other molecular mechanisms such as DNA methylation, histone modifications, histone variants, chromatin structure changes, and the actions of non-coding RNAs.

It should be emphasized that although epigenetic modifications can be inherited and thus somewhat stable, they are also reversible. In fact, an important point to keep in mind is that epigenetic information and its transmission are quite different from genetic information. In the classical genetic information, mutations acquired in the DNA can be irreversibly transmitted across all generations once they occur. Instead, in the epigenetic process, these epigenetic modifications can be added by specific enzymes but can also be removed. This is why the epigenetic mechanism and the duration of the epigenetic inheritance might vary making it a reversible process.

One of the key functions of epigenetic modifications is the regulation and maintenance of cell-type-specific gene expression. However, gene regulation is far from being their only cellular role. In fact, epigenetic modifications are also crucial for controlling the expression of transposable elements, silencing repetitive sequences, preventing non-homologous recombination, and ensuring a proper DNA structure.

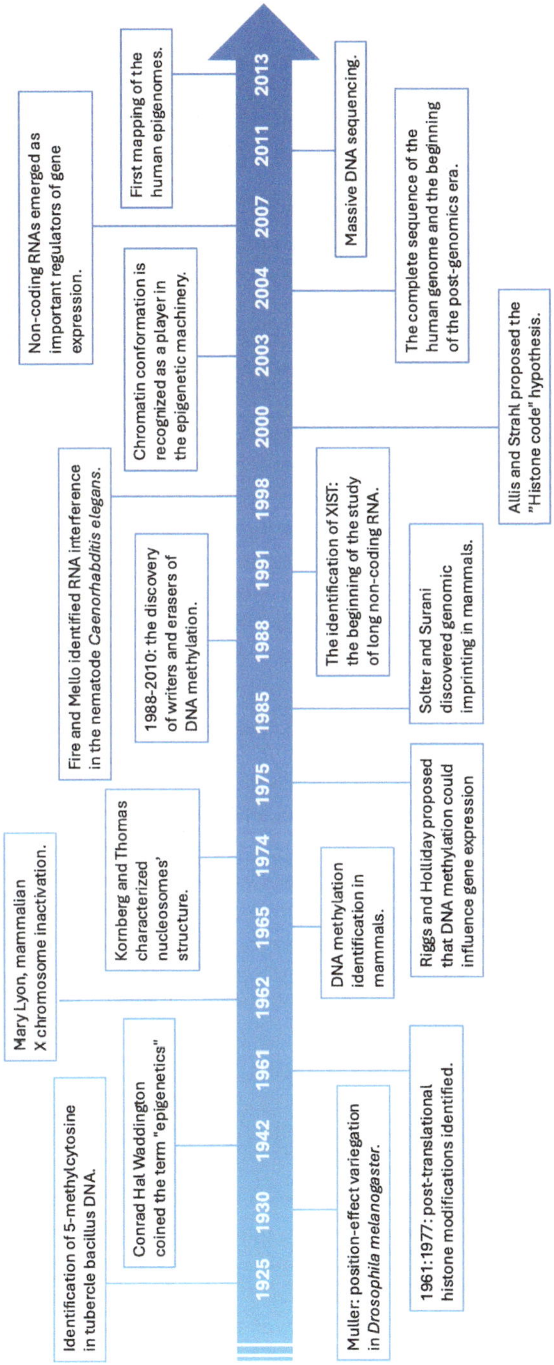

Fig. 1.2 The evolution of epigenetics

For example, elements such as centromeres and telomeres are epigenetically determined [7]. Therefore, a broader role of epigenetics is to ensure genomic stability. Indeed, the loss of epigenetic silencing, in particular the loss of DNA methylation at repetitive elements has been associated with various diseases' development.

Epigenetics encompasses the study of a wide range of biological phenomena, including cellular differentiation and development, metabolism, disease, phenotypic variability, inheritance and evolution, and environmental effects. It can be considered a natural extension of genetics because its goal is to understand the regulation of genes and other heritable elements in organisms.

Epigenetics is thus a discipline with the potential to explain biological events that cannot be accounted for soley by genetic mechanisms. In fact, the discovery of the mechanisms of inheritance, the decoding of the genetic code, and the understanding of how genes are organized didn't explain the mechanisms by which they are expressed in a coordinated manner. Moreover, in higher organisms, many genes can have a discontinuous structure, with alternative exons and multiple transcription start sites, and only a small portion of the genome is transcribed into proteins. For years, the non-coding DNA was referred to as "junk DNA" until its function and role in gene expression regulation was understood.

Epigenetics has gained increasing importance, becoming one of the fastest-growing fields in biology and biomedicine, with exponential growth in scientific publications, national and international funding, the development of research consortia, and the emergence of dedicated start-ups. It has become the discipline that studies the mechanisms of genome function: it is the set of processes that decode the information contained within the vast library of the genome, making it readable and interpretable, and determining how, where, and when a specific genetic information should be expressed by the cell within a given time and cellular district. The epigenetic marks thus act as a code "above" the sequence (from the Greek "ἐπί") and provide the tools necessary to convert the genotype into phenotype. We must imagine a book without punctuation or with paperclips closing some pages. We can think of epigenetics like punctuation, or as the ability to change the meaning of the story by adding or obscuring chapters with paperclips. But how is it possible to achieve this goal from a molecular point of view? What are the mechanisms underlying all of this?

1.3 Epigenetics: How Does It Work?

The first epigenetic modification to be identified was DNA methylation. The discovery of the presence of 5-methylcytosine (5mC) dates back to 1925, when it was found in tubercle bacillus DNA [8], and more than 20 years later in calf thymus DNA; however, its biological role remained uncharacterized for a long time, until 1975 when it was first demonstrated that 5mC could influence gene expression [9, 10], highlighting the significance of the "fifth nucleotide" in eukaryotic biology [11]. Since then, epigenetic research has moved its focus from the original developmental field toward the study of gene expression regulation mediated by epigenetic modifications.

DNA methylation is now recognized as a typical epigenetic mark and it satisfies the stringent definition proposed by Holliday and Riggs, being mitotically heritable. This signature in multicellular eukaryotes occurs predominantly but not exclusively at cytosine residues within CpG dinucleotides [12], and is mediated by DNA methyltransferases. The discovery of the "writers" and "erasers" of DNA methylation took decades: from 1988 with the cloning of the first writer enzyme, DNA methyltransferase 1 (DNMT1) [13], to at least 2010 with the identification of the 10–11 translocation cytosine dioxygenases (TET) family, which act as erasers of 5mC [14]. Today we know that the enzymes involved in DNA methylation belong to three classes: (i) writers: DNA methyltransferases that establish and maintain 5mC; (ii) erasers: TETs that oxidize 5mC to 5-hydroxymethylcytosine (5hmC), 5-formylcytosine (5fC) and 5-carboxylcytosine (5caC) that is finally replaced with cytosine; (iii) readers, such as methyl-CpG binding proteins (MeCPs), that recognize and bind methylated DNA.

In vertebrates, DNA methylation is tightly regulated during development, controlling tissue and differentiation states. It also regulates many cellular processes beyond embryonic development, including transcription, chromatin structure, X-chromosome inactivation, genomic imprinting, and chromosome stability.

DNA methylation plays central roles not only in chromatin-based transcriptional regulation but also in DNA replication and repair. Alterations in DNA methylation and its mediators may thus deregulate many genes involved in cell cycles, DNA repair, and chromosome stability, including tumor suppressor genes and oncogenes, contributing to the genome instability typical of various human diseases including cancer.

Alongside DNA methylation (5mC), post-translational histone modifications were rapidly identified following the resolution of the DNA double helix structure. Indeed, histone methylation, acetylation, phosphorylation, ubiquitylation, sumoylation, and ADP ribosylation were reported between 1962 and 1977, but the role of these modifications has been difficult to understand.

It must be noted that in eukaryotic cells, the DNA molecule is not "naked" inside the nucleus, but it is complexed with proteins to form a nucleoprotein structure called chromatin, discovered in 1974 [15]. The basic unit of chromatin is the nucleosome, which consists of a histone octamer formed by two units of each of the histones H2A, H2B, H3, and H4, and 147 base pairs of DNA wrapped around it. Chromatin undergoes additional condensation by winding in a polynucleosome fiber. The modulation of its structure is crucial for gene expression regulation, as well as for replication, recombination, and DNA repair, because this affects the accessibility of DNA to regulatory factors.

From the functional point of view, the chromatin structure is commonly divided into euchromatin and heterochromatin. Euchromatin generally refers to genome regions containing actively transcribed or potentially active genes; it remains decondensed during interphase. Nucleases have access to the regulatory sequences in these regions that typically contain unmethylated CpG islands. Additionally, N-terminal lysine residues of the core histones H3 and H4 are hyperacetylated. On the other hand, heterochromatin denotes the highly condensed and thus transcriptionally inactive regions of the genome. Within heterochromatin, nucleases have no access to DNA, which is usually methylated in the CpG dinucleotides and histones are markedly hypoacetylated. Whether heterochromatin is present in all cell types or limited to specific lineages and cell types determines the distinction between constitutive or facultative heterochromatin, respectively.

To date, many post-translational modifications have been identified in nucleosomal histones. While some remain largely uncharacterized, with the enzymes involved still unknown despite the scientist's efforts, for others, the mechanisms are clearer: specific histone modifications seem to be associated with particular genomic regions (e.g., H3K4me3 and H3K27Ac are typically found at the transcription start site (TSS) of active genes; H3K27me3 is typical at the TSS of inactive genes, whereas the simultaneous presence of the activation-associated histone modifications H3K4me3 and the repression-associated modification H3K27me3 is typical of poised genes) leading to the hypothesis that it may be possible to predict the transcriptional state of a specific region. Since multiple post-translational histone modifications can coexist on each nucleosome, around the year 2000, the "histone code" was proposed, meaning that "distinct histone amino-terminal modifications can generate synergistic or antagonistic interaction affinities for chromatin-associated proteins" (from Jenuwein and David [16]). This translates into the regulation of chromatin accessibility and, consequently, the functional state of that DNA segment. Currently, it is believed that this code follows specific rules, making it possible to decipher it once we have understood the underlying rules, as has been done with the genetic code.

Also, chromatin conformation has been recognized as a key player in the epigenetic machinery. Thanks to chromatin looping, it is indeed possible to bring together regions/sequences that may be widely separated along the linear DNA sequence. In this way, they can be brough into close proximity in three-dimensional space, allowing, for example, concerted regulation to be achieved. The most well-known example is the formation of the enhancesome, which allows distant enhancer regions to interact with promoter regions of a gene to regulate its expression. Finally, not only the state of chromatin (accessible or condensed) and its folding are important for the functional state of DNA, but it has recently been observed that the specific positioning of DNA within the nucleus also plays a functional role. In fact, more recently, increasing evidence suggests that DNA exhibits non-random positioning within the nucleus, where chromosomes, gene loci, and nuclear bodies adopt specific spatial and temporal arrangements. This organization into precise sub-compartments could serve as a foundation for regulating chromatin state and ensuring optimal transcriptional efficiency. However, a complete understanding of how chromosome positioning influences genomic function remains elusive [17].

Another paradigm shift of the modern days is the reevaluation of "junk DNA", once thought to be evolutionary remnants, but now recognized as functional. It is no coincidence that the non-coding regions proportionally increase as we climb up the evolutionary tree and clearly play a critical role in defining what makes us human compared to other species.

Thanks to the widespread adoption of RNA-sequencing techniques, it is now well established that the vast majority of human RNA transcripts do not encode proteins. In fact, although almost 80% of the human genome is transcribed, only about 2% of it, namely mRNA, codes for proteins. Over the past few decades, long and small non-coding RNAs have emerged as important regulators of gene expression at various levels, thus being involved in the regulation of cell physiology and in shaping cellular functions. This demonstrates that non-coding DNA is far from being junk DNA. Regulatory RNAs can be divided into two groups based on their size: small non-coding RNAs (sncRNAs), which include miRNAs, siRNAs, and piRNAs, and long noncoding RNAs (lncRNAs).

The first revolutionary discovery in the field of ncRNAs, which earned the Nobel Prize for Andrew Zachary Fire and Craig Cameron Mello was RNA interference (RNAi). This process was first identified in 1998 in the nematode *Caenorhabditis elegans* [18]. RNAi leads to the processing of double-stranded RNAs (dsRNAs) into small interfering RNAs (siRNAs), 21–22 nucleotide-long dsRNAs that target complementary mRNAs for degradation, leading to post-transcriptional gene silencing.

Since the discovery of eukaryotic small RNAs as the main effectors of RNA interference in the late 1990s, various other endogenous small RNAs have been characterized, particularly in vertebrates. Most notably microRNAs (miRNAs) and PIWI-interacting RNAs (piRNAs) have been identified. These small RNAs are associated with Argonaute proteins and influence almost all major biological processes through sequence-specific gene regulation. The discovery of RNAi revealed broad roles for small non-coding RNAs, with functions ranging from gene regulation to antiviral defense and transposon control [18].

The characterization of miRNAs highlighted that gene expression is controlled at an unexpected level [19], and that the inhibition of gene expression is a physiological mechanism in mammals.

In addition, the identification of the lncRNA XIST (1991) marked the beginning of the study of lncRNAs [20]. XIST plays a key role in a process called X inactivation, first described by Mary Lyon in 1961 [21], following the identification of the Barr body in 1948. Before the discovery of XIST, the molecular mechanisms underlying X inactivation remained unknown. Over the following decades, several other lncRNAs implicated in X inactivation emerged (such as Tsix, Ftx, Jpx), along with thousands of others involved in diverse gene regulation systems [22–24]. LncRNAs are also often involved in another biological phenomenon that appears to violate Mendel's laws: genomic imprinting, discovered by Davor Solter, Azim Surani, and co-workers in the mid-1980s [25, 26]. To ensure genomic imprinting (i.e., a monoallelic, parent-specific expression pattern) differential DNA methylation occurs during gametogenesis to distinguish the parental origin of the alleles; imprinted gene expression is then regulated through lncRNAs, histone modifications, insulators, and higher-order chromatin structure.

The general definition of lncRNAs, i.e., molecules longer than 200 nucleotides that are not translated into functional proteins, includes a heterogeneous collection of transcripts differing in biogenesis and genomic origin.

Genomes are extensively transcribed and give rise to thousands of long noncoding RNAs (estimates range from 16,000 to over 100,000), although the number of functional lncRNAs is still debated. While evidence is still lacking to support the functionality of most lncRNAs, thereby rendering them potential transcriptional by-products, it is well-documented that a growing number of lncRNAs have important cellular functions. It has now been demonstrated that lncRNAs can have multiple functions, including the regulation of transcription, proteins/RNA functioning, chromatin remodeling, and the genome 3D organization in the nucleus [27].

To add further complexity, non-coding RNAs can be post-transcriptionally modified (e.g., RNA methylation) and those modifications have been recently recognized as players in the epigenetic machinery.

1.4 From Epigenetics to Epigenomics

With the discovery of the molecular players (DNA methylation, chromatin modification and structure, ncRNAs), the focus of epigenetic research gradually moved from a conceptual framework to the analysis of individual epigenetic signatures and their consequential functional effect. Similar to the significant advances in genetics from DNA to the genome during the twentieth century, epigenetics also experienced significant development from the middle of the twentieth century until the end of the century, culminating in the publication of the first mapping of the human epigenome [28].

Obtaining the original sequence of the human genome was undoubtedly a landmark achievement. But this was only the beginning of the story. The decoding led to the understanding of only about 1–2% of the genome, containing protein-coding genes, but it was impossible to decode the functions of the other sequences. Only epigenomic mapping has provided the opportunity to tackle this new interpretative challenge. However, it should be emphasized that, unlike the genome, which is the same in every cell of an organism and remains constant over time, the epigenome differs between cell types and can change over time in response to environmental or physiological stimuli, and this remains a major challenge in epigenomic mapping. Furthermore, generating an epigenomic map requires the study of hundreds of modifications (DNA methylation, histone acetylation, histone phosphorylation, histone methylation, etc.). Not only that: for each histone modification, it is essential to determine its exact location (e.g., histone methylation H3K4me, H3K27me, H3K9me3 have completely different "meanings"). It is immediately clear that this is a big challenge, and the techniques required are far from simple and inexpensive. These technical difficulties, along with some reluctance to leave knowledge that seemed well established, may partially explain why epigenetics was slow to take off.

In the twenty-first century, numerous international consortia have emerged to address the fundamental challenges of epigenomics, fostering collaborations to share resources and protocols: Encyclopedia of DNA Elements (ENCODE) (2003), Roadmap Epigenomics Project (2008), and International Human Epigenome Consortium (IHEC) (2010). These initiatives have led to the annotation of thousands of genes and millions of candidate regulatory elements. However, how these elements mediate regulatory effects on specific target genes over distances of kilobases or sometimes megabases, remains largely unknown. More recently, the 4D Nucleome project addresses the role of nuclear organization in gene expression and cell function in both health and disease, exploring the DNA 3D organization in the nucleus and its changes over time (the fourth dimension).

Despite initial concerns about the value and utility of mapping epigenomes in human cells, driven by rapid technological advances, the field of epigenomics is experiencing unprecedented growth with no signs of deceleration.

1.5 Epigenetics: What's Next?

Despite significant progress, many questions remain unanswered, and heritability, especially meiotic, and the influence of the environment remain two areas of ongoing debate.

Over the years epigenetics has progressively focused on the issue of epigenetic inheritance.

Initially, mitotic heritability was proposed as a form of "cellular memory", to ensure tissue homogeneity: indeed, every cell type undergoing mitosis must maintain its own identity to form an organ or tissue. With the discovery of the molecular mechanisms underlying certain epigenetic markers, mitotic heritability has been demonstrated, but only for specific modifications such as DNA methylation, and certain histone methylations such as H3K27me3 and H3K9me3.

Growing evidence suggested that epigenetic information can not only be mitotically inherited but also meiotically passed from parents to offspring, in various organisms, including humans. Intergenerational and transgenerational epigenetic inheritance both involve the transmission of non-DNA sequence-base information across generations. However, in the latter case, effects are observed in generations not directly exposed to the initial signal or environmental factor that triggered the change. Nevertheless, evidence for transgenerational inheritance in humans via the germline, without direct exposure to the driving external stimulus, remains a subject of debate [29].

The epigenome is not stable throughout an individual's life, but the environment can act upon it, leading to changes. The environment thus can play a role in determining gene expression and phenotypic modifications through epigenetic mechanisms. Thus, epigenetics offers the potential to revolutionize our understanding of this interplay between genetics and the environment. In fact, the DNA, even when packaged within the nucleus, is exposed to various types of effector agents, and several studies have focused to understand if, how, when, and where there is an environmental effect on the epigenome, especially in animal models. Pathological conditions, aging, environmental exposures, and behaviors, such as lifestyle, drug use, physical and chemical agents, physical activity, stress, and diet, can continously send signals to our cells, often leading to epigenetic modifications that allow adaptation to specific circumstances through changes in gene expression, all without altering the DNA sequence. Therefore, epigenetics can investigate how exposure to environmental factors can modify gene expression without altering the DNA sequence. It has been demonstrated for example that air pollution can lead to changes in histone modifications, while cigarette smoke alters the methylation status, and therefore the expression, of specific genes. In the latter case, it is important to emphasize that the epigenetic state can return to normal once smoking is discontinued. Diet is among the most studied environmental factors that can influence epigenetic mechanisms of gene control: the foods we eat and even what is consumed by a mother during pregnancy are broken down within the body, and their metabolites can generate significant epigenetic effects, such as the donation of methyl groups, particularly during early stages of development when the epigenome is being established. Therefore, not only what we consume, especially in the early years of life, contributes to shaping our epigenetic profile, but it has also been proposed that parental diet may play a role.

Alterations in epigenetic signatures, in particular DNA methylation, have also been observed in patients with various chronic diseases. Given that epigenetic changes can be meiotically stable, it has been proposed that the gap in missing heritability observed in many chronic conditions could can be explained by epimutations inherited from the parents. By modulating immune and inflammatory response pathways, these epimutations may influence the risk of developing autoimmunity or chronic inflammation, in the same way genetic risk factors act.

Thus, in many diseases, genetic predisposition can increase susceptibility to a condition, but the expression or manifestation of the disease often also depends on interaction with the environment, where epigenetic factors seem to play a key role. However, contrary to genetic modifications, epigenetic marks can be transient, and they can be reversible [30]. This has been particularly attractive in biomedical research, especially in cancer research, in the past years because the enzymes responsible for adding or removing epigenetic tags have become targets for development of new drugs to try to restore the original setting of the genes through epigenetic editing.

This more comprehensive understanding of genetic and environmental influences affected by epigenetics has led to more advanced approaches in disease prevention and treatment, aiming to consider both genetic and environmental factors. Personalized medicine, which takes into account individual genetic characteristics, is an example of how research in this field is positively impacting health and disease approaches.

The reversibility of epigenetic modifications, as well as their hereditary transmission, opens further areas of investigation.

Thanks to epigenetics, it will be possible to study and understand both the origins of various pathologies and the influence of lifestyle on our genetic material. Deciphering the specific epigenetic changes associated with pathological states, or even with aging, could radically transform therapeutic strategies. If the correct epigenetic signals are identified, it may be possible to develop drugs aimed at restoring proper gene expression.

The integration of genetics and epigenetics will further enhance our knowledge about who we are, how we function in health and disease, and possibly solve some of the most devastating modern diseases.

References

1. Beadle GW, Tatum EL (1941) Genetic control of biochemical reactions in neurospora. Proc Natl Acad Sci USA 27(11):499–506
2. Finishing the euchromatic sequence of the human genome (2004) Nature 431(7011):931–945
3. Gurumurthy CB, Grati M, Ohtsuka M, Schilit SLP, Quadros RM, Liu XZ (2016) CRISPR: a versatile tool for both forward and reverse genetics research. Hum Genet 135(9):971–976
4. Waddington CH (1940) Organisers and genes. The Cambridge University Press, Cambridge
5. Nicoglou A, Merlin F (2017) Epigenetics: a way to bridge the gap between biological fields. Stud Hist Phil Biol Biomed Sci 66:73–82
6. Martienssen RA, Riggs AD, Russo VEA (1996) In: CSH (ed) Epigenetic mechanisms of gene regulation. Laboratory CSH, New York
7. Feng JX, Riddle NC (2020) Epigenetics and genome stability. Mamm Genome 31(5–6):181–195
8. Johnson TB, Coghill RD (1925) The discovery of 5-methyl-cytosine in tuberculinic acid, the nucleic acid of the tubercle bacillus. J Am Chem Soc 47:2838–2844
9. Riggs AD (1975) X inactivation, differentiation, and DNA methylation. Cytogenet Cell Genet 14(1):9–25
10. Holliday R, Pugh JE (1975) DNA modification mechanisms and gene activity during development. Science 187(4173):226–232
11. Doerfler W (2006) De novo methylation, long-term promoter silencing, methylation patterns in the human genome, and consequences of foreign DNA insertion. Curr Top Microbiol Immunol 301:125–175
12. Bestor TH (2000) The DNA methyltransferases of mammals. Hum Mol Genet 9(16):2395–2402
13. Bestor T, Laudano A, Mattaliano R, Ingram V (1988) Cloning and sequencing of a cDNA encoding DNA methyltransferase of mouse cells. The carboxyl-terminal domain of the mammalian enzymes is related to bacterial restriction methyltransferases. J Mol Biol 203(4):971–983
14. Zhang H, Zhang X, Clark E, Mulcahey M, Huang S, Shi YG (2010) TET1 is a DNA-binding protein that modulates DNA methylation and gene transcription via hydroxylation of 5-methylcytosine. Cell Res 20:1390–1393
15. Kornberg RD (1974) Chromatin structure: a repeating unit of histones and DNA. Science 184(4139):868–871
16. Jenuwein T, Allis CD (2001) Translating the histone code. Science (1979) [Internet]. 293(5532):1074–1080. Available from: https://doi.org/10.1126/science.1063127
17. Rajapakse I, Groudine M (2011) On emerging nuclear order. J Cell Biol 192(5):711–721
18. Fire A, Xu S, Montgomery MK, Kostas SA, Driver SE, Mello CC (1998) Potent and specific genetic interference by double-stranded RNA in Caenorhabditis elegans. Nature 391(6669):806–811
19. Lagos-Quintana M, Rauhut R, Lendeckel W, Tuschl T (2001) Identification of novel genes coding for small expressed RNAs. Science 294(5543):853–858
20. Brockdorff N, Ashworth A, Kay GF, Cooper P, Smith S, McCabe VM et al (1991) Conservation of position and exclusive expression of mouse Xist from the inactive X chromosome. Nature 351(6324):329–331
21. Lyon MF (1961) Gene action in the X-chromosome of the mouse (Mus musculus L.). Nature 190:372–373
22. Chureau C, Chantalat S, Romito A, Galvani A, Duret L, Avner P et al (2011) Ftx is a non-coding RNA which affects Xist expression and chromatin structure within the X-inactivation center region. Hum Mol Genet 20(4):705–718
23. Chureau C, Prissette M, Bourdet A, Barbe V, Cattolico L, Jones L et al (2002) Comparative sequence analysis of the X-inactivation center region in mouse, human, and bovine. Genome Res 12(6):894–908

24. Lee JT, Davidow LS, Warshawsky D (1999) Tsix, a gene antisense to Xist at the X-inactivation centre. Nat Genet 21(4):400–404
25. Surani MA, Barton SC, Norris ML (1984) Development of reconstituted mouse eggs suggests imprinting of the genome during gametogenesis. Nature 308(5959):548–550
26. McGrath J, Solter D (1984) Completion of mouse embryogenesis requires both the maternal and paternal genomes. Cell 37(1):179–183
27. Statello L, Guo CJ, Chen LL, Huarte M (2021) Gene regulation by long non-coding RNAs and its biological functions. Nat Rev Mol Cell Biol 22(2):96–118
28. Rivera CM, Ren B (2013) Mapping human epigenomes. Cell 155(1):39–55
29. Fitz-James MH, Cavalli G (2022) Molecular mechanisms of transgenerational epigenetic inheritance. Nat Rev Genet 23(6):325–341
30. John RM, Rougeulle C (2018) Developmental epigenetics: phenotype and the flexible epigenome. Front Cell Dev Biol 6:130

Molecular Basis of Epigenetic Mechanisms

2

Lucia Migliore and Marianna Giangreco

Contents

2.1 Introduction ... 16
2.2 Epigenetic Mechanisms: DNA Methylation ... 17
2.3 Chromatin Structure and Histone Modifications ... 20
 2.3.1 Post-translational Histone Modifications ... 21
 2.3.2 ATP-Dependent Chromatin-Remodeling Complexes ... 22
2.4 RNA-Mediated Gene Silencing ... 23
 2.4.1 miRNA ... 23
 2.4.2 lncRNA ... 25
 2.4.3 piRNA ... 25
2.5 RNA Methylation ... 26
2.6 Concluding Remarks ... 26
References ... 27

L. Migliore (✉)
Department of Translational Research and of New Surgical and Medical Technologies, University of Pisa, Medical School, Pisa, PI, Italy

Pisa University Hospital, Pisa, Italy
e-mail: lucia.migliore@med.unipi.it

M. Giangreco
Department of Translational Research and of New Surgical and Medical Technologies, University of Pisa, Medical School, Pisa, PI, Italy

L. Emmi, V. Fanos (eds.), *Epigenetics and Rare Diseases*, Rare Diseases of the Immune System, https://doi.org/10.1007/978-3-032-02902-7_2

2.1 Introduction

The term "epigenetics"refers to the study of modifications located "above" (from the Greek επι, epi) the DNA molecule, but which do not alter its sequence, and can directly influence the expression of a gene. There are several epigenetic mechanisms, the most studied are: DNA methylation, modifications of histone tails, and the interventions of non-coding RNAs (ncRNAs), such as microRNAs. Epigenetic mechanisms regulating gene expression are essential in various physiological processes including embryonic development, differentiation, and cell identity maintenance.

A growing number of studies have made it possible to understand that, thanks to complex enzymatic machinery capable of writing, removing, and interpreting epigenetic marks, the human genome is highly plastic and able to respond to numerous environmental stimuli, finely modulating gene expression. Enzymes and protein domains associated with epigenetic modifications are called writers, erasers, and readers. Writers are able to modify nucleotide bases on DNA and specific amino acid residues on histones. Erasers are a set of enzymes proficient in removing these marks. Finally, readers are a group of proteins that possess specialized domains capable of recognizing specific epigenetic marks in a locus.

Exposure to environmental factors such as chemical pollutants, dietary components, physical activities, and other external stresses can cause changes in the epigenome [1]. Unfortunately, when the environmental stimulus is early, very intense, or when the epigenetic machinery is deficient, perturbations of this plasticity can lead to many human diseases, helping us to understand the close relationship between genes and the environment in complex diseases. Therefore, epigenetic modifications have proven to be valuable biomarkers for the early diagnosis and prognosis of diseases and given their reversible condition, they are also considered potential therapeutic targets.

The effects of the environment on epigenome can be observed not only in the post-birth period but also in the prenatal period. This aspect is very important if we think that these modifications could influence cellular differentiation in fetus with the consequent development of various and severe malformations. In the 1980s and 1990s of the last century, the English pediatrician and epidemiologist David Barker established the time period from conception until the second year of life as a crucial phase of susceptibility for an individual's lifelong health [2]. He proposed "The Developmental Origins of Health and Disease" (DOHaD) hypothesis which holds that environment during the earliest stages of life—pre-conception, pregnancy, and infancy—shapes developmental trajectories and ultimately health outcomes including non-communicable diseases across the lifespan.

The developmental plasticity would be modulated by epigenetic changes to the DNA such as DNA methylation.

2.2 Epigenetic Mechanisms: DNA Methylation

DNA methylation is characterized by addition of a methyl group (-CH_3) deriving from S-adenosyl methionine (SAM) to cytosine (C) which is transformed into 5-methylcytosine (5mC). This addition is catalyzed by DNA methyltransferases (DNMTs).

C methylation occurs predominantly in CG dinucleotide sequences (called CpG), at the level of gene promoter sequences, but is also found less frequently in other non-CpG contexts (e.g., mCH, where H = A, T, or C) [3].

DNA methylation is an epigenetic mark conserved in living beings, allowing them to survive in different environments. DNA methylation does not exclusively affect C residues. N6-methyladenine (6 mA) residues have been found in various species, including humans. 6 mA methylation levels, reported in some studies, are lower than 5mC, and their biological role still remains to be clarified [4].

DNA methylation is a dynamic process that occurs during embryonic development and its basic function is the regulation of gene expression. It is also implicated in transcriptional repression, modification of chromatin architecture, genomic imprinting, embryogenesis, and inactivation of transposable genetic elements.

It is important to underline that promoter methylation is associated with transcriptional inhibition because it determines a change in the chromatin structure, preventing the binding of transcriptional factors to DNA.

In the vertebrate kingdom, approximately 80% of CpGs are primarily subject to methylation with a general hypermethylated status of the whole genome. The exception to this statement is the presence of particular DNA sequences known as CpG islands (CGIs), which tend to be hypomethylated relative to the overall genome. The condition of CGIs hypomethylation among gene promoters determines the transcription activation, allowing the binding between protein transcription factors and promoters. Even the methylation of non-CpG sequences in DNA could be involved in the regulation of gene expression, favoring or disfavoring, depending on the case, the direct binding with specific transcriptional factors. The methylation of gene regions beyond that of the promoter can also have a role in silencing repetitive DNA elements present within the gene [5]. Furthermore, important changes in methylation levels were found at intron-exon junctions, suggesting an association with splicing processes.

As previously reported, DNA methylation is catalyzed by DNMTs, which are divided into de novo DNA methyltransferases (DNMT3A, DNMT3B, DNMT3L) and maintenance DNA methyltransferase (DNMT1), responsible for maintaining or copying methylation marks after DNA replication. These enzymes play critical roles in setting DNA methylation patterns genome-wide at specific developmental time points, particularly during early development and in specific tissues undergoing dynamic methylation in living beings. DNMT1 is the methyltransferase with the highest abundance in somatic cells. This enzyme, localized at the level of replication forks during DNA duplication, is responsible for the preservation of methylation patterns in the new DNA strand after replication, having a high affinity for hemimethylated substrates (Fig. 2.1a, b). It is also required for correct embryonic

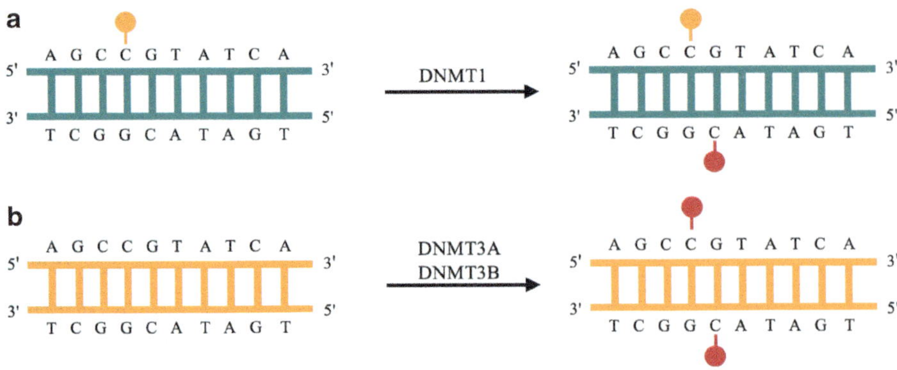

Fig. 2.1 DNA methylation pathways. DNA methylation is catalyzed by DNMTs, which are divided into maintenance DNA methyltransferase (DNMT1) and de novo DNA methyltransferases (DNMT3A, DNMT3b)

development, imprinting, and X inactivation. DNMT3A and DNMT3B, together with DNMT3L, which do not have their own methyltransferase activity, but modulate their activity (depending on the cell type and developmental stage), catalyze de novo methylation during gametogenesis and embryonic development in mammals and aberrant methylation patterns in carcinogenesis. DNMT2 is the most widely distributed DNMT in living beings. It has RNA methylation activity, and it is considered the first DNMT in life. However, several studies demonstrate that all three DNMTs not only cooperate, but can also participate in both de novo methylation and methylation maintenance [6].

Three reader protein families recognize and have the ability to bind methylated CpGs: the methyl-CPG-binding domain (MBD) protein, the ubiquitin-like with PHD and RING finger domains (UHRF) protein, and the zinc finger proteins. MBD protein contains a conserved methyl-CPG-binding domain (MBD), which has a high affinity for a single methylated CpG site.

The MBD family includes MBD1, MBD2, MBD3, MBD4, as well as MeCP2, the first protein of this family to be identified, and a mutation in its gene causes the majority of Rett syndrome cases.

The UHRF family of proteins, through binding to DNMT1 and targeting semi-methylated DNA, has the task of maintaining DNA methylation. The zinc finger protein family, which includes the Kaiso, ZBTB4, and ZBTB38 proteins, inhibits transcription mainly in a DNA methylation-dependent manner.

Until not long ago, it was believed that DNA methylation was an irreversible epigenetic event, associated with the repression of gene expression, having the possibility of modifications only through DNA replication. Accumulating evidence indicated the existence of a DNA demethylation machinery. In fact, studies have discovered several pathways leading to 5hmC-mediated DNA demethylation. Recently TET1 protein (Ten Eleven Translocation protein 1) was discovered. It is capable of modifying methylcytosines and potentially erasing DNA methylation [7]. TET1, together with the TET2 and TET3 proteins, promotes the conversion of

5-methylcytosine (5mC) to 5-hydroxymethylcytosine (5hmC), which in turn can be deaminated (by AID/APOBEC) or oxidized (by the action of TET) to 5hmU (5-hydroxymethyluracil) or 5fC (5-formyl-cytosine).

In addition, the amino group of 5mC can be deaminated by AID/APOBEC (activation-induced cytidine deaminase/apolipoprotein B mRNA-editing enzyme complex) into a carbonyl group, resulting in thymine (T). This mutation will lead to the formation of an incorrect G/T pairing, which can be corrected thanks to the DNA Base Excision Repair (BER) system.

The resulting products can be further converted to cytosine (C) by the enzyme thymine DNA glycosylase (TDG) [8] (Fig. 2.2).

Hydroxymethylation at the promoter level, within the gene and in transcription termination regions, showed a strong positive correlation with gene expression both within a tissue and among different tissues, suggesting that the presence of 5hmC could characterize active genes and play a role in the regulation of gene expression, via DNA demethylation.

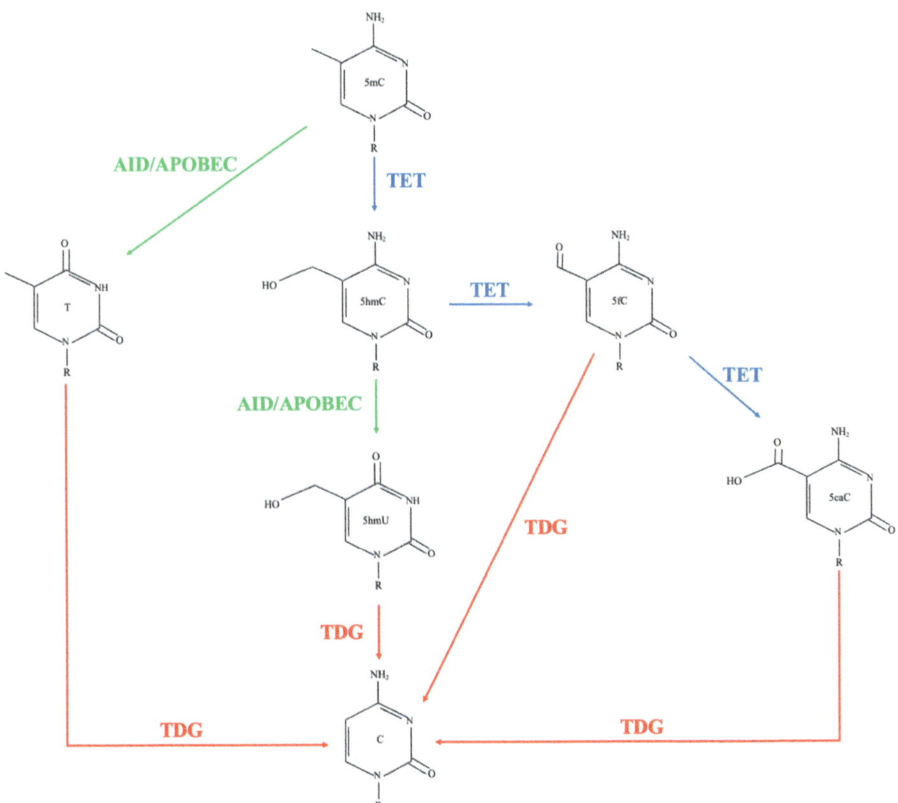

Fig. 2.2 Schematic of DNA demethylation pathways. (See the text for details)

DNA demethylation and TET proteins as well as implicated in embryonic development and hematopoietic differentiation, are also involved in other differentiation and maturation process.

For example, loss of TET activity is associated with altered neuronal functioning and has been linked to cognitive processes related to learning and memory as well as cocaine addiction [9]. Therefore, the activity of TETs is probably an important factor in numerous stages of development, and their functional deficiency can lead to the failure of specific cell type functions.

Methylation patterns are dynamically remodeled during reprogramming that occurs at specific stages throughout the life of an organism [10].

Variations in global methylation have frequently been found in many pathologies, in particular in many tumor forms, where hypermethylation of tumor suppressor genes is often observed, compared to global hypomethylation, especially of non-coding sequences, correlated with greater genomic instability [11].

Methylations are also inherited and some methylation patterns have transgenerational effects [12].

2.3 Chromatin Structure and Histone Modifications

The modifications of histone tails (such as methylation, acetylation, ubiquitination, adenylation, phosphorylation reactions) are catalyzed by numerous enzymes. These changes have a substantial effect on the activity of transcription, DNA repair and the condensation of chromosomes.

DNA, inside eukaryotic cells, is packaged in a dynamic structure called chromatin. The basic units of chromatin are nucleosomes, which are composed of four core histone proteins (H2A, H2B, H3, H4) present in duplicate forming an octamer, around which is wrapped a 145–150 base-pairs of DNA.

The linker histone H1 joins the nucleosomes and promotes the formation of the immediately higher structural level, consisting of the 30 nm chromatin fiber. Compaction, however, can limit the accessibility of DNA by the transcription apparatus and represents a further level of regulation of the transcription process. There are two states of chromatin compaction. The heterochromatic state, tightly condensed, is characterized by strong interactions between DNA and histone proteins, which inactivate the transcriptional process. On the contrary, in the euchromatic state, the interactions between DNA and histone proteins become weaker and therefore the chromatin is less condensed and consequently transcriptionally active.

Histone proteins are made up of a globular domain and an N-terminal, "tail"-shaped domain. The "histone tails" have an external position, compared to the rest of the histone molecules, so they have been attributed a role as "sensors" toward the surrounding environment. The different states of chromatin compaction largely depend on the N-terminal modifications of the histone tails, which protrude from the nucleosomes.

Following these modifications, the chromatin could be in a condensed or relaxed state, transcriptionally inactive or active, respectively.

The regulation of chromatin can take place through post-translational histone modifications or through ATP-dependent remodeling.

2.3.1 Post-translational Histone Modifications

The most commonly studied post-translational histone modifications are *acetylation* and *methylation*. In histone acetylation, an acetyl (-$COCH_3$) group is transferred from an acetyl-CoA molecule to a lysine of a histone tail. This transfer neutralizes a positively charged histone protein and weakens its affinity for negatively charged DNA, promoting the formation of euchromatin. Histone acetyltransferases (HATs), which include GCN5, p300/CBP, PCAF, Tip 60, catalyze the acetylation of histones and stimulate transcription [13], while histone deacetylases (HDACs) are enzymes that remove the acetyl group from histone tails and repress transcription (Fig. 2.3).

In mammals, there are 18 HDAC genes that are grouped into four classes: class I (HDAC1, 2, 3), class II (HDAC4, 5, 7, 9 in IIa and HDAC6 in IIb, class III (sirtuins 1–7), and class IV (HDAC11) [14]. The methylation of arginine (R) and lysine (K) residues on histone tails is regulated by histone methyltransferases (HMT). K residues can be mono-, di-, and trimethylated, while R residues can be mono- and di-methylated. Also, in this case, methylation can inhibit or stimulate the transcriptional process. It has been shown that methylation of lysine 4 (K4), K36, and K79 within histone 3 (H3) determines transcriptional activation. On the other hand, the methylation of K9H3, K20H3, and K27H3 causes a condensation of the chromatin, which is transcriptionally inactive. Instead, arginine methylation is associated with transcriptional activation. During methylation, a methyl group is transferred from S-adenosyl methionine. Histone methylation, unlike acetylation, does not

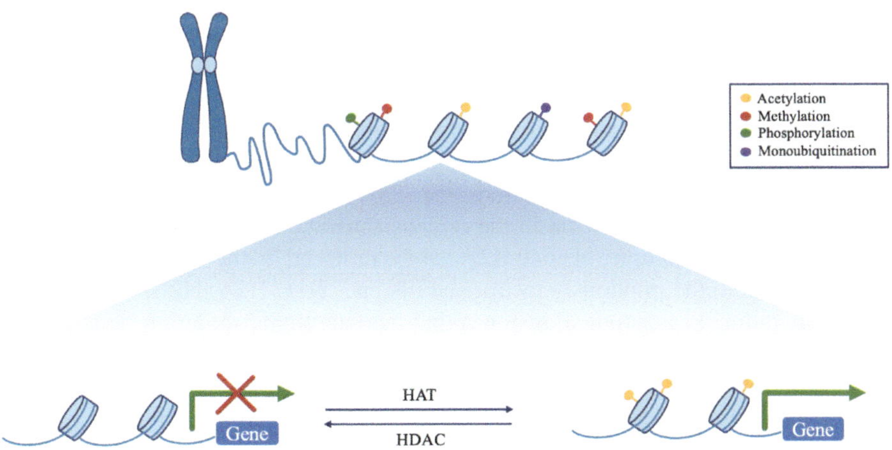

Fig. 2.3 Histone modifications. For example, histone acetylation stimulates gene transcription, while deacetylation represses transcription

necessarily cause changes in chromatin structure. This is because methylation does not neutralize a positive charge on the histone protein. It is thought that methylation, through the binding of effector proteins, can lead to chromatin changes.

In addition to acetylation and methylation, there are other histone modifications that are less characterized, such as *phosphorylation* and *monoubiquitination*. Phosphorylation, as well as acetylation, by adding a negative charge to the histone tails neutralizes the positive charge of the tails themselves. This results in a reduction of histone-DNA interactions leading to a transcriptionally active chromatin conformation. Monoubiquitination, which mainly involves histones H2A and H2B, may be associated with transcriptional activation or repression [15].

Therefore, depending on the modifications of the histone tails, the chromatin itself can be more or less condensed and consequently relatively accessible by the transcriptional apparatus, thus influencing gene expression.

All these histone modifications can interact with each other and we speak of crosstalk between histone modifications. For example, the monoubiquitination of histone H2B is essential for the methylation of H3K4 [16].

2.3.2 ATP-Dependent Chromatin-Remodeling Complexes

Because transcription cannot take place unless nucleosome elimination or displacement occurs first, the action of certain protein complexes called remodeling or remodeling complexes, which are ATP-dependent and thus require energy in the form of ATP to perform their function, is crucial. Therefore, they require energy in the form of ATP in order to perform their function. Once the remodeler has eliminated or displaced the nucleosomes, it can be released. In this way, the promoter of the target gene is recognized and bound by transcription factors that can initiate the transcriptional process.

Chromatin remodelers are the imitation switch (ISWI), chromodomain helicase DNA-binding (CHD), Switch/sucrose non-fermentable (SWI/SNF), and INO80. ISWI and CHD promote histone positioning, maturation, and nucleosome spacing. The SWI/SNF subfamily is involved in "chromatin access". These remodelers facilitate the binding of transcription factors to chromatin by causing nucleosomes sliding along the DNA, eliminating nucleosome components (such as the dimers H2A and H2B) or ejecting the entire nucleosome. This ensures the transcription process. INO80 remodelers are involved in the replacement of canonical histones with histone variants. For example, histone H2A can be replaced by the variant H2A.Z [17].

In addition to the previously mentioned histones (H1, H2A, H2B, H3, and H4), which are defined as canonical histones that mediate higher-order chromatin organization, mammals possess replication-independent, noncanonical variant histones. Thus, one difference that distinguishes histone variants from canonical histones is that the latter are only incorporated into nucleosomes during the replication phase, a limitation that doesn't exist for histone variants. These variants contribute to creating distinct nucleosome structures and functions. Variants include H2AZ and H3.3, which are variants of histones H2 and H3, respectively, that together destabilize H1, making chromatin looser and more accessible to transcription factors [18]. Histone variants interfere not only with normal physiological functions, but also with ageing, senescence, and various pathogenic conditions. Many studies have shown that replacing canonical histones with their variants disrupts chromatin dynamics and reprograms the transcriptome and the epigenome of the cell [19].

2.4 RNA-Mediated Gene Silencing

Although not universally considered part of the epigenetic apparatus, non-coding RNA molecules, which can be located inside and outside the nucleus, are believed to play an essential role in epigenetic processes.

Most of the mammalian genome, approximately 98%, is transcribed into ncRNAs, functional RNA molecules, not translated into proteins. While less than 2% is transcribed into proteins, ncRNAs have multiple functions. Those involved in gene regulation include short-interfering RNA (siRNAs), PIWI-interacting RNAs (piRNAs), long non-coding RNAs (lncRNAs), and microRNAs (miRNAs), which are the most widely studied class of non-coding RNAs [20]. It is now established the essential importance of their function in evolutionary processes, in apoptosis, and cell proliferation.

2.4.1 miRNA

miRNAs regulate the expression of target genes at the post-transcriptional level, pairing by complementarity with the 5′ region of target messenger RNAs, preventing their translation or causing their degradation. The production of a functional miRNA consists in the transformation of the primary miRNA (pri-miRNA) into a mature miRNA of approximately 22 nucleotides (nt).

MicroRNA genes, included within intronic regions or within the exons of protein-coding genes, are transcribed by RNA polymerase II as pri-miRNAs, which have stem-loop structures (i.e., a structure with double-strand RNA regions and 3′ and 5′ extensions to single-strand RNA). Subsequently, enzymes Drosha (type III RNase) and DGCR8 (an RNA-binding protein) process pri-miRNAs, in the nucleus, cutting at the 3′ and 5′ extensions and generating the miRNA precursors (pre-miRNA). These are transported into the cytoplasm by Exportin5 and undergo a second endonuclease cleavage reaction by the enzyme Dicer (class III RNase): this cut generates a miRNA duplex (~22 nt).

The Argonauta protein (AGO2) binds to the miRNA duplex and this causes the detachment of the passenger filament, which is degraded, while the guide filament remains attached to Argonauta. This complex is called miRNA-induced silencing complex (miRISC), which mediates the binding of miRNA to mRNA target, inhibiting the translation of the mRNA or determining its degradation (Fig. 2.4). Since most protein-coding genes, in mammals, are subject to regulation by at least one miRNA, and considering the presence of multiple non-conserved sites, it is

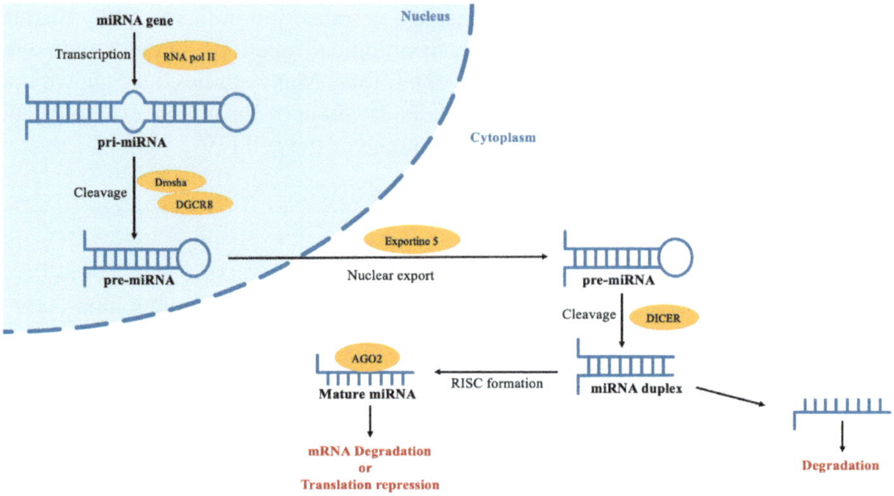

Fig. 2.4 Biogenesis and function of miRNA. (See the text for details)

unsurprising that the biogenesis and the function of these genes are meticulously controlled. The regulation of this mechanism is justified by the fact that any alteration has been associated with a wide range of human diseases, such as cancer.

MiRNAs can be regulated both at the transcriptional level, thanks to the presence of canonical promoters and enhancers in their regulatory regions, and at the post-transcriptional level. Therefore, their expression is closely related to the action of transcription factors, such as c-myc and p53, or to epigenetic modifications, including DNA methylation and alterations of histone tails.

The mechanism described, based on the intervention of miRNAs, is the best characterized, compared to many other molecular mechanisms that are being identified in recent years, which involve a growing number of regulatory circuits based on interactions among messenger RNAs (mRNA) and other types of ncRNA.

Small interfering RNAs (siRNAs) are a class of double-stranded RNAs, generally 20–24 bp long, with 5′ phosphorylated and 3′ hydroxylated ends with two prominent nucleotides.

The biogenesis of siRNAs is very similar to that of miRNAs. Long dsRNAs are processed by Dicer, resulting in the formation of siRNAs. The Argonauta protein binds to the siRNA leader strand, forming the siRISC complex, while the passenger strand is excreted and degraded. The siRNAs thus induce gene silencing by degrading the target mRNA. Their action is similar to that of miRNAs in the RNA interference (RNAi) pathway. While siRNAs are 100% complementary to the mRNA target, miRNAs have a broader specificity and can have more than 100 targets simultaneously. In addition, siRNAs promote the silencing of gene expression by endonucleocytic cleavage of the target mRNA, whereas miRNAs inhibit translation microRNAs (miRNAs) by physically binding to the mRNA target or by degrading the mRNA target itself [21].

2.4.2 lncRNA

More and more evidence suggest that long non-coding RNAs (lncRNAs) also have an important role in the regulation of gene expression. Unlike small noncoding RNAs, lncRNAs are molecules larger than 200 nt and are generally divided into five groups based on their position and orientation relative to nearby protein-coding genes: sense, antisense, bidirectional, intronic, or intergenic lncRNAs. Many lncRNAs are transcribed by RNA polymerase II and presumably undergo 5′ capping, 3′ polyadenylation, and splicing, but unlike mRNAs they are not thought to code for proteins. lncRNAs are able to interact with proteins, RNA and DNA, forming complexes capable of regulating gene expression through multiple mechanisms, including the recruitment of transcription factors or complexes that modify chromatin, the formation of heterogeneous complexes of nuclear ribonucleoproteins, acting as a decoy to sequester miRNAs and RNA-binding proteins, or by interacting directly with RNA and DNA through base complementarity. In this way, lncRNAs are involved in many biological processes, such as the maintenance of genome integrity and imprinting, the regulation of the cell cycle, development, and differentiation. As with small non-coding RNAs, deregulation of lncRNAs is also associated with an ever-increasing number of human pathologies [22].

2.4.3 piRNA

PIWI-interacting RNAs (piRNAs), discovered in germ cells, are single-stranded ncRNAs (26–31 nucleotides long) with non-conserved nucleotide sequences. piRNAs are named for their interaction with PIWI (P-element-induced wimpy testis) proteins, a family of germ-line-specific Argonaute nucleoproteins. This binding results in translation inhibition.

The biogenesis of piRNAs is still poorly understood and the information we have comes from studies in Drosophila. Precursor piRNAs are derived from specific genetic regions called piRNA clusters, localized mainly in relatively short genomic loci on chromosomes 17, 5, 4, and 2.

Transcribed by RNA polymerase II, piRNAs are transported from the nucleus to the cytoplasm where they are processed into mature piRNAs in an unknown manner, probably independent of the action of Dicer. To date, piRNAs have been implicated in genomic rearrangement, epigenetic regulation, protein regulation, transposon silencing, and maintenance of spermatogenesis [21]. piRNAs have been shown to be implicated in the silencing of retrotransposons, as well as the silencing of other genetic elements in germ lines, particularly during spermatogenesis.

Furthermore, the piRNA/PIWI complex, through the recruitment of DNMT3a and DNMT3b, promotes DNA and histone methylation, leading to transcriptional gene silencing. Recent studies have shown that dysregulation of piRNAs, which contributes to abnormal gene expression, can promote or prevent the development of various diseases such as cancer [23].

2.5 RNA Methylation

Like DNA, RNA can also be modified. In mammals, the most common modification of mRNA appears to be N^6-methyladenosine (m^6A). It can regulate mRNA stability, splicing, translation, and nuclear export. Unlike DNA or histone methylation, knowledge about N6-methyladenosine (m6A) modification of RNA is a little cloudy.

In recent years, based on whole transcriptome mapping approaches, six additional internal modifications have been reported to be widespread on mammalian mRNA: pseudouridine (Ψ), N^1-methyladenosine (m^1A), ribose methylations (N_m), 5-methylcytidine (m^5C), N^7-methylguanosine (m^7G), and N^4-acetylcytidine (ac^4C).

N^6-methyladenosine is co-transcribed by the m^6A methyltransferase complex, which consists of the catalytic subunit METTL3 and other accessory subunits. Wilms' tumor 1-associating protein (WTAP) promotes the formation of the METTL3-METTL14 heterodimer, with METTL3 catalyzing the transfer at the N^6 position of adenosine.

In addition, there are several WTAP interactors that are important for methylation, such as WIRMA, ZC3H13, RBM15, and RBM15B (two paralogous RNA-binding proteins). Depletion of WTPA or WTPA interactors results in a reduction of m^6A in mRNA.

This process is dynamic and reversible, as there are demethylases (FTO and ALKBH5) that remove the m6A residues present in the RNA. Both FTO and ALKBH5 are iron- and α-ketoglutarate-dependent dioxygenases that mediate the oxidative demethylation of m^6A.

The recruitment of the writer complex to specific promoters could be induced by the intervention of transcription factors. In human stem cells, the SMAD2/3 transcription factors bind METTL3, METTL14, and WTAP by promoting transcript methylation. The position of m^6A can also be determined by histone modifications. Several studies have shown that m^6A is mainly found within exons that, compared to introns, have higher levels of histone modifications such as H3K36me3, which is associated with m^6A formation in mRNA. Thus, chromatin marks may guide mRNA methylation.

More recently, this methylation has also been linked to various diseases and cancers [24].

2.6 Concluding Remarks

A deregulation of epigenetic profiles has been described in several human pathologies, including genomic imprinting disorders and many complex diseases. Cancer was the first complex human disease associated with epigenetic defects, but the list of diseases exhibiting epigenetic modifications has grown rapidly in recent years. There are many studies on complex diseases, such as obesity, type 2 diabetes mellitus, cardiovascular diseases, neurodegenerative diseases, neurodevelopmental, behavioral, psychiatric disorders, and many others, in which alterations in epigenetic marks have been demonstrated in many regions of the genome or even in specific genes. The pathogenetic mechanisms of these diseases are particularly interesting, because it has been shown that epigenetic modifications can also be influenced by environmental factors such as diet and lifestyle. Deleterious epigenetic profiles can also be a consequence of alterations in genes due to factors that read, write, modify, and erase epigenetic marks, implicated in many serious genetic disorders in humans [25]. Epigenetic markers are therefore beginning to find their application in clinical practice as diagnostic, prognostic, or response to pharmacological treatment indicators and, at the same time, the technologies for analyzing the epigenome with rapid and reproducible methods, and at reasonable costs, are increasing. The tissue specificity of epigenetic modifications and the not always easy correlation between modifications observed in the patient and the environmental factors that induced them, perhaps a few decades before the observation, make the study of animal and cellular models of these diseases indispensable. These models, given the reversible nature of these markings, also lend themselves well to the testing of drugs or natural compounds with epigenetic properties capable of counteracting pathological progression. Many epigenetic drugs have been developed and already approved by the US Food and Drug Administration: inhibitors for targeting epigenetic writers, such as DNA methyltransferases and histone methyltransferases, epigenetic erasers, such as histone demethylases and histone deacetylases, and epigenetic readers, such as bromodomain and extra-terminal domain proteins [26]. They are already used in the therapy of myelodysplastic syndromes and cutaneous T-cell lymphoma, and many studies are underway to evaluate their effectiveness in other tumor and non-tumor pathologies [27].

References

1. Feil R, Fraga MF (2012) Epigenetics and the environment: emerging patterns and implications. Nat Rev Genet 13(2):97–109
2. Barker DJ, Eriksson JG, Forsén T, Osmond C (2002) Fetal origins of adult disease: strength of effects and biological basis. Int J Epidemiol 31(6):1235–1239
3. Jang HS, Shin WJ, Lee JE, Do JT (2017) CpG and non-CpG methylation in epigenetic gene regulation and brain function. Genes (Basel) 8(6):148
4. Li Z, Zhao P, Xia Q (2019) Epigenetic methylations on N6-adenine and N6-adenosine with the same input but different output. Int J Mol Sci 20(12):2931
5. Pappalardo XG, Barra V (2021) Losing DNA methylation at repetitive elements and breaking bad. Epigenetics Chromatin 14(1):25
6. Del Castillo Falconi VM, Torres-Arciga K, Matus-Ortega G, Díaz-Chávez J, Herrera LA (2022) DNA Methyltransferases: from evolution to clinical applications. Int J Mol Sci 23:8994
7. Rasmussen KD, Helin K (2016) Role of TET enzymes in DNA methylation, development, and cancer. Genes Dev 30:733–750
8. Joshi K, Liu S, Breslin SJP, Zhang J (2022) Mechanisms that regulate the activities of TET proteins. Cell Mol Life Sci 79(7):363
9. Feng J, Shao N, Szulwach KE, Vialou V, Huynh J, Zhong C, Le T, Ferguson D, Cahill ME, Li Y, Koo JW, Ribeiro E, Labonte B, Laitman BM, Estey D, Stockman V, Kennedy P, Couroussé T, Mensah I, Turecki G, Faull KF, Ming GL, Song H, Fan G, Casaccia P, Shen L, Jin P, Nestler EJ (2015) Role of Tet1 and 5-hydroxymethylcytosine in cocaine action. Nat Neurosci 18(4):536–544

10. Zeng Y, Chen T (2019) DNA methylation reprogramming during mammalian development. Genes (Basel) 10(4):257
11. Jin Z, Liu Y (2018) DNA methylation in human diseases. Genes Dis 5(1):1–8. https://doi.org/10.1016/j.gendis.2018.01.002
12. Fitz-James M, Cavalli G (2022) Molecular mechanisms of transgenerational epigenetic inheritance. Nat Rev Genet 23(6):325–341
13. Ramaiah MJ, Tangutur AD, Manyam RR (2021) Epigenetic modulation and understanding of HDAC inhibitors in cancer therapy. Life Sci 277:119504
14. Haberland M, Montgomery RL, Olson EN (2009) The many roles of histone deacetylases in development and physiology: implications for disease and therapy. Nat Rev Genet 10(1):32–42
15. Hyun K, Jeon J, Park K, Kim J (2017) Writing, erasing and reading histone lysine methylations. Exp Mol Med 49:e324
16. Lee J-S, Smith E, Shilatifard A (2010) The language of histone crosstalk. Cell 142:682–685
17. Clapier CR, Iwasa J, Cairns BR, Peterson CL (2017) Mechanisms of action and regulation of ATP-dependent chromatin-remodelling complexes. Nat Rev Mol Cell Biol 18:407–422
18. Talbert PB, Henikoff S (2021) Histone variants at a glance. J Cell Sci 134:jcs244749
19. Dubey SK, Dubey R, Kleinman ME (2024) Unraveling histone loss in aging and senescence. Cells 13(4):320
20. Chen LL (2016) Linking long noncoding RNA localization and function. Trends Biochem Sci 41(9):761–772
21. Beňačka R, Szabóová D, Guľašová Z, Hertelyová Z, Radoňak J (2023) Non-coding RNAs in human cancer and other diseases: overview of the diagnostic potential. Int J Mol Sci 24:16213
22. Weidle UH, Birzele F, Kollmorgen G, Rüger R (2017) Long non-coding RNAs and their role in metastasis. Cancer Genomics Proteomics 14(3):143–160
23. Zhang Q, Zhu Y, Cao X, Tan W, Yu J, Lu Y, Kang R, Wang X, Li E (2023) The epigenetic regulatory mechanism of PIWI/piRNAs in human cancers. Mol Cancer 22:45
24. Jiang X, Liu B, Nie Z, Duan L, Xiong Q, Jin Z, Yang C, Chen Y (2021) The role of m6A modification in the biological functions and diseases. Signal Transduct Target Ther 6:74
25. Bjornsson HT (2015) The Mendelian disorders of the epigenetic machinery. Genomc Res 25:1473–1481
26. Davalos V, Esteller M (2023) Cancer epigenetics in clinical practice. CA Cancer J Clin 73(4):376–424
27. Berdasco M, Esteller M (2019) Clinical epigenetics: seizing opportunities for translation. Nat Rev Genet 20:109–127

Epigenetics and Endocrine Disruptors 3

Mauro Bologna and Ilaria Bologna

Contents

3.1 Introduction 30
3.2 What Is Epigenetics and What Are the Endocrine Disruptors? 30
3.2.1 Bisphenols 33
3.2.2 DDT 35
3.2.3 Dioxins 36
3.2.4 Parabens 38
3.2.5 PFAS 38
3.2.6 Phtalates 40
3.2.7 Pesticides 40
3.2.8 Polychlorinated Biphenyls 41
3.2.9 Triclosan 42
3.3 Microplastics and Other ED Compounds 42
3.4 Conclusion 43
References 44

M. Bologna (✉)
Department of Life, Health and Environmental Sciences, University of L'Aquila, L'Aquila, Italy

Italian Society of Psycho-Neuro-Endocrine-Immunology, Rome, Italy
e-mail: mauro.bologna@univaq.it

I. Bologna
Department of Prevention, Local Health Authority n.1, L'Aquila, Italy

Department of Life, Health and Environmental Sciences, University of L'Aquila, L'Aquila, Italy

L. Emmi, V. Fanos (eds.), *Epigenetics and Rare Diseases*, Rare Diseases of the Immune System, https://doi.org/10.1007/978-3-032-02902-7_3

Abbreviations

BPA	Bisphenol A
CBZ	Carbendazim
DLCs	Dioxin-like compounds
DDT	Dichlorodiphenyltrichloroethane
EDs	Endocrine disruptors
EFSA	European Food Safety Authority
GDM	Gestational Diabetes Mellitus
IARC	International Agency for Research on Cancer
PCBs	Poly-chloro-biphenyls
PBB	Poly-bromo-biphenyls
PCDD	Polychlorinated dibenzo-p-dioxins
PCDF	Polychlorinated dibenzofurans
PFAS	Per-fluoro-alkyl-substances
PFDA	Perfluorodecanoic acid
PFHxS	Perfluorohexane sulfonic acid
PFNA	Perfluorononancanoic acid
PFOA	Perfluorooctanoic acid
PFOS	8-carbon Perfluoroctane sulfonic acid
PFOSA	Perfluorooctanesulfonamide
POPs	Persistent organic pollutants
PVC	Polyvinyl chloride
SVHC	Substances of very high concern
TCDDs	2,3,7,8-tetrachlorodibenzodioxin
TPO	thyroid peroxidase
THT	thyroid hormone transporter

3.1 Introduction

We open this chapter with some essential definitions.

3.2 What Is Epigenetics and What Are the Endocrine Disruptors?

With the term Epigenetics we indicate, since the original description by Conrad Waddington in the early 1940s, the interaction between the genes of a live being and its environment of life [1, 2]. This interaction leads to the non-Mendelian genetic inheritance of phenotypic traits in many species. The concepts regarding Epigenetics are extensively discussed in other chapters of this volume and should focus particularly our attention toward protection of the embryonic and fetal stages in all the developing organisms (mostly human but also in other species).

With the term Endocrine Disruptors (EDs) we indicate all the chemical compounds, both synthetic and natural, able to interfere with the endocrine messages and receptors regulating the physiological mechanisms of a live being [3]. A recent Expert Consensus Statement indicated the Key Characteristics of EDs in biomedical fields and represents a very important document summarizing the fundamental principles necessary to identify the corresponding hazards for Human and Environmental Health [4]. Further evidences have been collected from experiments regarding mitochondrial dysfunction, which is implicated in development and progression of many chronic diseases, through the effects of environmentally derived oxidative stress in which mitochondrial dysfunction is involved [5].

EDs may contribute to the observed declines of wildlife populations and of the reproductive health of humans. Several different classes of chemicals have been demonstrated to have endocrine-disrupting properties, often joined to transgenerational epigenetic effects and also to carcinogenic pathways. Epigenetic histone modifications include acetylation, deacetylation, phosphorylation, and ubiquitination. Therefore, the possible influences of epigenetically active compounds are very wide and still incompletely studied. Even paternal exposure to environmental risk factors may have influences on the offspring health through methylation of sperm DNA [6]. Other data showing negative effects on male fertility are emerging, with particular regard to Bisphenol A [18]. The fundamental goal of this chapter is to promote disease prevention toward the most vulnerable parts of the population, by health professionals and health organizations, with special attention to environmental and occupational exposures. But due to the reversibility of epigenetic modifications, it may be possible to reverse diseases related to epigenetic changes by the development of new molecularly targeted drugs (like DNA methylation and histone deacetylation inhibitors) [15].

Examples of ED compounds having epigenetic actions are already very numerous and documented. EDs range from naturally occurring plant hormones to synthetic hormones and to persistent and non-persistent chemicals mimicking the hormone actions through a variety of mechanisms. Therefore, for the present work, we decided to proceed by illustrating and screening the major classes and examples of ED compounds with proven action in the epigenetic sense. Still limited are the studies concerning the so-called "emerging EDs of interest" [16]. Moreover, the probability exists about the simultaneous effects of exposure to multiple compounds. A major factor of relevance is the age at which the exposure of the individual occurs. Chemical exposure during the fetal or neonatal life may indeed cause irreversible effects even at lower doses and at short-term exposure [10, 11, 16].

Of particular note is the fact that the brain is incredibly sensitive to small changes in the hormonal environment during early development, with natural sex differences in gonadal hormones shaping the organization of sex-specific neural circuits early in life. ED exposure during these periods can lead to lifelong impacts on neurobiological health outcomes [11].

In Table 3.1, we list the compounds that we will discuss in this chapter, indicating and referencing for each group of compounds the demonstrated actions as Epigenetic effects (EPIG) and Endocrine Disruptor effects (ED).

The fundamental key characteristics (KCs) of compounds classified as Endocrine Disruptors (EDs) are illustrated in Fig. 3.1 (Reproduction Permission in Progress from the Publisher) [RPPP] (Ref [4]).

Such key characteristics are listed below, mentioning the ability of the compound to:

Table 3.1 Major classes of compounds described in this chapter, with mention of their corresponding epigenetic (EPIG) and Endocrine Disruption (ED) effects (with bibliographic references)

Class of compounds	EPIG	ED
Bisphenols	Hypomethylation and/or hypermethylation; histone modifications and altered expression of non-coding RNAs [13, 23, 25]	Estrogenic, androgenic, and obesogenic effects; cardiovascular disorders; thyroid alterations; reproductive and metabolic effects [16, 18, 24]
DDT	Methylation, non-coding RNAs and histone alterations [15]	Reproductive effects; metabolic effects [16]
Dioxins	Methylation alterations [15]	Antiestrogenic and antiandrogenic effects; liver damage; metabolic effects; thyroid, reproductive, and developmental effects [16, 29]
Parabens	Epigenetic mechanisms demonstrated in GDM	Increasing GDM in women at high risk [21]
PFAS	Histone alterations; methylation alterations [9, 10, 11, 12]	Reproductive and developmental effects; hepatotoxic effects; thyroid function disruption, metabolic effects [27]
Phthalates	Hypermethylation, altered expression of non-coding RNAs, histone modifications [15]	Anti-androgenic and metabolic effects; thyroid alterations; reproductive and developmental effects; liver damage [22]
Pesticides	DNA hypomethylation [11]	Metabolic effects, thyroid, reproductive and developmental effects [29, 31]
Polychlorinated biphenyls	Hypermethylation and altered non-coding RNAs [11]	Thyroid effects; reproductive effects; liver damage; metabolic effects [26]
Micro- and nano-plastics	Several examples: See text	Several examples: See text

KC1: Interact with or activate hormone receptors.
KC2: Antagonize hormone receptors.
KC3: Alter hormone receptor expression.
KC4: Alter signal transduction in hormone-responsive cells.
KC5: Induce epigenetic modifications in hormone-producing or hormone-responsive cells.
KC6: Alter hormone synthesis.
KC7: Alter hormone transport across cell membranes.
KC8: Alter hormone distribution or circulating levels of hormones.
KC9: Alter hormone metabolism or clearance.
KC10: Alter the fate of hormone-producing or hormone-responsive cells.

The following paragraphs contain a short discussion of the scientific evidences regarding each of the major classes of compounds listed in Table 3.1.

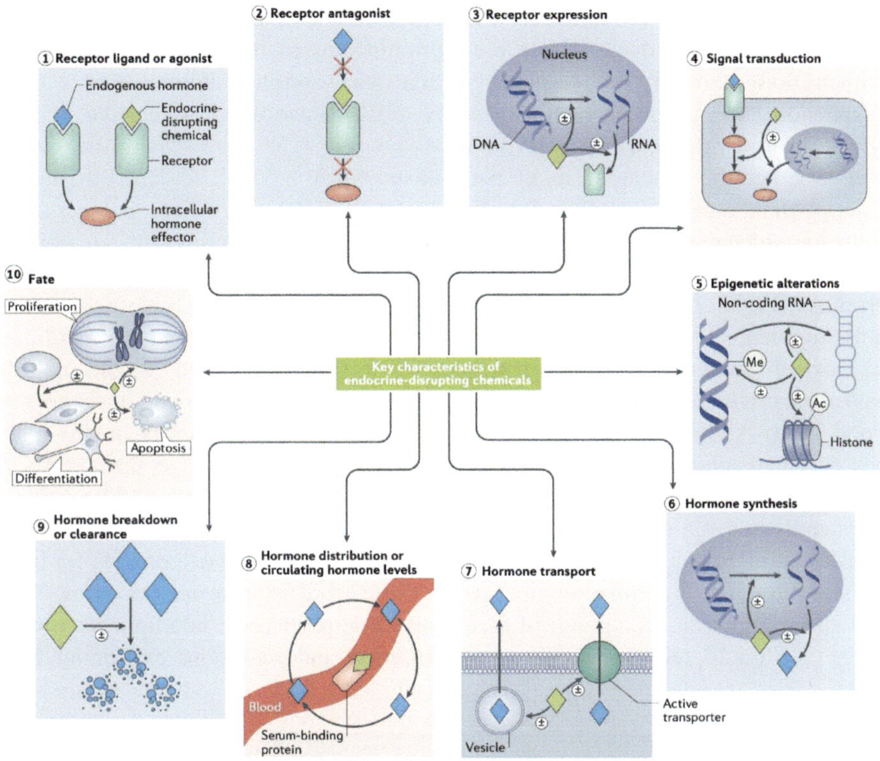

Fig. 3.1 The key characteristics of endocrine-disrupting chemicals. Arrows identify the ten specific key characteristics (KCs) of endocrine-disrupting chemicals (EDCs). The ± symbol indicates that an EDC can increase or decrease processes and effects. KC1 states that an EDC can interact with or activate hormone receptors. KC2 states that an EDC can antagonize hormone receptors. KC3 states that an EDC can alter hormone receptor expression. KC4 states that an EDC can alter signal transduction (including changes in protein or RNA expression, post-translational modifications, and/or ion flux) in hormone-responsive cells. KC5 states that an EDC can induce epigenetic modifications in hormone-producing or hormone-responsive cells. KC6 states that an EDC can alter hormone synthesis. KC7 states that an EDC can alter hormone transport across cell membranes. KC8 states that an EDC can alter hormone distribution or circulating hormone levels. KC9 states that an EDC can alter hormone metabolism or clearance. KC10 states that an EDC can alter the fate of hormone-producing or hormone-responsive cells. Depicted EDC actions include amplification and attenuation of effects. Ac, acetyl group; Me, methyl group. (Reproduction Permission obtained from the Publisher) Ref [4]

3.2.1 Bisphenols

Bisphenols are a group of chemical compounds widely used in manufacturing of plastics, in particular of polycarbonates. Human exposure derives from several sources, but one of the major ones is through the canned foods. Since many metal cans are lined internally with plastics containing bisphenols, a diet largely involving canned foods may imply a significant exposure to this class of compounds. They are also contained in epoxy resins, food storage containers, dental sealants [13], thermal paper, compact discs, electronic equipments, and toys; BPA can enter the human body through the respiratory, digestive, and transdermal routes [21]. Bisphenols, and particularly bisphenol A (BPA), have an estrogen-like activity which is largely documented [7].

Public health agencies in many parts of the world have all screened the health risks of BPA. But, due to the scientific uncertainty, many jurisdictions have taken steps to reduce exposure on a precautionary basis. In particular, infants are considered to be at greater risk, since BPA is present in plastic baby bottles and related products (Fig. 3.2). Bottle producers have largely switched from polycarbonate to polypropylene. The European Chemicals Agency has added BPA to the Candidate List of substances of very high concern (SVHC), procedure that should make it easier to restrict or ban their use in the future. In 2023 after the EFSA (European Food Safety Authority) reported about the toxicity of BPA, the European Union has passed the resolution in early 2024 to ban BPA in all the food contact material including plastic and coated packaging and it said it would also address other bisphenols to avoid replacing with other harmful substances.

BPA and DES (diethylstilbestrol) were shown in vitro to be responsible for insulin secretion and hyperinsulinemia favoring insulin resistance in adipocytes, liver, and skeletal muscle, as a result of their estrogenic properties; in adipocytes moreover, they could reduce adiponectin production and secretion, enhancing lipid

Fig. 3.2 Bisphenols in canned foods; baby bottles; cashier receipts. (Reproduction Permission obtained from the Publisher) (Nature 17apr 2014, vol 507, page 306, Josie Glausiusz)

accumulation. Exposure to BPA was shown to induce hyperinsulinemia and to reduce insulin sensitivity in animal studies in vivo. The epigenetic mechanisms seem to include hypomethylation and/or hypermethylation, histone modifications, and dysregulation of non-coding RNA expression (microRNAs and long non-coding RNA) [13, 15, 23].

BPA is involved in impaired fertility, premature births, in vitro fertilization failure, and sperm morphology and motility [21, 24]. Due to its antiandrogenic effect, exposure to BPA is also associated in males with declined semen quality, altered reproductive hormone levels, and adverse reproductive outcome [18].

There is a negative association between BPA exposure and semen motility, linked to a dose-dependent decrease in mitochondrial membrane potential, which suggests to investigate people and particularly workers at major risk of exposure compared with general population [24].

As well, BPA impacts on the placenta by reducing global methylation [15] and it is also linked to thyroid hormone alteration [11, 16, 17]. Some data link BPA with polycystic ovary syndrome [28].

3.2.2 DDT

The dichlorodiphenyltrichloroethane (DDT) is a chemical compound in the category of organochlorides: originally, it is a colorless, tasteless, and almost odorless crystalline powder. It was developed as an insecticide, but soon it became infamous for its environmental impacts. DDT was first synthesized in 1874 with its insecticidal action being discovered only afterward, in 1939. DDT was used in the second half of World War II to limit the spread of the insect-borne diseases, mostly malaria and typhus, among civilians and troops. Antimalaric campaigns were applied intensively in geographically humid areas where mosquitoes lived and spread malaria contagion. The WHO's anti-malaria campaign of the 1950s and 1960s relied heavily on DDT and the results were promising, though there was a resurgence of the epidemics in developing countries afterward.

DDT was promoted by government and industry for use as an agricultural and household pesticide, but there were also concerns about its use from the beginning. Opposition to DDT was focused by the 1962 publication of Rachel Carson's book "Silent Spring". The book talked about environmental impacts that correlated with the widespread use of DDT in agriculture in the United States, and it questioned the logic of extensive introduction of potentially dangerous chemicals into the environment with little prior investigation of their environmental and health effects. The book cited claims that DDT and other pesticides caused cancer and that their agricultural use was a threat to wildlife, particularly birds. The Carson book was a seminal event for the environmental movement and resulted in a large public movement that eventually led, in 1972, to a ban on DDT's agricultural use in the United States. Along with the passage of the Endangered Species Act, the United States ban on DDT was a major factor in the comeback of the bald eagle (the national bird of the United States) and the peregrine falcon from near-extinction in the United States and elsewhere. DDT is now included in the IARC list of carcinogens (International Agency for Research on Cancer. Volume 113: DDT, lindane and 2,4-D. IARC Working Group. Lyon; 2–9 June 2015). Concerning DDT's endocrine-disrupting effects, there is strong evidence that DDT is an androgen receptor antagonist and moreover it can induce changes in DNA methylation, non-coding RNA, and histones within sperm [15]. DDT effects on female fertility showed a reduced ovarian reserve [16]. Diabetes is also linked to pesticide exposure, with a remarkable amount of data [31].

3.2.3 Dioxins

Dioxins and dioxin-like compounds (DLCs) are a very large group of chemical compounds **that are persistent organic pollutants (POPs) in the environment. They develop mostly as by-products of burning or various industrial processes or, in the case of dioxin-like PCBs and** PBBs**, unwanted minor components of intentionally produced mixtures.**

Some of them are highly toxic, but the toxicity among them varies extensively (up to 30,000-fold). They are grouped together because their mechanism of action is the same. They activate the aryl hydrocarbon receptor (AH receptor), although with very different binding affinities, leading to high differences in toxicity and other effects. The dioxin group of compounds includes:

(a) Polychlorinated dibenzo-p-dioxins (PCDDs), or simply dioxins, which are derivatives of dibenzo-p dioxin. There are 75 PCDD congeners, differing in the number and location of chlorine atoms, and 7 of them are specifically toxic, the most toxic being 2,3,7,8-tetrachlorodibenzodioxin (TCDD).
(b) Polychlorinated dibenzofurans (PCDFs), or furans, which are derivatives of dibenzofuran. There are 135 isomers; 10 have dioxin-like properties.
(c) Polychlorinated biphenyls (PCBs), derived from biphenyl, of which 12 are "dioxin-like"; under certain conditions PCBs may form dibenzofurans through partial oxidation.
(d) Polybrominated analogs of the above classes may have similar effects.

Almost all such chemical species, included in the generic term "dioxin", have ED properties, epigenetic effects, and carcinogenic action. They are now included in the IARC classifications of carcinogenic compounds. (Polychlorinated Dibenzo-para-Dioxins and Polychlorinated Dibenzofurans, IARC Monographs on the Evaluation of Carcinogenic Risks to Humans Volume 69, Lyon; 1997).

Since most dioxins are generated by burning plastic waste products, or by burning undifferentiated house wastes, one of the most important measures to reduce dioxin production is the differentiated collection of plastic waste products and their careful management that should follow. In the household, as a mandatory rule, we should all do the following things: buy less plastic (mostly single-use plastics like food wrappings and food packages), carefully collect all sorts of plastic waste products and discard them in separate, differentiated waste collection initiatives. And remember, NEVER BURN plastics (at room temperature and with limited oxygen atmosphere). Only highly technological waste burners (functioning at very high temperatures and at very high oxygen concentrations) may burn plastics with limited development of dioxins. Dioxins are among the most toxic substances ever generated by human activities, since they ARE NOT NATURAL PRODUCTS and have started appearing in the environment only after the 1950s (years of introduction of many varieties of plastic polymers) and with the practice of plastic waste burning, fostered by the "use and discard" practice involving plastic materials and goods (Fig. 3.3). In some in vitro studies, several dioxins were demonstrated to impair glucose uptake by the adipose tissue and the pancreas and to influence negatively the insulin secretion.

TCDD)modifies DNA methylation in gametes both in vivo and in vitro [15]. Adult mice exposed to TCDD reduced glucose-stimulated insulin release.

Human accidental exposure resulting from a chemical plant explosion in Seveso, Italy in 1976, showed an increased risk of diabetes, especially in women. Military personnel's exposure to "orange agent" herbicide in Vietnam war revealed an association between serum TCDD concentrations and the prevalence of type-2 diabetes

Fig. 3.3 Figure 6. The Figure very well illustrates the historical success of the Single Use Plastics (SUP) in the american life, followed by the worldwide diffusion of this pervasive habit. A *Life* magazine cover celebrating "Throwaway Living" in 1955. (From: Rochman, C.M. 2020. The story of plastic pollution: From the distant ocean gyres to the global policy stage. Oceanography 33(3):60–70. CC-BY)

[25]. In the literature are also reported reproductive, developmental effects, and thyroid effects [16].

3.2.4 Parabens

Parabens are chemicals that are widely used as additives in cosmetic and pharmaceutical products. Parabens are esters of para-hydroxybenzoic acid, from which the name is derived. Common parabens include methylparaben (E number E218), ethylparaben (E214), propylparaben (E216), butylparaben and heptylparaben (E209). Parabens are effective preservatives in many types of formulas. These compounds, and their salts, are used primarily for their bactericidal and fungicidal properties. They are found in shampoos, commercial moisturizers, shaving gels, personal lubricants, topical/parenteral pharmaceuticals, sun-tan products, makeup, and toothpaste. They are also used as food preservatives. Parabens are additionally found in pharmaceutical products such as topical treatments for wounds. These treatments help heal wounds by keeping the skin moist and preventing infection. The antimicrobial properties of parabens play a role in the effectiveness of the treatment. This application is helpful for those who have chronic wounds and need to prevent infections as much as possible. Disposal of such items realizes a wide diffusion of parabens in the environment. Parabens are significantly associated with increasing GDM (Gestational Diabetes Mellitus) prevalence in women at high risk [21].

On these compounds, however, more studies are needed and some are indeed in progress.

3.2.5 PFAS

Perfluoroalkyl substances (PFASs), which include PFOA (perfluorooctanoic acid), PFOS (8-carbon perfluoroctane sulfonic acid), PFNA (perfluorononancanoic acid), PFDA (perfluorodecanoic acid), PFHxS (perfluorohexane sulfonic acid), PFOSA (perfluorooctanesulfonamide), Gen X, and many others are a numerous class of man-made chemicals composed of completely fluorinated carbon chains, with different functional head groups. PFASs are widely used in industrial products since 1940s and in everyday use objects, because of their water, oil, stain and heat resistant properties, and they are contained in food wrappers, non-stick cookware, fire-extinguishing foam, water repellent fabrics, painting, personal care products, etc., and in contaminated water and soil, since they are persistent pollutants. Due to the strong carbon-fluorine bond, PFASs are stable and persistent in the environment, such as air dust, water, soil, and in many species of animals, plants, and humans. For such high persistence properties, they are also named "chemicals of eternity". The mean elimination half-lives of serum PFASs in humans are 5.4 years for perfluorooctyl sulfonate (PFOS) and 3.8 years for perfluorooctanoic acid (PFOA) [9]. PFOA is also classified by IARC (International Agency for Research on Cancer) 2B (possibly carcinogenic to human) (Ref. Some Chemicals Used as Solvents and in

Polymer Manufacture. IARC Monographs on the Evaluation of Carcinogenic Risks to Humans Volume 110, Lyon, 2016).

Human exposure to PFAS is thought to be through contaminated water, food, indoor and outdoor air, maternal transfer in utero, and breastfeeding for neonates. PFAS do not accumulate in adipose tissue unlike other contaminants but primarily can be detected in serum, lungs, kidney, liver, and brain [9, 10].

The placenta is a target of PFAS, where the effects include increased time to pregnancy, pregnancy-induced hypertension and preeclampsia, and low birth weight of the infant, which is the most frequently reported adverse outcome of pregnancy in human and animal studies. OAT4 (organic anion transporter 4) is the protein involved in the placental transport of PFAS [10]. Several studies showed in the exposed animals, a reduced fetal and placental weight, and enzyme, hormone, signaling, and apoptosis dysregulation [10]. In humans, cells derived from first-trimester trophoblast (HTR-8/SV neo cells) contain an increased level of microRNAs, have global hypomethylation and hyperacetylation of protein, and an increased production of reactive oxidative species.

PFASs have also been shown to induce biological effects that dysregulate glucose homeostasis in vivo animal studies; exposure to PFOS during gestation, and early postnatal period showed a glucose intolerance and insulin resistance in animal models [11, 25]. There is evidence, through human, animal, and in vitro studies, suggesting that these chemicals disrupt metabolic function by damaging the liver. In the literature are also reported positive associations between exposure to these chemicals and total cholesterol and triglyceride blood levels. In animal studies, PFAS were found to induce expression of hepatic genes involved in lipid synthesis potentially causing steatosis [10].

Another target of PFASs is the thyroid gland and its homeostasis: here they seem to act by reducing the circulating levels of thyroxine (T4) by competitive binding to THT (thyroid hormone transporter) proteins, by increasing T4 metabolism in the thyroid and in liver, by reducing production of T4 or reducing TPO (thyroid peroxidase) activity, and by disrupting the placental capacity to balance fetal thyroid hormone and being positively associated with raising of TSH in pregnant women [10].

PFASs exposure has also been associated with reduced kidney function, compensatory increase in glomerular filtration rate and chronic kidney disease, and consequent increased cardiovascular risk.

In recent studies, PFAS-exposed animals show fertility or fecundity decline, menstrual cycle irregularity and an earlier age at menopause or premature ovarian insufficiency/premature ovarian failure, due to the accelerated follicles loss and endocrine disorder; the serum levels of E2 (estradiol) and P4 (progesterone) were reduced in association with reduction of histone acetylation and serum levels of FSH and LH in exposed animals [11, 12].

In human studies, exposure to PFASs was associated with increased time to pregnancy, poor response to ovarian stimulation, and significantly different in vitro fertilization outcome [11, 12]. Prenatal exposure is of particular relevance and its consequences on childhood obesity were recently the subject of systematic review and meta-analysis [27].

3.2.6 Phtalates

Phthalates, or phthalate esters, are esters of phthalic acid. They are mainly used as plasticizers, i.e., substances added to plastics to increase their flexibility, transparency, durability, and longevity. They are used primarily to soften polyvinyl chloride (PVC) and sometimes found in fragrances included in cosmetics. Note that while phthalates are usually plasticizers, not all plasticizers are phthalates. The two terms are specific and unique and cannot be used interchangeably.

Lower-molecular-weight phthalates are being replaced in many products in many parts of the world (mostly United States and Europe) over health concerns. They are being replaced by higher molecular-weight phthalates as well as non-phthalic plasticizers. Their biological effects as EDs are well documented [13, 15]. Phthalates exposure has a positive correlation with diabetes [13] and metabolic syndrome [30]; it causes glucose metabolism dysregulation, insulin resistance, and adipogenesis. Moreover, prenatal exposure increased preterm birth risk and decreased gestational age [22]. They have a strong antiandrogenic and weak estrogenic effect, which may reduce in males the amount of semen and sperm counts, sperm motility, and often also altered sperm morphology, while in females ovarian cell capacity decline and infertility are observed [15].

3.2.7 Pesticides

Pesticides are a very large and heterogeneous series of substances that are used to control pests. They include herbicides, insecticides, nematicides, fungicides, and many others. The most common of these are herbicides, which account for approximately 50% of all pesticide use globally. Most pesticides are used as plant protection products (also known as crop protection products), which in general protect plants from weeds, fungi, or insects. In general, a pesticide is a chemical or biological agent (such as a virus, bacterium, or fungus) that deters, incapacitates, kills, or otherwise discourages pests. Unfortunately, many pesticides have health-damaging properties, both in humans and in environmental ecological equilibria. Pests include insects, plant pathogens, weeds, mollusks, birds, mammals, fish, nematodes (roundworms), and microbes that destroy property, cause nuisance, or spread disease, or are disease vectors. Along with these benefits, pesticides also have drawbacks, such as potential toxicity to humans and other species.

Several pesticides have been demonstrated to have ED actions. Therefore, this subject is extremely complex and difficult to investigate, since human exposure may derive from several compounds encountered at the same time, each at low to medium concentrations. The effects of such mixtures are far from being known in precise toxicological and scientific terms. Most of the toxicological science describes the effects of single substances at high concentrations, while environmental exposures today imply mostly multiple exposures to different toxicants, each at low to medium concentrations. A general mitigation strategy is to reduce, or better to eliminate, pesticide use, but the issue is very debated and complex. There is only a slow solution to all this, if any.

In in vivo animal studies, atrazin promotes insulin resistance and malathion increases glucose and insulin levels in rats. In in vitro experimental studies, organochlorine pesticides are shown to impair the response to insulin downregulating lipid metabolism genes [25].

CBZ (carbendazim) is known to cause diabetes and hepatic and reproductive toxicity in experimental animals.

Glyphosate is a very widely used herbicide and represents a highly debated subject, with demonstrated effects on human carcinogenicity (lymphomas) and child neurobehavioral implications (autism spectrum disorders, through perturbations of the intestinal microbiome) [14].

Organophosphates exposure has also shown a remarkable association with diabetes prevalence in humans [31] and with androgenic functions [29].

3.2.8 Polychlorinated Biphenyls

Polychlorinated biphenyls (PCBs) are highly carcinogenic chemical compounds, formerly used in industrial and consumer products, whose production was banned in the United States by the Toxic Substances Control Act in 1976 and internationally by the Stockholm Convention on Persistent Organic Pollutants in 2001.

PCBs are organochlorine compounds. They were once widely used in the manufacture of carbonless copy paper, as heat transfer fluids, and as dielectric and coolant fluids for electrical equipment.

Because of their longevity, PCBs are still widely in use, even though their manufacture has declined drastically since the 1960s, when a series of relevant problems was identified. With the discovery of PCBs'environmental toxicity, and classification as persistent organic pollutants, their production was banned by the United States in 1978. Polychlorinated biphenyls are now regulated by the 1978 legislation: "no person may manufacture, process or distribute in commerce or use any PCB in any manner other than in a totally enclosed manner (…)". EPA may also regulate PCB disposal.

The International Agency for Research on Cancer (IARC) classified PCBs as definite carcinogens in humans. According to the U.S. Environmental Protection Agency (EPA), PCBs cause cancer in animals and are probable human carcinogens. Many rivers and buildings, including schools, parks, and other sites, were contaminated with PCBs and there has also been contamination of food supplies with these substances. Moreover, because of their use as a coolant in electric transformers, PCBs still persist in built environments.

Some PCBs share a structural similarity and toxic mode of action with dioxins. Other toxic effects such as endocrine disruption (notably blocks of thyroid system) and neurotoxicity are known. The bromine analogues of PCBs are polybrominated biphenyls (PBBs), which have analogous applications and environmental concerns.

An estimated 1.2 million tons have been produced globally. Though the EPA enforced the federal ban as of 1978, PCBs continued to create health problems in later years through their continued presence in soil and sediment, and from products which were made before 1979. In 1988, Japanese scientists estimated that 370,000 tons were in the environment globally, and 780,000 tons were present in products, landfills, and dumps or kept in storage.

As for endocrine-disrupting effects, PCBs are associated with diabetes [25], with female fertility showing a reduced ovarian reserve and a declined fertility as increased time to pregnancy, poor response to ovarian stimulation, altered cycle length, and bleeding [16]. Moreover, mention is deserved for associations with gestational diabetes mellitus (GDM) [26] and with diabetes [31].

3.2.9 Triclosan

Triclosan is used in industry as a disinfectant and antiseptic agent and is contained in personal care products, toothpaste, toys, bedding, and clothes. It's an agent disrupting thyroid function, and reproductive and developmental function [11]. More evidences are needed indeed for these compounds.

3.3 Microplastics and Other ED Compounds

Plastic materials have allowed extraordinary progresses in every area of human activities and also in medicine. Now, however, we have evidence that plastics have many drawbacks and are dangerous for environmental and human health [7, 8, 20]. Plastics are subject to mincing and crumbling operated by the physical forces in the environment (UV rays, ice, etc.) and are therefore reduced in small pieces (*microplastics* —of size between 5 mm to 1 nanometer [nm]—and *nanoplastics* —size below 1 nm) that are easily dispersed in the waters and in the atmosphere, creating a planetary problem. Air and water pollutants at the molecular level can adhere to plastic fragments, concentrate onto them, and being further dispersed in the environment. The finding of microplastics and nanoplastics in human blood and tissues has been ascertained (Fig. 3.4).

The wide health consequences of that are under extensive investigation [20]. We have already mentioned this issue (see above, paragraph on Dioxins) also in relation to the problem of plastic waste collection and recycling. The issue, however, is so enormous and pervasive that microplastic particles have been found everywhere in the human body: lungs, liver, blood, heart, and (very recently) also in the brain [32].

Some of the toxicants illustrated in this chapter (and summarized in Table 3.1) are derived from industrial production of plastics and from their degradation in biological systems. Therefore, this subject constitutes a very prevalent environmental and health concern. Plastic pollution is everywhere and microplastics can be found extensively in the human and animal bodies, not to mention all other forms of life both on land and in water.

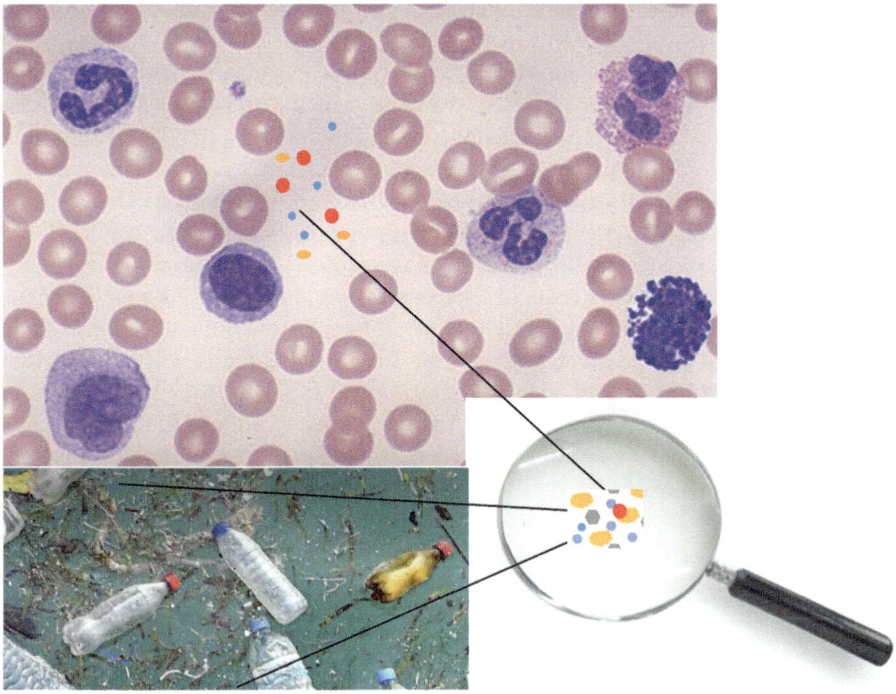

Fig. 3.4 Plastics in the body. In the blood and in many internal organs, particles of micro and nano-plastics can be found (color dots), detectable and quantifiable through specific analytical methods (magnifying lens). These particles are introduced in the body through environmental exposures (water, food, air inspired) and reach all the internal organs as already evident in many recent studies (see text)

Extensive research is being pursued and results are emerging constantly. Health issues and ecological issues are involved, as it appears already evident, in many observations.

3.4 Conclusion

The environmental problem involving EDs intermingles with the widespread presence of micro- and nano-plastics and results to be of huge proportions. The spectrum of diseases involved is also very wide and relevant, spanning from endocrine perturbations to immunological perturbations to metabolic consequences such as obesity and diabetes and the related events involving the cardiovascular and neoplastic diseases. Further concerns are regarding developmental effects in humans and in wildlife [19]. However, it is not always possible to determine the single effect of each chemical compound when people are often nowadays exposed to a mixture of chemicals deriving from several sources, which may have an additive action, together with genetic predisposition and factors such as sedentariness and overnutrition.

Some epigenetic modifications may be reversed by nutrients (nutritional epigenetics), which may either alter the availability of the substrates for enzymatic reactions that catalyze DNA or histone modifications or directly inhibit these enzymes. Nutritional epigenetics knowledge is limited since nutrients interact with other nutrients, genes, lifestyle factors, etc. Therefore, it is hard to identify each nutrient's individual and precise effect on every epigenetic modification.

Several major endocrine disruptors are plasticizers: avoiding or minimizing their usage will be very helpful in reducing epigenetic modulations. Saying no to plastics may improve health. Finding safe alternatives for plastic containers, cans, toothbrushes, plastic bags, etc., and staying away from canned foods will help to evade endocrine disruptions in our bodies. Phthalates are typically found in fragrances, and staying away from them minimizes the toxic effects of these EDCs. Pesticides are EDCs that can cause hormone disruption. Switching to organic food is one way of avoiding the intake of toxic pesticides.

Water pollution by toxic chemicals is widespread, and soil and groundwater contamination follow. Even with successive purification, water will contain trace amounts of many EDCs: therefore, filtering tap water through effective water filters can decrease EDC intake/exposure.

The cosmetic industry uses largely these dangerous chemicals for the manufacturing of several products, through compounds that can make the items smell good and look irresistible, but many toxic chemicals hide behind them. Switching to home-grown, organic brands is a cheaper and better alternative. Avoiding the usage of all these products lessens our exposure to EDCs and thus minimizes the risk of epigenetic modulations caused by them.

This brief and very partial discussion of the main issues of this very wide subject and of the related scientific evidences may invite the interested readers to go deeper into the bibliography and to extend the current investigations to seek for further and wider environmental and human data.

The essential meaning of all this for the readers of this book and of this chapter in particular, mainly physicians and health professionals, is to guide their patients and families to an effective cooperation in the adoption of healthy life behaviors and in controlling essential steps of the problem, like purchase of goods (with minimal packaging disposal), waste production and waste management with better knowledge and the most accurate procedures. The environment and human health may both improve, this way.

References

1. Waddington CH (1940) Organisers and Genes. Cambridge University Press, Cambridge
2. Van Speybroeck L (2002) From epigenesis to epigenetics, the case of C, H. Waddington. Ann NY Acad Sci 981:61–81

3. Colborn T et al (2003) (2003). Developmental effects of endocrine-disrupting chemicals in wildlife and humans. Environ Health Perspect 101:378–384. https://doi.org/10.1289/ehp.93101378
4. La Merrill MA et al (2022) Consensus on the key characteristics of endocrine-disrupting chemicals as a basis for hazard identification. Nature Rev Endocrinol 16:45–xx. https://doi.org/10.1038/s41574-019-0273-8
5. Zhou Z et al (2020) Mitochondrial epigenetics and environmental health: making a case for endocrine disrupting chemicals. Toxicol Sci 178(1):16–25. https://doi.org/10.1093/toxsci/kfaa129
6. Cheng Y et al (2024) Alterations in sperm DNA methylation may as a mediator of paternal air pollution exposure and offspring birth outcomes: insight from a birth cohort study. Environ Res:119941. https://doi.org/10.1016/j.envres.2023.117941
7. MacKay H, Abizaid A (2018) A plurality of molecular targets: the receptor ecosystem for bisphenol-a (BPA). Horm Behav 101:59–67. https://doi.org/10.1016/j.yhbeh.2017.11.001
8. Landrigan P (2024) Plastics, fossil carbon, and the heart. N Engl J Med. https://doi.org/10.1056/NEJMe2400683
9. Chen-Yu L et al (2018) Prenatal Perfluorooctyl sulfonate exposure and Alu DNA Hypomethylation in cord blood. Int J Environ Res Public Health 15:1066. https://doi.org/10.3390/ijerph15061066
10. Blake BE, Fenton SE (2020) Early life exposure to per- and polyfluoroalkyl substances (PFAS) and latent health outcomes: a review including the placenta as a target tissue and possible driver of peri- and postnatal effects. Toxicology 443:152565. https://doi.org/10.1016/j.tox.2020.152565
11. Kirtana A, Seetharaman B (2022) Comprehending the role of endocrine disruptors in inducing epigenetic toxicity. Endocr Metab Immune Disord Drug Targets 22:1059–1072. https://doi.org/10.2174/1871530322666220411082656
12. Thorson JLM, Skinner MK (2023) Endocrine disrupting chemicals induce the epigenetic transgenerational inheritance of disease. In: Carnevali O, Hardiman G (eds) Chapter 2.4 in environmental contaminants and endocrine health. Academic Press. https://doi.org/10.1016/B978-0-12-824464-7.00020-9
13. Chevalier N, Fénichel P (2015) Endocrine disruptors: new players in the pathophysiology of type 2 diabetes? Diabetes Metab 41(2015):107–115. https://doi.org/10.1016/j.diabet.2014.09.005
14. Pu Y et al (2020) Maternal glyphosate exposure causes autism-like behaviors in offspring through increased expression of soluble epoxide hydrolase. PNAS 117(21):11753–11759. https://doi.org/10.1073/pnas.1922287117
15. Shi Y et al (2021) The role of epigenetic in the reproductive toxicity of environmental endocrine disruptors. Environ Mol Mutagen 62:78–88. https://doi.org/10.1002/em.22414
16. Ding T et al (2022) Endocrine disrupting chemicals impact on ovarian aging: evidence from epidemiological and experimental evidence. Environ Pollut 305:119269. https://doi.org/10.1016/j.envpol.2022.119269
17. Streiter M, Gore AC (2021) Epigenetics, estrogenic endocrine-disrupting chemicals (EDCs), and the brain. Adv Pharmacol 92:73–99. https://doi.org/10.1016/bs.apha.2021.03.006
18. Deng X et al (2024) Adverse effects of bisphenol a and its analogues on male fertility: an epigenetic perspective. Environ Pollut 345:15. https://doi.org/10.1016/j.envpol.2024.123393
19. Colborn T et al (1993) Developmental effects of endocrine-disrupting Chemicals in Wildlife and Humans. Environ Health Perspect 101:378–384
20. Leslie HA et al (2022) Discovery and quantifications of plastic particle pollution in human blood. Environ Int 163:107199. https://doi.org/10.1016/j.envint.2022.107199
21. Kunisz M et al (2021) Epigenetic modifications associated with exposure to endocrine disrupting Chemicals in Patients with gestational diabetes mellitus. Int J Molecul Science 22:4693. https://doi.org/10.3390/ijms22094696

22. Trasande L et al (2024) Prenatal phthalate exposure and adverse birth outcomes in the USA: a prospective analysis of births and estimates of attributable burden and costs. www.thelancet.com/planetary-health. Lancet Planet Health 8:e74–e85
23. Besaratinia A (2023) The state of research and weight of evidence on the epigenetic effects of bisphenol a. Int J Mol Sci 24:7951. https://doi.org/10.3390/ijms24097951
24. Castellini C et al (2022) Association between urinary bisphenol a concentrations and semen quality: a meta-analytic study. Biochem Pharmacol 197:114896. https://doi.org/10.1016/j.bcp.2021.114896
25. Fénichel P, Chevalier N (2017) Environmental endocrine disruptors: New diabetogens? C R Biol 340(2017):446–452. https://doi.org/10.1016/j.crvi.2017.07.003
26. Yan D, Jiao Y, Yan H, Liu T, Yan H, Yuan J (2022) Endocrine-disrupting chemicals and the risk of gestational diabetes mellitus: a systematic review and meta-analysis. Environ Health 21(1):53. https://doi.org/10.1186/s12940-022-00858-8
27. Frigerio G, Ferrari CM (2023) Fustinoni: Prenatal and childhood exposure to per-/polyfluoroalkyl substances (PFASs) and its associations with childhood overweight and/or obesity: a systematic review with meta-analyses. S Environ Health 22(1):56. https://doi.org/10.1186/s12940-023-01006-6
28. Srnovršnik T, Virant-Klun I, Pinter B (2023) Polycystic ovary syndrome and endocrine disruptors (Bisphenols, Parabens, and Triclosan). Sys Rev Life (Basel) 13(1):138. https://doi.org/10.3390/life13010138
29. Hamed MA, Akhigbe TM, Adeogun AE, Adesoye OB, Akhigbe RE (2023) Impact of organophosphate pesticides exposure on human semen parameters and testosterone: a systematic review and meta-analysis. Front Endocrinol (Lausanne) 14:1227836. https://doi.org/10.3389/fendo.2023.1227836. eCollection 2023
30. Mérida DM, Moreno-Franco B, Marquès M, León-Latre M, Laclaustra M, Guallar-Castillón P (2023) Phthalate exposure and the metabolic syndrome: A systematic review and meta-analysis. Environ Pollut 333:121957. https://doi.org/10.1016/j.envpol.2023.121957. Epub 2023 Jun 14
31. Evangelou E, Ntritsos G, Chondrogiorgi M, Kavvoura FK, Hernández AF, Ntzani EE, Tzoulaki I (2016) Exposure to pesticides and diabetes: a systematic review and meta-analysis. Environ Int 91:60–68. https://doi.org/10.1016/j.envint.2016.02.013. Epub 2016 Feb 22
32. Amato-Lourenco LF et al (2024) Microplastics in the olfactory bulb of the human brain. JAMA Netw Open 7(9):e2440018. https://doi.org/10.1001/jamanetworkopen.2024.40018

4 New Insights into the Fetal Programming Hypothesis of the Linkage Between Neurodevelopmental Disorders and Impaired Neurogenesis

G. Faa, P. Paribello, M. Fraschini, M. Castagnola, and M. Manchia

Contents

4.1 Introduction 48
4.2 Neurogenesis in the Human Brain 48
4.3 Molecular Pathways Involved in Neurogenesis and Brain Development 50
4.4 Fetal Programming of Neurodevelopmental Disorders 52
4.5 Conclusions 54
References 54

G. Faa (✉)
Department of Medical Sciences and Public Health, University of Cagliari, Cagliari, Italy

Department of Biology, College of Science and Technology, Temple University, Philadelphia, PA, USA

P. Paribello
Unit of Psychiatry, Department of Medical Sciences and Public Health, University of Cagliari, Cagliari, Italy

Unit of Clinical Psychiatry, University Hospital Agency of Cagliari, Cagliari, Italy

M. Fraschini
Department of Electrical and Electronic Engineering, University of Cagliari, Cagliari, Italy

M. Castagnola
Proteomics Laboratory, European Center for Brain Research, (IRCCS) Santa Lucia Foundation, Rome, Italy

M. Manchia
Unit of Psychiatry, Department of Medical Sciences and Public Health, University of Cagliari, Cagliari, Italy

Unit of Clinical Psychiatry, University Hospital Agency of Cagliari, Cagliari, Italy

Department of Pharmacology, Dalhousie University, Halifax, NS, Canada

L. Emmi, V. Fanos (eds.), *Epigenetics and Rare Diseases*, Rare Diseases of the Immune System, https://doi.org/10.1007/978-3-032-02902-7_4

4.1 Introduction

The development of the central nervous system (CNS) is a very complex event occurring during embryogenesis, which leads to the complex organization of the human brain and, specifically, of the cerebral cortex, whose composite architecture distinguishes humans from other species. The term neurodevelopment refers to the development of multiple neurological molecular pathways that are at the basis of different important cognitive functions, including intelligence, language, consciousness, motor abilities, memory, and sensory perceptions [1]. The acquisition of intellectual functioning, reading and listening comprehension abilities, social skills, memory, and attention is strictly related to optimal prenatal and postnatal neurodevelopment [2]. The objective of physiological neurodevelopment is to achieve the highest levels of neuropsychological performance for each individual.

One of the most relevant challenges for researchers in neuroscience is the identification of the embryonic origin of the different types of cortical neurons that characterize the adult neocortex of human beings. The main goal for neurogenesis researchers is to discover the molecular pathways involved in the emergence of all the cortical neuronal subtypes. The identification of neural stem/progenitor cells in the subventricular zone of the lateral ventricles of the adult brain represents a new challenge for researchers involved in the study of neurogenesis. Indeed, studies on embryology may help elucidate the mechanisms that maintain neural stem cells in adulthood, and that regulate their mobilization under pathological conditions [3].

In this chapter, we will describe the most important phases of human neurogenesis and brain development, focusing on the role played by neural stem/progenitor cells in brain development during intrauterine life. In the second part, according to the Barker hypothesis of fetal programming of adult diseases, we will discuss the factors which may interfere with physiological neurodevelopment during gestation, causing a derangement in neurogenesis, with important influences in neuropsychological performances later in life [4].

4.2 Neurogenesis in the Human Brain

The development of appropriate neuronal and glial cell diversity in the embryonic brain represents a major question for researchers involved in neuroscience. During development, apical radial glial progenitors represent the most important neural stem cells that, following a well-orchestrated developmental program, originate neurons and glial cells of the mature brain. The search for the molecular mechanisms at the basis of this program, which regulates a radial glial progenitor progression, is the focus of multiple recent research programs in neuroscience [5].

In the developing human brain, neurogenesis is regulated by a complex interplay that controls stem/progenitor cell survival, renewal, and differentiation into multiple cell types that characterize the mature brain tissue. Neurogenesis occurs in two main neurogenic areas: i) the subventricular zone (SVZ), localized in the areas lining the lateral ventricles of the embryonic telencephalon [6]; ii) the subgranular zone (SGZ), localized in the dentate gyrus of the hippocampus [7]. Neurogenesis continues throughout childhood and adulthood in these two specialized cerebral zones, underlying, at the cell level, the peculiar plasticity of the human brain [8].

In this context, the cerebrospinal fluid (CSF) should be considered a member of the neurogenic niche, being involved in regulating the proliferation and migration of the stem/progenitors of the human brain. The contact between CSF and neural progenitors is mediated by neural stem cell protrusions, which extend through the ependymal layer into the CSF of the lateral ventricles. During embryogenesis, CSF increases the neurogenic activity of stem/progenitor cells localized in the subventricular zone [9].

Neural stem/progenitor cells of the subependymal zone of the lateral ventricles show three main characteristics: (1) self-maintenance for long periods, probably lifelong; (2) high regenerative capability; (3) ability to differentiate into multiple specific cell lineages, some of which migrate toward the subpial region. In recent years, the apical radial glia (aRG) cells have been identified as the neural stem/progenitor cells, being able to give rise to all the neuronal subtypes of the six-layered human cerebral cortex, as well as to glial cells [10].

Neurons and glial cells represent the principal differentiated cell types of the human CNS. Neurons are deputed to transmit information to neighboring neurons through action potentials and neurotransmitters. Glial cells provide critical support for optimal neuronal function and survival. Neural stem cells of the subventricular zone may undergo both symmetrical and asymmetrical division. The former underlies their ability to self-renew and expand the stem cell pool of the subventricular zone. The asymmetric division produces one neural stem cell and one neural progenitor cell, characterized by a limited capacity of self-renewal but with the ability to differentiate toward neurons or glial cells.

Apical radial glial cells are bipolar-shaped cells that play a crucial role in the development of the cerebral cortex of the human embryo, being able to differentiate into neuronal and glial cells [11]. These cells are characterized by long cytoplasmic extensions extending from the periventricular zone toward the brain cortex, through which they maintain contact with both the ventricular and the pial surfaces [12]. These cytoplasmic extensions are not evidenced by routine stains, such as hematoxylin-eosin, but they are well evidenced by immunostaining for Wilm's tumor-1 (WT1), appearing as long lines extending from the ventricular to the pial zone and arranged in a parallel pattern [13].

Of great interest is also the genomic regulation of the development of the CNS. The seminal work of Edoardo Boncinelli showed that specific conserved DNA sequences of 183 nucleotides (homoeo box) are expressed during embryogenesis in humans [14]. Subsequently, a series of homeobox genes were identified in the spinal cord, backbone rudiments, limb buds (or limbs), heart, and skin of human embryos and early fetuses in the fifth- to ninth-week post-fertilization period, indicating their crucial role in neurodevelopment [15]. In addition, the analysis of mouse Hox mutants has clearly demonstrated that Homeobox genes have important roles in normal regionalization of the hindbrain and branchial arches, and this has led to interest in how their early patterns are established in the CNS [16]. Of great interest is the finding that Hox genes Emx 1 and 2 are key for brain development. For instance, Emx2 homozygous mice embryos are not vial due to severe urogenital alterations and show cerebral hemispheres with a reduced size and specific morphological alterations in allocortical structures of the medial wall of the brain [17]. For instance, the dentate gyrus is missing and the hippocampus proper is reduced [17]. Finally, the analysis of the phenotype of Emx2 null embryos has shown that this transcription factor plays important roles in neuroblast proliferation, migration and differentiation, as well as in the development of the diencephalon, where it has been shown to cooperate with another homeobox gene (Otx2) [18].

Concerning embryological development, multiple sequential processes occur in the neuroepithelial cell progenitors, namely, i) cell proliferation, ii) migration, and iii) differentiation. The cortical neurogenesis is characterized by the persistence of the newly formed neuronal progenitors in the ventricular and subventricular zones, which represent the proliferative zone of the developing cortex. Progenitor cells differentiating into neurons migrate radially toward the pial zone of the developing neocortex, following the ways indicated by the extensions of radial glial cells.

In humans, at six weeks of gestation, in the thin developing cerebral cortex, two layers may be identified: (i) a thick proliferative periventricular zone, formed by densely packed undifferentiated small stem/progenitor cells; (ii) a marginal zone with few scattered cells. At the gestational age of 7 weeks, a third zone develops between the previous zones: the subventricular zone. At gestational age 12, the following zones may be identified in the developing cortex: (a) the ventricular zone (VZ) formed by undifferentiated stem/progenitors; (b) the subventricular zone (SVZ), formed by 10–15 layers of proliferating neuronal and glial precursors; (c) the intermediate zone (IZ), formed by a loose fibrillary network embedding scattered migrating cells; (d) the subplate zone (SPZ); (e) the cortical plate (CP), a strip of post-mitotic neurons which will originate the mature brain cortex; (f) the pial zone, a thin area intermingled between the cortical plate and the pia, formed by a loose fibril network [13].

The cortical neurogenesis is completed in about 108 embryonic days. In the developing nervous system, particularly in the developing brain cortex, radial glial cells play a fundamental role in forming the cortical layers that characterize the mammalian brain. During development, newborn neurons are "guided" by radial glia during their migration from the SBZ toward the subpial zone [19]. Recently, these cells, or closely related cell types called radial glia-like cells, have been described in the adult CNS, functioning as neurogenic progenitors in normal homeostasis and in response to injury [20].

4.3 Molecular Pathways Involved in Neurogenesis and Brain Development

In the human embryo, neural stem/progenitor cells lining the lateral ventricles must adapt their transcriptional activity to the changing intrauterine environment, which is influenced by maternal diet and, generally, by maternal life conditions [2]. Neural stem cells, called apical radial glial (aRG) progenitors, expand the progenitor pool through symmetric self-renewing divisions and switch to asymmetric divisions, generating specific neuronal subtypes. Multiple molecular signaling complexes may influence aRG proliferative behavior, favoring the self-renewing symmetrical division or a neurogenic asymmetric division within the cortical proliferative zone.

The differentiation of neural stem cells in the subventricular zone is under the control of multiple genes, as well as of numerous external epigenetic factors, including fibroblast growth factor (FGF) [21], epidermal growth factor (EGF) [22], Wnt [23], and Sonic Hedgehog [24]. More recently, other epigenetic mechanisms, including DNA methylation, chromatin remodeling, and noncoding RNAs (ncRNAs), have been found to play critical regulatory roles in human neurogenesis. The dynamic nature of epigenetic mechanisms represents a very important field of research, given their role in gene regulation, and particularly in the control of the multiple steps of neurogenesis in response to environmental signals [25, 26].

Neural stem cells must be able to adapt rapidly to their transcriptional activity according to the changing intrauterine environment. The highly apical-basally polarized radial glial morphology is a prerequisite for cortical development, regulating neural migration during neurogenesis. Cadherin-based adherens junctions play a crucial role in the maintenance of radial glial apicobasal polarity [27].

Sonic hedgehog (Shh) and Notch signaling have been indicated as crucial factors in regulating neurogenic divisions of apical radial glial cells, representing the principal neocortical progenitor cell. Indeed, Shh may activate progenitor cells of the ventricular zone, inducing upregulation of Blbp and Hes1, two downstream targets of the Notch signaling pathway. The activation by Shh of Notch signaling pathways plays a key role in the maintenance of neural stem/progenitors, favoring the symmetric divisions of apical radial glial cells [28]. The interplay between Shh and Notch signaling pathways on neural progenitor cells plays a key role in neurogenesis, participating in the correct balance between asymmetric and symmetric divisions of the neural stem/progenitor cells during neurogenesis [29].

Another novel mechanism involved in controlling the balance between asymmetric and symmetric divisions of radial glial progenitors in the developing brain cortex is regulated by two kinesins, Kif13 B and Kif1A, which play opposing roles in neurogenesis. Kif13B promotes non-neurogenic symmetric division, whereas Kif1A promotes asymmetric division and neurogenesis [30].

The up- and downregulation of the paired box transcription factor Pax6 has been recently indicated as a key factor in the regulation of the transcriptional activity of neural stem/progenitor cells during development [31]. Pax6 plays a relevant role in the patterning of the developing cortex, regulating the expansion of the neural stem cell pool and the differentiation of neuronal precursors into the neuronal and glial populations [32]. Recently, a molecular pathway has been identified in the negative regulation of Pax6 activity. The histone acetyltransferase Kat2a mediates lysine acetylation on Pax6, leading to its ubiquitination, followed by its degradation via the proteasome [33]. The final member of the molecular pathway leading to Pax6 degradation is represented by the E3 ligase Ring finger protein 8 (Rnf8), the ubiquitin ligase responsible for targeting Pax6 for proteasome degradation.

An intriguing question arises from the discovery of these molecular pathways involved in neurogenesis: how many other transcription factors are involved in the Kat2a-Rfn8 degradation pathway? And does this pathway belong to a more complex mechanism through which all the transcription factors involved in neurogenesis may be up- or downregulated in response to changing environmental conditions during gestation?

4.4 Fetal Programming of Neurodevelopmental Disorders

The theory of fetal programming of diseases presenting later in life, in childhood or adulthood, was first proposed by Barker in the previous century [34]. According to this hypothesis, poor maternal living conditions in the prenatal period may represent a risk factor for multiple diseases. First related to atherosclerosis, over the years, this hypothesis has become the focus of various studies aimed at investigating the link between changes in neurogenesis during early brain development and the onset of neurodegenerative disorders later in life [35–37].

Neurodevelopmental disorders are characterized by developmental deficits or delays, usually manifest in early childhood and may persist throughout the individual's lifetime. Neurodevelopmental disorders include autism spectrum disorders (ASD), attention deficit hyperactivity disorder (ADHD), intellectual developmental disorder, communication disorder, motor disorders, and specific learning disorders [38]. The role of fetal programming has been highlighted in a study on the influence of modified neurogenesis and the development of ASD [39]. Evidence deriving from the field of neuroimaging genetics appears particularly explicatory in demonstrating the close relationship between anatomical development and underlying genetic architecture for neurodevelopment itself. Such an approach helps to identify and elucidate biological mechanisms and pathways that may relate to the underlying genetic risk. Through the integration of genetic and neuroimaging data with a sophisticated machine-learning approach, it may aid clinicians in substantiating a diagnosis of complex neurodevelopmental disorders, such as ASD. Indeed, in a subset of children with ASD, abnormal neurogenesis has been reported to be responsible for an abnormal increase in the number of neurons in the prefrontal cortex, probably due to an imbalance between symmetrical and asymmetrical divisions of the neural stem cells during brain cortex development [39]. Recently, de novo mutations in the transcriptional co-factor Zmiz1 or its regulatory regions have been identified in unrelated patients with syndromic intellectual disabilities and ASD [40]. Importantly, Zmiz1 is highly expressed during embryonic brain development in mice and humans, particularly in areas prominently impacted by intellectual disability and ASD, such as the cortex, hippocampus, and cerebellum [40]. Approximately 25% of individuals with constitutional mutations in the PTEN gene satisfy the criteria for ASD. The PTEN gene is involved in cell growth and survival regulation and specific mutations such as the D252G and H93R have been associated with increased head circumference and macrocephaly in ASD. Mutations in the PTEN, SHANK3, and SYNGAP1 genes have all been associated with white matter

(WM), grey matter (GM), and ventricular anomalies. Further, the CNTNAP2 gene has been reported to be associated with WM and GM anomalies and is clinically associated with speech-language delay. The specific variant rs2538991 in this gene has also been reported with WM anomalies located in the right anterior cingulate gyrus in ASD. Anecdotally, de novo Chromodomain Helicase DNA-binding Protein 8 (CHD8) mutations have also been reported in ASD subjects featuring intellectual disability, macrocephaly, and craniofacial anomalies. MECP2 mutations appear to be associated with impaired neuronal development, possibly leading to ASD, while SHANK3 mutations appear to be associated with reduced brain volume and altered connectivity, impacting social and communication behaviors. Neurexins and synaptic function neurexins are essential for synapse formation. Mutations in neurexin genes such as NRXN1-α and NRXN2-α have been associated with anomalies in neuronal action potential maturation and with altered brain microstructures. This affects social behavior and structural connectivity in brain regions associated with social processing. SNPs for CD38 and Oxytocin Regulation CD38, a transmembrane protein involved in oxytocin secretion, have also been described in the context of ASD. Lower plasma oxytocin levels are observed in individuals with the R140W allele, and treatment with intranasal oxytocin has been postulated to impact social behaviors. Additionally, oxytocin receptor (OXTR) gene polymorphisms influence social behavior and brain activation patterns during emotion recognition tasks, further lending support for their role in influencing ASD phenotype. Arginine Vasopressin Receptor 1A (AVPR1A) gene variants are linked to differential amygdala activation and verbal memory performance. Certain lines of evidence suggest that certain AVPR1A polymorphisms may be more frequently observed in families with ASD individuals and have been hypothesized to affect social behavior and contribute to ASD pathophysiology. Further, Zmiz1 regulates multiple developmental processes, including neurogenesis, neuron connectivity, and synaptic signaling [40]. In another study [41], Le Duc et al. used exome and genome sequencing as well as array analysis of 13 individuals that identified pathogenic, heterozygous, mostly de novo variants in the WDFY3 gene as a monogenic cause of mild and non-specific neurodevelopmental delay. The WDFY3 gene encodes for the autophagy-linked FYVE protein that is implicated in the selective degradation of ubiquitinated protein aggregates via autophagy [42]. This gene has previously been implicated in cognitive function and in humans WDFY3 loss-of-function variants lead to macrocephaly via downregulation of the Wnt pathway [41]. Another study explored the role of BRINP1, a gene that appears to be a key regulator of the physiological functions of the CNS. Using BRINP1-deficient (KO) mice, the authors showed that neurogenesis in the subgranular zone of dentate gyrus was increased in BRINP1-KO mice, creating a more immature neuronal population in the granule cell layer [43]. Interestingly, BRINP1-KO mice showed abnormal behaviors with an increase in locomotor activity, reduced anxiety-like behavior, poor social interaction, and slight impairment of working memory, all of which resemble symptoms of human psychiatric disorders such as ADHD [43].

Finally, another study explored the role of loss-of-function mutations in the UPF3B gene resulting in variable clinical presentations including intellectual disability (ID, syndromic and non-syndromic), autism, childhood-onset schizophrenia, and attention deficit hyperactivity disorder [44]. UPF3B is a core member of the nonsense-mediated mRNA decay (NMD) pathway that functions to rapidly degrade transcripts with premature termination codons (PTCs). Thus, UPF3B mutations appeared to partly explain the clinical phenotype of UPF3B patients, consisting of developmental delays including intellectual disability [44].

4.5 Conclusions

In this chapter, we have recapitulated the biological underpinnings of neurodevelopment attempting to correlate genomic variation and alterations of embryological events with clinical presentations. Data consistently show that biological regulatory systems finely tune the development of the CNS during embryogenesis and fetal growth. The presence of specific mutations or of alterations in embryogenesis might predispose to the development of neurodevelopmental disorders but probably also of psychiatric disorders later in life.

References

1. Ouhaz Z, Fleming H, Mitchell AS (2018) Cognitive functions and neurodevelopmental disorders involving the prefrontal cortex and Mediodorsal thalamus. Front Neurosci 12:33
2. Cohen Kadosh K, Muhardi L, Parikh P, Basso M, Jan Mohamed HJ, Prawitasari T et al (2021) Nutritional support of neurodevelopment and cognitive function in infants and young children-an update and novel insights. Nutrients 13(1):199
3. Meng H, Huan Y, Zhang K, Yi X, Meng X, Kang E et al (2024) Quiescent adult neural stem cells: developmental origin and regulatory mechanisms. Neurosci Bull 40:1353
4. Banik A, Kandilya D, Ramya S, Stünkel W, Chong YS, Dheen ST (2017) Maternal factors that induce epigenetic changes contribute to neurological disorders in offspring. Genes 8(6):150
5. Hippenmeyer S (2023) Principles of neural stem cell lineage progression: insights from developing cerebral cortex. Curr Opin Neurobiol 79:102695
6. de Sonnaville SFAM, van Strien ME, Middeldorp J, Sluijs JA, van den Berge SA, Moeton M et al (2020) The adult human subventricular zone: partial ependymal coverage and proliferative capacity of cerebrospinal fluid. Brain Commun 2(2):fcaa150
7. Eriksson PS, Perfilieva E, Björk-Eriksson T, Alborn AM, Nordborg C, Peterson DA et al (1998) Neurogenesis in the adult human hippocampus. Nat Med 4(11):1313–1317
8. Alshebib Y, Hori T, Goel A, Fauzi AA, Kashiwagi T (2023) Adult human neurogenesis: a view from two schools of thought. IBRO Neurosci Rep 15:342–347
9. Alonso MI, Lamus F, Carnicero E, Moro JA, de la Mano A, Fernández JMF et al (2017) Embryonic cerebrospinal fluid increases neurogenic activity in the brain ventricular-subventricular zone of adult mice. Front Neuroanat 11:124
10. Shohayeb B, Muzar Z, Cooper HM (2021) Conservation of neural progenitor identity and the emergence of neocortical neuronal diversity. Semin Cell Dev Biol 118:4–13
11. Campbell K, Götz M (2002) Radial glia: multi-purpose cells for vertebrate brain development. Trends Neurosci 25(5):235–238
12. Malatesta P, Appolloni I, Calzolari F (2008) Radial glia and neural stem cells. Cell Tissue Res 331(1):165–178

13. Vinci L, Ravarino A, Gerosa C, Pintus MC, Marcialis MA, Marinelli V et al (2016) Stem/progenitor cells in the cerebral cortex of the human preterm: a resource for an endogenous regenerative neuronal medicine? J Pediatr Neonatal Individ Med JPNIM 5(1):e050121
14. Simeone A, Mavilio F, Bottero L, Giampaolo A, Russo G, Faiella A et al (1986) A human homoeo box gene specifically expressed in spinal cord during embryonic development. Nature 320(6064):763–765
15. Simeone A, Mavilio F, Acampora D, Giampaolo A, Faiella A, Zappavigna V et al (1987) Two human homeobox genes, c1 and c8: structure analysis and expression in embryonic development. Proc Natl Acad Sci USA 84(14):4914–4918
16. Krumlauf R, Marshall H, Studer M, Nonchev S, Sham MH, Lumsden A (1993) Hox homeobox genes and regionalisation of the nervous system. J Neurobiol 24(10):1328–1340
17. Pellegrini M, Mansouri A, Simeone A, Boncinelli E, Gruss P (1996) Dentate gyrus formation requires Emx2. Dev Camb Engl 122(12):3893–3898
18. Cecchi C (2002) Emx2: a gene responsible for cortical development, regionalization and area specification. Gene 291(1–2):1–9
19. Rakic P (2009) Evolution of the neocortex: a perspective from developmental biology. Nat Rev Neurosci 10(10):724–735
20. Miranda-Negrón Y, García-Arrarás JE (2022) Radial glia and radial glia-like cells: their role in neurogenesis and regeneration. Front Neurosci 16:1006037
21. Gremo F, Presta M (2000) Role of fibroblast growth factor-2 in human brain: a focus on development. Int J Dev Neurosci Off J Int Soc Dev Neurosci 18(2–3):271–279
22. Garcez RC, Teixeira BL, dos Santos Schmitt S, Alvarez-Silva M, Trentin AG (2009) Epidermal growth factor (EGF) promotes the in vitro differentiation of neural crest cells to neurons and melanocytes. Cell Mol Neurobiol 29(8):1087–1091
23. Kalani MYS, Cheshier SH, Cord BJ, Bababeygy SR, Vogel H, Weissman IL et al (2008) Wnt-mediated self-renewal of neural stem/progenitor cells. Proc Natl Acad Sci USA 105(44):16970–16975
24. Ho KS, Scott MP (2002) Sonic hedgehog in the nervous system: functions, modifications and mechanisms. Curr Opin Neurobiol 12(1):57–63
25. Hsieh J, Zhao X (2016) Genetics and epigenetics in adult neurogenesis. Cold Spring Harb Perspect Biol 8(6):a018911
26. Jakovcevski M, Zimmer-Bensch G (2022) Epigenetic function in neurodevelopment and cognitive impairment. e-Neuroforum 28(1):41–53
27. Veeraval L, O'Leary CJ, Cooper HM (2020) Adherens junctions: guardians of cortical development. Front Cell Dev Biol 8:6
28. Dave RK, Ellis T, Toumpas MC, Robson JP, Julian E, Adolphe C et al (2011) Sonic hedgehog and notch signaling can cooperate to regulate neurogenic divisions of neocortical progenitors. PLoS One 6(2):e14680
29. Kong JH, Yang L, Dessaud E, Chuang K, Moore DM, Rohatgi R et al (2015) Notch activity modulates the responsiveness of neural progenitors to sonic hedgehog signaling. Dev Cell 33(4):373–387
30. Helmer P, Vallee RB (2023) A two-kinesin mechanism controls neurogenesis in the developing brain. Commun Biol 6(1):1219
31. Shohayeb B, Cooper HM (2023) The ups and downs of Pax6 in neural stem cells. J Biol Chem 299(5):104680
32. Ochi S, Manabe S, Kikkawa T, Osumi N (2022) Thirty years' history since the discovery of Pax6: from central nervous system development to neurodevelopmental disorders. Int J Mol Sci 23(11):6115
33. Dong Z, He W, Lin G, Chen X, Cao S, Guan T et al (2023 Mar) Histone acetyltransferase KAT2A modulates neural stem cell differentiation and proliferation by inducing degradation of the transcription factor PAX6. J Biol Chem 299(3):103020
34. Faa G, Fanos V, Manchia M, Van Eyken P, Suri JS, Saba L (2024) The fascinating theory of fetal programming of adult diseases: a review of the fundamentals of the barker hypothesis. J Public Health Res 13(1):22799036241226817

35. Faa G, Marcialis MA, Ravarino A, Piras M, Pintus MC, Fanos V (2014) Fetal programming of the human brain: is there a link with insurgence of neurodegenerative disorders in adulthood? Curr Med Chem 21(33):3854–3876
36. Piras M, Fanos V, Ravarino A, Marcialis MA, Vinci L, Pintus MC et al (2014) Fetal programming of Parkinson's and Alzheimer's diseases: the role of epigenetic factors. J Pediatr Neonatal Individ Med JPNIM. 3(2):e030270–e030270
37. Faa G, Manchia M, Pintus R, Gerosa C, Marcialis MA, Fanos V (2016) Fetal programming of neuropsychiatric disorders. Birth Defects Res Part C Embryo Today Rev 108(3):207–223
38. Cainelli E, Bisiacchi P (2022) Neurodevelopmental disorders: past, present, and future. Child Basel Switz 10(1):31
39. Kaushik G, Zarbalis KS (2016) Prenatal Neurogenesis in Autism Spectrum Disorders. Front Chem 4:12
40. Rajan KC, Tiemroth AS, Thurmon AN, Meadows SM, Galazo MJ (2024) Zmiz1 is a novel regulator of brain development associated with autism and intellectual disability. Front. Psychiatry 15:1375492
41. Le Duc D, Giulivi C, Hiatt SM, Napoli E, Panoutsopoulos A, Harlan De Crescenzo A et al (2019) Pathogenic WDFY3 variants cause neurodevelopmental disorders and opposing effects on brain size. Brain. J Neurol 142(9):2617–2630
42. Filimonenko M, Isakson P, Finley KD, Anderson M, Jeong H, Melia TJ et al (2010) The selective macroautophagic degradation of aggregated proteins requires the PI3P-binding protein Alfy. Mol Cell 38(2):265–279
43. Kobayashi M, Nakatani T, Koda T, Matsumoto KI, Ozaki R, Mochida N et al (2014) Absence of BRINP1 in mice causes increase of hippocampal neurogenesis and behavioral alterations relevant to human psychiatric disorders. Mol Brain 14(7):12
44. Jolly LA, Homan CC, Jacob R, Barry S, Gecz J (2013) The UPF3B gene, implicated in intellectual disability, autism, ADHD and childhood onset schizophrenia regulates neural progenitor cell behaviour and neuronal outgrowth. Hum Mol Genet 22(23):4673–4687

The Placenta and the Development of Human Health

5

Antonio Ragusa, Antonino Lo Re, Maria Segata, Denise Rinaldo, Giovanni la Motta, Marika De Vito, Caterina De Luca, and Alessandro Svelato

Contents

5.1 The Placenta: A Key Element 58
5.2 Details of the Individual Mechanistic Phenomena 63
5.3 Maternal BMI 64
5.4 Maternal Diabetes 64
5.5 Maternal Nutrition 65
5.6 Maternal Smoking 65
5.7 Stress and Maternal Health 66
5.8 Endocrine Disruptor and Environmental Toxins 67
5.9 Placenta as a Mediator of Intergenerational Epigenetic Effect 67
5.10 Implications for Public Health 67
5.11 Plastic in the Placenta and the Effects on Human Health as an Example of Environmental Interference 67
5.12 Conclusion 70
References 72

A. Ragusa · A. L. Re
Istituto per la Sicurezza Sociale, Reparto di Ostetricia e Ginecologia, Repubblica di San Marino, San Marino, Italy

M. Segata · G. la Motta · M. De Vito
U.O.C. Ostetricia e Ginecologia, Ospedale Maggiore Carlo Alberto Pizzardi, Bologna, Italy

D. Rinaldo
Department of Obstetrics -ASST Grande Ospedale Metropolitano Niguarda, Milan, MI, Italy

C. De Luca (✉)
Department of Gynecology and Obstetrics, University of Catanzaro "Magna Graecia", Catanzaro, Italy

A. Svelato
Department of Experimental and Clinical Medicine, University of Catanzaro "Magna Graecia", Catanzaro, Italy

L. Emmi, V. Fanos (eds.), *Epigenetics and Rare Diseases*, Rare Diseases of the Immune System, https://doi.org/10.1007/978-3-032-02902-7_5

5.1 The Placenta: A Key Element

During pregnancy, the placenta develops from the trophoblast and works as an interface between the mother and her fetus. While many aspects of the placenta are still to be explored, two elements are well understood: its structure and the fundamental stages of its development. At birth, the placenta weighs just over half a kilogram and consists of two distinct sides: one attached to the maternal uterine walls, which appears like a blood-soaked sponge, and the other facing the fetus, characterized by a complex network of blood vessels connected to the umbilical cord.

The placenta must develop rapidly to fulfill the functions of other fetal organs until they become fully operational. It performs the roles that will later be taken over by the liver, such as nutrient metabolism, by the lungs, such as oxygen and carbon dioxide exchange and by the kidneys, such as waste products elimination. Within the first week after fertilization, specialized cells called trophoblasts emerge on the embryo surface. These cells not only produce hormones that signal the presence of the embryo to the maternal body but also have the crucial task of penetrating the uterine wall. There, they rapidly multiply or, alternatively, regulate the amount of blood they can transport. By the end of the first trimester, the arteries open into the spaces between the chorionic villi, supplying the fetus with maternal blood rich in nutrients and oxygen. The cytotrophoblasts also invade the uterine veins, allowing blood to flow from the placenta back to the maternal body, thereby completing the exchange circuit and removing carbon dioxide and other fetal waste products.

Blood from the maternal arteries bathes the placenta surface just a few cell layers away from the fetal blood vessels, optimizing the exchange of nutrients, gases, and metabolites. In recent years, research has revealed that the placenta releases large quantities of fetal DNA into the maternal bloodstream, making prenatal genetic testing possible, with just a simple maternal blood sample. These tests are gradually replacing more invasive procedures, such as chorionic villus sampling and amniocentesis. While fetal genes direct much of the placenta's development, the surrounding microenvironment plays an equally crucial role. Research over the past two decades highlighted how the success of pregnancy depends on the interaction between maternal cells in the uterine tissue and the invasive projections of the placenta [1]. The areas of contact between the placenta and the uterus, consisting of projections extending into the uterine tissue, include multiple layers. One layer is made up of cells called cytotrophoblasts, while another layer, composed of fused cells known as syncytiotrophoblasts, forms the surface of the placenta. Over time, the placenta takes the shape of a disk, anchored to the uterine wall through branching structures.

During the second and third weeks after fertilization, these branches fill with support cells and blood vessels. By the time a woman discovers pregnancy, the mature configuration of the chorionic villi has already been established. The placenta's ability to self-redistribute maternal blood flow is essential for placental rapid development in early pregnancy. This process can occur thanks to the unique journey of the cytotrophoblasts. Initially, these cells attach to the surface of the uterine wall and then migrate progressively deeper. More than twenty years ago, Fisher discovered that during this process, cytotrophoblasts undergo a transformation, adopting the same characteristics of the cells that normally line blood vessels. This ability to mimic allows cytotrophoblasts to penetrate the oxygen-rich maternal arteries, climbing the inner lining of the vessels and gradually replacing it [2].

Alterations in epigenetic regulation can lead to adverse pregnancy outcomes, such as preeclampsia, gestational trophoblastic disease, and fetal growth restriction. These changes can be influenced by environmental factors, including maternal diet, stress, and exposure to toxins, which can disrupt the epigenome and affect placental function [3–5].

The placenta's unique epigenetic features, some of which resemble those found in malignancies, highlight its sensitivity to environmental influences [3].

Given the rapid development of the placenta and the variety of cells involved, it is not surprising that errors can occur in the formation of the maternal-fetal interface. Such anomalies can lead to pregnancy complications, including:

- Preterm birth (defined as delivery of the baby before 37 weeks).
- Intrauterine growth restriction (the fetus is smaller than its expected growth potential).
- Preeclampsia (a condition characterized by the sudden onset of maternal hypertension and vascular damage).

Research has identified potential epigenetic biomarkers in the placenta that could predict pregnancy complications and fetal development issues. For instance, circulating epigenetic markers in maternal blood may serve as indicators for the onset and severity of preeclampsia. These biomarkers offer promising avenues for early diagnosis and intervention, potentially improving pregnancy outcomes [6, 7].

The placenta plays a fundamental role in mediating prenatal programming through its functions of nutrient transfer, waste elimination, and hormonal regulation. Emerging evidence indicates that the placenta is not merely a passive conduit for maternal-fetal exchange but an active participant that can influence the epigenetic landscape of the developing fetus.

There are several mechanisms through which the placenta mediates these complex processes.

Nutrient Perception and Metabolism the placenta actively detects nutrient availability and responds accordingly. Factors such as maternal diet can induce changes in placental gene expression, influencing fetal growth and development.

Specifically:

- *Regulation of nutrient transport*: placental transporters for glucose, amino acids, and lipids can be modulated in response to maternal diet and other metabolic stimuli. For instance, high-fat diets or deficiencies in essential nutrients can alter the expression of placental genes involved in energy metabolism, influencing the risk of developing obesity and metabolic diseases in the fetus [8];

- *Insulin and fetal growth*: alterations in placental insulin signaling (3), as seen in gestational diabetes, can impair glucose transport, contributing to both macrosomia and intrauterine growth restriction (IUGR). Additionally, these changes can have long-lasting epigenetic effects, modulating the expression of genes associated with insulin sensitivity and fetal energy metabolism after birth [9].

Hormone Production the placenta is a key source of steroid and protein hormones essential for maintaining pregnancy and regulating fetal development [10].

- *Steroid hormones*: progesterone and estrogens regulate placental function and fetal growth. For example, estrogen modulates the expression of genes related to placental angiogenesis, ensuring adequate blood flow [11].
- *hCG (human chorionic gonadotropin)*: this hormone, produced by the placenta, not only supports corpus luteum function but also regulates the expression of genes affecting the central nervous system and fetal metabolism [12].
- *Interaction with maternal-fetal systems*: placental hormones regulate crucial epigenetic processes, influencing the programming of hypothalamic-pituitary-adrenal (HPA) axis, fundamental for stress response and metabolic regulation in postnatal life.

Exposure to Environmental Factors the placenta is highly sensitive to a wide range of environmental stimuli capable of altering fetal epigenetic programming [13].

- *Maternal psychosocial stress*: chronic stress exposure during pregnancy is associated with changes in placental gene expression. For example, maternal cortisol can alter the activity of placental 11β-HSD enzymes, modulating glucocorticoid levels reaching the fetus [14].
- *Environmental toxins*: cigarette smoke, air pollution, and heavy metals can induce epigenetic modifications in the placenta, including DNA methylation and histone modifications, with a negative impact on fetal development [15].
- *Placental microbiota*: recent studies suggest that placental microbiota may also influence fetal metabolism and immune response through epigenetic mechanisms [16].

Role in Prenatal Programming growing evidence suggests that the placenta plays a central role in fetal programming.

It can influence:

- *Growth trajectories*: alterations in placental function can lead to accelerated or restricted fetal growth, with long-term effects on obesity, diabetes, and cardiovascular disease risk [17, 18].
- *Immunity and inflammation*: activation of placental inflammatory processes can predispose the fetus to chronic inflammatory diseases [19].
- *Neuropsychiatry*: placental epigenetic modifications in response to maternal stress or environmental exposure have been linked to an increased risk of autism spectrum disorders and anxiety in offspring [20, 21].

A clear example of this phenomenon is seen in maternal anemia, which has long been known to predispose neonates to autism and other neuropsychiatric disorders [22, 23].

In summary, the placenta is not merely a biological filter, but an active regulator of fetal growth and development. Its metabolic, endocrine, and immune functions, combined with its ability to adapt to environmental factors, make it a crucial determinant of long-term offspring health.

The placenta is a complex interface between intrauterine life and the world that the fetus will encounter after birth.

Epigenetics and the Placenta

Epigenetics refers to the study of heritable modifications in gene expression that do not involve changes to the underlying DNA sequence. In recent years, epigenetics has received considerable attention, particularly in the context of prenatal development and placental function [24].

Epigenetics examines how environmental factors, such as maternal nutrition, stress, and pollution, can influence fetal gene expression. The intrauterine environment does not only determine immediate fetal development but can also condition future health and disease susceptibility. For example, maternal malnutrition during pregnancy can induce alterations in DNA methylation profiles, affecting brain development and increasing the risk of metabolic and neurological disorders [20, 21]

.

The placenta is a highly dynamic tissue, undergoing rapid changes throughout pregnancy. During this process epigenetic mechanisms are essential for regulating placental gene expression. Notably, DNA methylation and histone modifications play a key role in placental development and function. Aberrations in these epigenetic processes can contribute to pregnancy complications, including preeclampsia, intrauterine growth restriction (IUGR), and gestational diabetes [9, 25].

Chronic diseases such as obesity, type 2 diabetes, and cardiovascular diseases are increasingly recognized not only as the result of genetic and behavioral factors but also of prenatal exposure. Evidence suggests that epigenetic alterations induced by environmental factors during pregnancy can predispose individuals to these conditions. For example, pregnant women exposed to high levels of air pollution may give birth to children with altered epigenetic profiles and with an increased risk of developing respiratory disorders, allergies, and other diseases [26]. Low birth weight, on the other hand, has been linked to an increased risk of developing metabolic diseases and cardiovascular mortality in adulthood.

As a divergent developmental lineage, placental cells exhibit epigenetic characteristics not observed in embryonic tissues, likely contributing to their unique biology. For example, human placenta has globally reduced DNA methylation levels compared to embryonic or adult somatic tissues, along with a set of placenta-specific imprinting loci, especially in cases of miscarriage [27]. However, many aspects of the human placental epigenome remain poorly defined, including how interactions between different epigenetic features collectively are able to influence placental development and function.

Although the precise mechanisms are not yet fully understood, it is hypothesized that certain intrauterine stress factors act on early developmental pathways to enhance adaptation to postnatal environment as part of an adaptive response. A striking example of this is the so-called "thrifty phenotype hypothesis": poor maternal nutrition during pregnancy may trigger physiological reprogramming to better adapt the fetus to an environment with limited nutrient availability. However, if the offspring is born within a nutrient-rich environment, this adaptation may increase the risk of developing metabolic diseases [28, 29].

In a recent study, Zhang and colleagues mapped histone modification and DNA methylation profiles in normal human placental cells, specifically cytotrophoblasts, during mid-to-late gestation and in cytotrophoblasts from patients with preeclampsia. The authors explain that regulatory state is characterized by extreme DNA hypomethylation and high levels of the repressive histone modification H3K9me3. These shifts in histone modification could explain epigenetic changes observed in preeclampsia and placental senescence.

Finally, the authors identified altered transcriptional and epigenetic profiles in cytotrophoblasts from preeclamptic patients. This study demonstrated very high levels of the histone marks H3K9me3, H3K27me3, and H3K27ac in cytotrophoblasts from preeclamptic patients [30].

Epigenetic modifications include DNA methylation, histone modifications, and RNA-based processes. DNA methylation involves the addition of a methyl group to a cytosine base, most commonly when cytosine is adjacent to guanine, forming a "CpG site." CpG methylation at gene promoters typically leads to gene silencing, although its effects on transcription vary depending on the genomic context.

Histone proteins act to compact DNA by wrapping it into units called "nucleosomes." Histone modifications, including methylation, acetylation, and phosphorylation of histone "tails," can influence gene expression by altering DNA accessibility and its packaging into transcriptionally active or inactive regions. Increasing evidence suggests that DNA methylation and histone modifications can also function in concert with RNA to regulate gene expression [31, 32].

Research linking epigenetic changes to fetal programming in humans focused on DNA methylation due to its stability in isolated DNA.

DNA methylation plays a crucial role in embryonic and fetal development. Among its key functions are establishment and maintenance of cellular identity, X-chromosome inactivation in females, and regulation of genomic imprinting [33].

The first days after conception represent a particularly dynamic phase of epigenetic reprogramming, during which the epigenetic characteristics of parental gametes are largely erased and cells are reset to a "naive" state, ready to acquire cell- and tissue-specific characteristics during gastrulation and subsequent stages.

This cascade of early developmental processes highlights the periconceptional period as a "critical window" for epigenetically mediated effects on fetal programming [34]. Epigenetic regulation is fundamental to placental development and function, with significant implications for maternal and fetal health. In fact, disruptions in these processes can lead to various pregnancy complications. It is therefore important to understand and potentially manipulate epigenetic mechanisms. Research in this field will probably help to enhance our ability to predict, diagnose, and treat pregnancy-related disorders, ultimately improving outcomes for both mothers and their children.

In summary, the placenta plays a crucial role in fetal development and maternal health, with epigenetic regulation being a key factor in its proper function. Epigenetic mechanisms, such as DNA methylation, histone modifications, and non-coding RNAs, are essential for controlling gene expression patterns that are vital for placental and fetal development. These mechanisms are involved in genomic imprinting, which is critical for parent-of-origin-dependent gene expression [3, 6, 7, 35].

5.2 Details of the Individual Mechanistic Phenomena

Methylation the addition of a methyl group to the 5′ position of cytosine residues in CpG dinucleotides is one of the most studied epigenetic mechanisms in the placenta. This modification is crucial for regulating gene expression during placental development. Studies have shown that differential methylation patterns in the placenta are associated with pregnancy-related conditions, including abnormal placental function in preeclampsia and IUGR [36]. Methylation signatures in the placenta also reflect maternal environmental exposures, such as diet, smoking, and stress, which can affect fetal development and long-term health outcomes [37].

Histone Modifications histone proteins undergo a variety of modifications, including acetylation, methylation, and phosphorylation, which influence chromatin structure and gene expression. In the placenta, these modifications are involved in regulating genes that control trophoblast differentiation, placental vascularization, and nutrient transport. Research demonstrated that maternal environmental factors, such as obesity and diabetes, may alter histone modifications in placental cells, leading to impaired placental function and adverse pregnancy outcomes [38].

Non-coding RNAs non-coding RNAs (ncRNAs), including microRNAs (miRNAs) and long non-coding RNAs (lncRNAs), are crucial regulators of gene expression in the placenta. miRNAs have been shown to modulate trophoblast differentiation, placental development, and immune responses. Dysregulation of miRNA expression in the placenta is linked to conditions like preeclampsia and fetal growth restriction [39]. lncRNAs also play roles in placental function, including regulation of trophoblast invasion and balance between immune tolerance and immune activation during pregnancy [40].

5.3 Maternal BMI

An association has been demonstrated between maternal body mass index (BMI) and epigenetic modifications in offspring, particularly regarding DNA methylation of key genes involved in energy metabolism, including PPARG and leptin. For example, a positive correlation exists between maternal BMI and the methylation of the "PPARGC1A" gene promoter in umbilical cord blood, suggesting that a high maternal BMI may influence the expression of this gene in offspring [41]. Other research found an association between a high pre-pregnancy maternal BMI and a reduced "LEP" gene methylation in neonatal cord blood. This hypomethylation could mediate the relationship between maternal obesity and neonatal adiposity, influencing leptin levels in offspring [42].

Studies on children born to mothers who underwent bariatric surgery have shown improved cardiovascular risk profiles compared to their unexposed siblings [43]. This suggests that maternal weight loss before pregnancy may positively impact offspring health, potentially through epigenetic modifications in genes involved in glucose and leptin regulation.

These findings highlight the importance of maternal BMI as a factor able to influence epigenetic programming in offspring, with potential implications for the risk of developing obesity and other metabolic conditions.

5.4 Maternal Diabetes

Maternal diabetes, particularly gestational diabetes, can induce epigenetic modifications in offspring, affecting DNA methylation in key genes involved in energy metabolism, such as leptin. These epigenetic alterations may mediate the relationship between maternal hyperglycemia and neonatal leptin levels, with potential implications for the offspring's future metabolic health. Maternal hyperglycemia is associated with reduced methylation of the leptin gene in umbilical cord blood. This hypomethylation can lead to increased leptin expression in offspring [44], appetite regulation, and energy metabolism, predisposing the fetus to greater adiposity accumulation.

However, epigenetic modifications induced by maternal hyperglycemia are not limited to the leptin gene. Other DNA methylation changes have been observed in genes involved in insulin secretion and glucose metabolism, increasing the risk of obesity and type 2 diabetes in adulthood [45, 46].

These findings highlight the importance of optimal glycemic control during pregnancy to prevent potentially harmful epigenetic modifications in offspring. Targeted interventions to maintain adequate blood glucose levels in mothers with gestational diabetes could therefore reduce the risk of metabolic diseases in offspring, promoting better long-term health.

5.5 Maternal Nutrition

Maternal nutrition plays a crucial role in fetal programming, influencing DNA methylation through the one-carbon cycle pathway and associated metabolites such as folate, riboflavin, choline, and betaine. These nutrients are essential for providing methyl groups required for DNA methylation reactions, which regulate gene expression during fetal development.

Epidemiological studies have shown that fetal exposure to maternal undernutrition is associated with changes in DNA methylation at specific loci, affecting offspring metabolism in adulthood. For example, the Dutch famine of 1944–1945 provided significant data: in utero famine exposure increased risk of developing obesity, hypertension, and type 2 diabetes in adulthood. These observations suggest that maternal malnutrition can induce epigenetic alterations that influence offspring metabolism [47]. Epigenetic modifications induced by maternal malnutrition affect key genes involved in energy metabolism, including those related to glycolysis and adipogenesis. For instance, alterations in the methylation of genes involved in glycolysis can impact the efficiency of cells to utilize glucose, while changes in adipogenesis-related genes may lead to variations in fat tissue formation and accumulation.

Maternal nutrition during pregnancy is one of the most studied environmental factors influencing placental epigenetics. Nutrients such as folate, choline, and vitamin D play a role in one-carbon metabolism, directly affecting DNA methylation patterns. Inadequate maternal nutrition can lead to altered placental gene expression, potentially contributing to adverse pregnancy outcomes and to an increased risk of developing diseases such as hypertension, diabetes, and obesity in offspring [48].

These findings highlight the importance of adequate maternal nutrition during pregnancy to ensure a proper epigenetic programming of the fetus. A balanced intake of essential nutrients can prevent alterations in DNA methylation, which may predispose the offspring to metabolic disorders in adulthood.

5.6 Maternal Smoking

Different studies demonstrated that maternal smoking during pregnancy induces changes in specific epigenetics in DNA of the offspring, representing one of the most robust epigenetic biomarkers for early exposure to environmental factors [49].

These alterations in DNA methylation have been observed in umbilical cord blood at birth and seem to persist at least until adolescence [50].

Research identified over 6000 CpG (cytosine-phosphate-guanine dinucleotide) sites in the fetal genome, whose methylation is affected by maternal smoking [51].

These changes involve genes associated with lung development, metabolism, and inflammatory processes. The epigenetic alterations induced by smoking during pregnancy have been linked to several negative outcomes for children's health. These include:

- **Inflammatory bowel diseases and psychiatric morbidity:** some methylation loci associated with maternal smoking in fact show a causal link with an increased risk of developing inflammatory bowel diseases and psychiatric disorders in the offspring.
- **Reduced lung function:** epigenetic changes can affect the expression of genes crucial for lung development, leading to a decrease in respiratory function in children exposed to smoking in utero.
- **Cardiovascular risk factors:** maternal smoking has been associated with epigenetic changes that may predispose offspring to develop cardiovascular risk factors, such as hypertension and lipid abnormalities.

It is important to emphasize that these changes in DNA methylation are not transient; evidence suggests that they persist from birth until adolescence, indicating a long-term impact of prenatal exposure to smoking on the offspring health.

This evidence reinforces the importance of awareness campaigns and interventions aimed at reducing smoking during pregnancy, in order to prevent potentially harmful epigenetic changes and to promote better long-term health for future generations.

5.7 Stress and Maternal Health

Maternal stress, both physical and psychological, can have profound effects on placental function and fetal development. Chronic stress activates the hypothalamic-pituitary-adrenal (HPA) axis, leading to changes in placental gene expression through epigenetic mechanisms, including DNA methylation and histone modifications. Studies have shown that maternal stress during pregnancy can lead to alterations in the placental expression of genes involved in immune regulation, potentially increasing the risk for preterm birth, low birth weight, and developmental disorders [52].

5.8 Endocrine Disruptor and Environmental Toxins

Exposure to environmental toxins, such as endocrine-disrupting chemicals (EDCs), has been implicated in changes to placental epigenetics. EDCs, such as bisphenol A (BPA), phthalates, and pesticides, can interfere with hormonal regulation during pregnancy, leading to altered DNA methylation patterns in the placenta. These changes can influence the expression of genes related to cell proliferation, apoptosis, and immune response, contributing to pregnancy complications and affecting fetal health [53, 54].

5.9 Placenta as a Mediator of Intergenerational Epigenetic Effect

One of the most compelling aspects of placental epigenetics is its potential role in mediating intergenerational effects. The placenta not only regulates fetal development but also transmits epigenetic modifications that may persist across generations. For instance, maternal nutrition and environmental exposures can lead to epigenetic changes in the placenta, which are then passed on to the offspring, influencing their health in later life. This phenomenon is consistent with the concept of "epigenetic programming" where early-life exposures shape the development of chronic diseases such as cardiovascular disease, metabolic disorders, and mental health conditions [53].

5.10 Implications for Public Health

Understanding the epigenetic influences of prenatal planning highlights the importance of maternal health and environmental conditions during pregnancy. Public health interventions focused on maternal nutrition, stress management, and reduced exposure to harmful factors could be critical to promote healthier generations.

The effects of epigenetic changes can extend to several generations, giving rise to discussions on the intergenerational transmission of health risks. Policies of prenatal care and targeted interventions could not only improve immediate outcomes for mother and children, but also reduce the incidence of future chronic diseases.

5.11 Plastic in the Placenta and the Effects on Human Health as an Example of Environmental Interference

Micro- and nanoplastics (MP and NP) are ubiquitous and enter the human body through respiration, skin, and gastrointestinal tract (Fig. 5.1). MP/NPs can alter different cellular regulation pathways in the placenta and breast milk, such as immunity, oxidation, metabolism, storage and production of lipids, and mechanism of

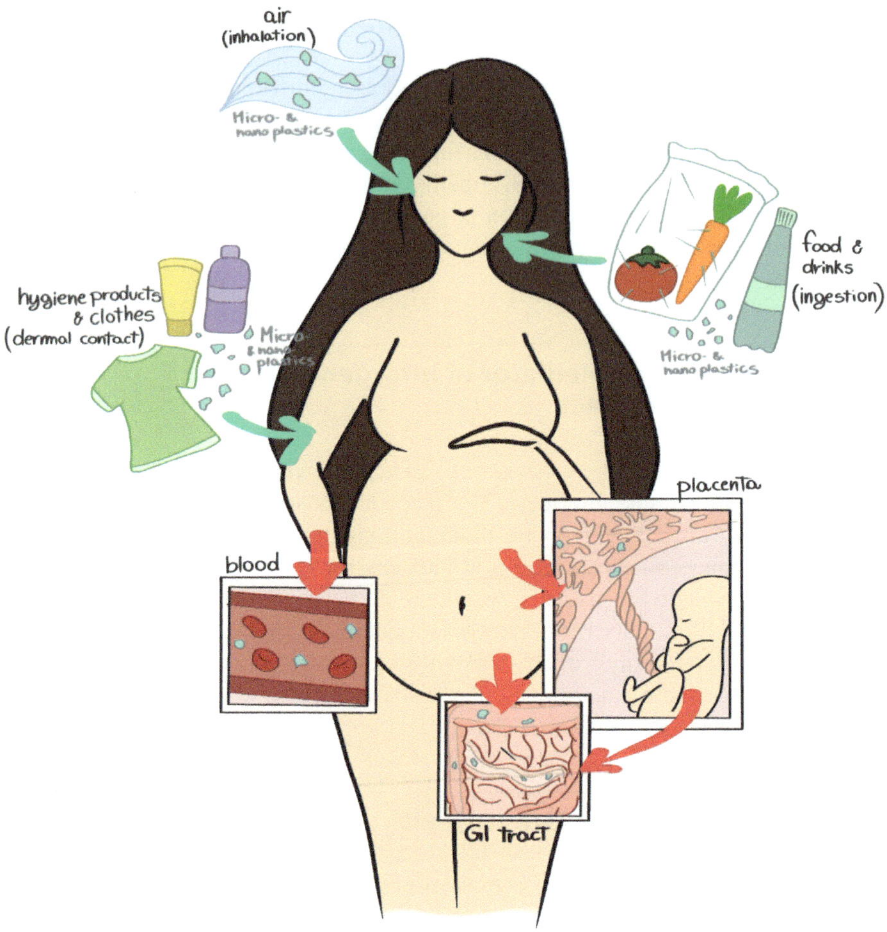

Fig. 5.1 Representation of micro- and nanoplastic contamination by inhalation, ingestion, and dermal contact. Consequent findings of MPs/NPs in blood, inside the placenta, and in the gastrointestinal tract of the pregnant mother and fetus. (From: Antonio Ragusa, Giulia Principi, Maria Matta. Pregnancy in the Era of the Environmental Crisis: Plastic and Pollution. *Clin. Exp. Obstet. Gynecol.* 2022, 49 (10), 216. https://doi.org/10.31083/j.ceog4910216. Image 606 Reprinted under terms of the CC-BY license [68])

action of some important proteins. These changes can lead to diseases in the short term (infants) and long term (adult life).

MP and NP have been detected in human placenta, meconium, and feces of infants, raising significant concerns about the effects of their exposure at an early stage of development [55–57].

However, the impact of plastic particle exposure during critical windows of vulnerability, such as pregnancy and early life years, remains largely unknown. Lack of knowledge about the health effects of nano- and microplastics, and of the chemicals they contain, prevents a scientifically based assessment and effective management of potential health risks. Crucial questions remain, such as the extent of human exposure to these particles, especially during pregnancy, and what the long-term consequences may be for children's well-being and development.

The discovery of microplastics in the human placenta allowed scientists to identify, in some cases, the type of plastic present. However, the wide range of chemical additives used in plastics makes it difficult to accurately identify particles, even with advanced tools such as Raman microscopy. These additives may interfere with analysis, complicating the determination of specific health effects. Moreover, it is not only plastic itself but also additives to influence the epigenetic expression of many proteins involved in processes critical to fetal growth and development. The most worrying chemicals are phthalates, bisphenol A (BPA) and polyvinyl alcohol, known to act as endocrine disruptors, affecting the normal functioning of the endocrine system since they have a chemical structure similar to natural sex hormones. In infants and growing children, this could have important implications for hormonal and metabolic development [58].

After the first study that confirmed the presence of microplastics in the placenta, new research focused on their location within cell compartments. Microplastics were found in all compartments of the placental villi and their presence is associated with significant alterations of intracellular organelles. In particular, morphological changes with a dose-response effect were observed in mitochondria and in endoplasmic reticulum, two cellular structures fundamental for metabolism and protein synthesis [59]. The results of this in vivo study were subsequently confirmed by a further in vitro study [60].

The proximity of microplastic particles to these organelles could impair their function, altering energy production and protein synthesis, crucial for proper fetal development. Studies in animal models, such as mice, have shown that exposure to microplastics can lead to abnormal accumulation of lipids, changes in protein synthesis processes, essential for neurological development, and imbalances in redox reactions, with increased production of free radicals. These phenomena can contribute to the so-called "perfect storm", that is, a set of factors that could negatively affect children's health not only at birth, but also later, in adulthood.

Microplastics could disrupt the immunological self-tolerance mechanism, an essential process that allows the mother to tolerate the fetus without rejection reactions. Once in the human body, microplastics are able to accumulate and cause localized toxicity, inducing or amplifying immune responses that could reduce the body's ability to defend against external pathogens and alter energy homeostasis. In addition, the alterations of lipid metabolism, induced by these materials, could have repercussions even in adulthood, increasing the risk of metabolic diseases through epigenetic mechanisms.

A significant study by Amereh et al. analyzed placentas of 43 pregnant women, using advanced digital microscopy and Raman microspectroscopy techniques to detect the presence of microplastics smaller than 5 mm [61]. Microplastics were detected in all placentas of pregnancies complicated by intrauterine growth restriction (IUGR), with concentrations ranging from 2 to 38 particles per placenta. In contrast, in uneventful pregnancies the presence of microplastics was much lower and was not detectable in most cases, with the exception of three out of thirty subjects. The data analysis showed a negative correlation between the presence of microplastics and neonatal anthropometric parameters, such as birth weight and length, cranial circumference, and one-minute Apgar score, indicating that exposure to microplastics could adversely affect fetal growth. However, it is important to note that while our study analyzed children born after normal pregnancies, Amereh survey focused on IUGR fetuses, suggesting that the effects of microplastics may be more apparent in situations already prone to complications.

Recent evidence seems to confirm that MPs and NPs play a crucial role in the development of pregnancy diseases, for instance, in preterm labor genesis [62].

The morphological alterations observed in placental cells could represent an attempt to remove plastic particles, which leads to a build-up of toxic substances and a prolonged cellular alarm state known as Cellular Danger Response (CDR) [63]. This persistent condition may have long-term consequences, affecting the fetal phenotype and predisposing the newborn to future diseases. In addition, endocrine disruptors contained in microplastics, such as phthalates and BPA, could further alter normal hormonal and metabolic development in children [64].

In summary, the presence of microplastics in the human placenta represents a risk to fetal and neonatal health, with implications that could extend well beyond the pregnancy period [65]. The presence of microplastics may trigger cellular changes and long-term consequences. Therefore, it is essential to conduct further research to better understand the mechanisms of interaction between MPs/NPs and placental tissues. It's mandatory to identify prevention strategies to reduce exposure to these particles specially during pregnancy, thus protecting future generations health.

5.12 Conclusion

The relationship between genes and environment is crucial to prenatal programming [66].

The placenta, an interface between the mother and her fetus, reveals a complex network of influences on fetal development and long-term health, since it is nothing but fitness. As research in this field progresses, it becomes increasingly clear that priority must be given to improving the environment where mothers carry out gestation, in order to promote optimal health outcomes for future generations.

The placenta is emerging as a key player in the epigenetic regulation of pregnancy and fetal development. Epigenetic modifications in the placenta are influenced by a variety of maternal environmental factors, including nutrition, stress, and exposure to toxins, which can have long-lasting effects on both maternal and fetal health. However, there is still a lot to understand about the specific molecular mechanisms that lead to these changes and how they contribute to pregnancy-related disorders and to long-term health of offspring.

Future research should focus on uncovering the precise epigenetic pathways that regulate placental function and its impact on pregnancy outcomes. Longitudinal studies that examine placental epigenome across different maternal health conditions and environmental exposures will be essential to understand how placental epigenetic can be leveraged for early interventions, thus improving pregnancy outcomes and reducing the risk of chronic diseases in future generations. Additionally, advancements in non-invasive prenatal testing could provide opportunities for monitoring placental epigenetic changes during pregnancy, enabling personalized medicine approaches for expectant mothers.

Investing in understanding and addressing these factors can pave the way for innovative strategies in non-communicable disease prevention and management, conditions rooted in the early stages of human development.

We should reflect on the origin of the information carried by the newborn, even before birth, recognizing the operation of natural selection, which, generation after generation, changed the shape and function of the placenta, allowing this vital organ to adapt to the mold provided by the structure of the outside world. The placenta is the depository of a deep phylogenetic memory, in which information, when vital, is transformed into a transmissible and heritable code, used by new generations to better adapt to the outside world (Fig. 5.2). We need to fight for this environmental

Fig. 5.2 Placenta represented as a metaphor for the intersection between intrauterine life and the outside world. (Image created with https://www.craiyon.com/)

justice, since the alterations described in this chapter, mainly affect the economically less well-off social classes [67, 68].

References

1. Olmos-Ortiz A, Flores-Espinosa P, Mancilla-Herrera I, Vega-Sánchez R, Díaz L, Zaga-Clavellina V (2019) Innate immune cells and toll-like receptor–dependent responses at the maternal–fetal interface. IJMS 20(15):3654
2. Logan SK, Fisher SJ, Damsky CH (1992) Human placental cells transformed with temperature-sensitive simian virus 40 are immortalized and mimic the phenotype of invasive cytotrophoblasts at both permissive and nonpermissive temperatures. Cancer Res 52(21):6001–6009
3. Nelissen ECM, Van Montfoort APA, Dumoulin JCM, Evers JLH (2011) Epigenetics and the placenta. Hum Reprod Update 17(3):397–417
4. Novakovic B, Saffery R (2012) The ever growing complexity of placental epigenetics—role in adverse pregnancy outcomes and fetal programming. Placenta 33(12):959–970
5. Hemberger M, Dean W (2023) The placenta: epigenetic insights into trophoblast developmental models of a generation-bridging organ with long-lasting impact on lifelong health. Physiol Rev 103(4):2523–2560
6. Januar V, Desoye G, Novakovic B, Cvitic S, Saffery R (2015) Epigenetic regulation of human placental function and pregnancy outcome: considerations for causal inference. Amer J Obs Gynecol 213(4):S182–S196
7. Apicella C, Ruano CSM, Méhats C, Miralles F, Vaiman D (2019) The role of epigenetics in placental development and the etiology of preeclampsia. IJMS 20(11):2837
8. Yu X, Li Q, Shao X, Sferruzzi-Perri AN, Wang YL (2025) Fine-tuned programming of placenta trophoblast determines optimal maternal–fetal nutrient allocation. Curr Opinion Genetics Develop 91:102305
9. Tong W, Allison BJ, Brain KL, Patey OV, Niu Y, Botting KJ et al (2025) Placental mitochondrial metabolic adaptation maintains cellular energy balance in pregnancy complicated by gestational hypoxia. J Physiol:JP287897
10. Pepe GJ, Albrecht ED (2009) Steroid endocrinology of pregnancy. GLOWM [Internet]. [citato 16 febbraio 2025]; Disponibile su: https://www.glowm.com/section-view/item/310
11. Ghosh D (2023) Structures and functions of human placental aromatase and steroid sulfatase, two key enzymes in estrogen biosynthesis. Steroids agosto 196:109249
12. Henke A, Gromoll J (2008) New insights into the evolution of chorionic gonadotrophin. Molecul Cell Endocrinol 291(1–2):11–19
13. Pino O, Di Pietro S, Poli D (2023) Effect of musical stimulation on placental programming and neurodevelopment outcome of preterm infants: a systematic review. IJERPH 20(3):2718
14. Frasch MG, Wakefield C, Janoschek B, Frank YS, Karp F, Reyes N, et al (2025) [citato 11 febbraio 2025] Perinatal psychoneuroimmunology of prenatal stress and its effects on fetal and postnatal Brain development. In: Yan Q. Psychoneuroimmunology [internet]. Springer, New York, NY. pp 303–332. (Methods in Molecular Biology; vol. 2868). Disponibile su: https://link.springer.com/10.1007/978-1-0716-4200-9_16
15. Green BB, Marsit CJ (2015) Select prenatal environmental exposures and subsequent alterations of gene-specific and repetitive element DNA methylation in fetal tissues. Curr Envir Health Rpt giugno 2(2):126–136
16. Du Q, Liu X, Zhang R, Hu G, Liu Q, Wang R et al (2025) Placental and fetal microbiota in rhesus macaque: a case study using metagenomic sequencing. American J Primatol gennaio 87(1):e23718
17. Dumolt JH, Rosario FJ, Barentsen K, Urschitz J, Powell TL, Jansson T (2024) Trophoblast-specific overexpression of adiponectin receptor 2 causes fetal growth restriction in pregnant mice. FASEB J 38(19):e70100

18. Mikkola TS, Ylikorkala O (2024) Pregnancy-associated risk factors for future cardiovascular disease—early prevention strategies warranted. Climacteric 27(1):41–46
19. Feist H, Schaumann N (2024) Occurrence of eosinophilic granulocytes in the decidua of placentas with perinatal clinical findings. Pediatr Dev Pathol:10935266241295341
20. Rees S, Harding R, Walker D (2008) An adverse intrauterine environment: implications for injury and altered development of the brain. Intl J Devlp Neurosci 26(1):3–11
21. Osman HC, Moreno R, Rose D, Rowland ME, Ciernia AV, Ashwood P (2024) Impact of maternal immune activation and sex on placental and fetal brain cytokine and gene expression profiles in a preclinical model of neurodevelopmental disorders. J Neuroinflammation 21(1):118
22. Georgieff MK (2020) Iron deficiency in pregnancy. Ame J Obs Gynecol 223(4):516–524
23. Orlandini C, Torricelli M, Spirito N, Alaimo L, Di Tommaso M, Severi FM et al (2017) Maternal anemia effects during pregnancy on male and female fetuses: are there any differences? J Matern Fetal Neonatal Med luglio 30(14):1704–1708
24. King SE, Schatz NA, Babenko O, Ilnytskyy Y, Kovalchuk I, Metz GAS (2025) Prenatal maternal stress in rats alters the epigenetic and transcriptomic landscape of the maternal-fetal interface across four generations. Commun Biol 8(1):38
25. Jiang Y, Zhou X, He Z, Wei L, Zhu S, Zhang H et al (2025) DHA improves neurodevelopmental abnormalities in offspring of gestational diabetes mellitus patients via the PPAR-γ/FATP4 pathway. Biochem Pharmacol 232:116726
26. Shiroshita A, Kataoka Y, Wang Q, Kajita N, Anan K, Tajima T et al (2024) Joint associations of air pollutants during pregnancy, infancy, and childhood with childhood persistent asthma: Nationwide database study in Japan. Ecotoxicol Environ Safety 281:116626
27. Matsumoto Y, Shinjo K, Mase S, Fukuyo M, Aoki K, Ozawa F et al (2022) Characteristic DNA methylation profiles of chorionic villi in recurrent miscarriage. Sci Rep 12(1):11673
28. Alabduljabbar S, Zaidan SA, Lakshmanan AP, Terranegra A (2021) Personalized nutrition approach in pregnancy and early life to tackle childhood and adult non-communicable diseases. Life 11(6):467
29. Fall CHD, Kumaran K (2019) Metabolic programming in early life in humans. Phil Trans R Soc B. 374(1770):20180123
30. Zhang B, Kim MY, Elliot G, Zhou Y, Zhao G, Li D et al (2021) Human placental cytotrophoblast epigenome dynamics over gestation and alterations in placental disease. Developmental Cell 56(9):1238–1252.e5
31. Toh H, Okae H, Shirane K, Sato T, Hamada H, Kikutake C et al (2024) Epigenetic dynamics of partially methylated domains in human placenta and trophoblast stem cells. BMC Genomics 25(1):1050
32. Bai J, Jiang P, Ji L, Lam WKJ, Zhou Q, Ma MJL et al (2024) Histone modifications of circulating nucleosomes are associated with changes in cell-free DNA fragmentation patterns. Proc Natl Acad Sci USA 121(42):e2404058121
33. Wu W, Zhang B, Zhao J, Hu W, Li Y, Feng Y et al (2025) Relationship between apoptosis gene DNA methylation and fetal growth and development. Gene 941:149224
34. Sánchez-Garrido MA, García-Galiano D, Tena-Sempere M (2022) Early programming of reproductive health and fertility: novel neuroendocrine mechanisms and implications in reproductive medicine. Hum Reprod Update 28(3):346–375
35. Maccani MA, Marsit CJ (2009) REVIEW ARTICLE: epigenetics in the placenta. American J Rep Immunol 62(2):78–89
36. Workalemahu T, Ouidir M, Shrestha D, Wu J, Grantz KL, Tekola-Ayele F (2020) Differential DNA methylation in placenta associated with maternal blood pressure during pregnancy. Hypertension 75(4):1117–1124
37. Heijmans BT, Tobi EW, Stein AD, Putter H, Blauw GJ, Susser ES et al (2008) Persistent epigenetic differences associated with prenatal exposure to famine in humans. Proc Natl Acad Sci USA 105(44):17046–17049
38. Bi S, Tu Z, Chen D, Zhang S (2023) Histone modifications in embryo implantation and placentation: insights from mouse models. Front Endocrinol 14:1229862

39. Zhao Y, Wu J, Liangpunsakul S, Wang L (2017) Long non-coding RNA in liver metabolism and disease: current status. Liver Research 1(3):163–167
40. Adu-Gyamfi EA, Cheeran EA, Salamah J, Enabulele DB, Tahir A, Lee BK (2024) Long non-coding RNAs: a summary of their roles in placenta development and pathology. Biol Reprod 110(3):431–449
41. Chavira-Suárez E, Ramírez-Mendieta AJ, Martínez-Gutiérrez S, Zárate-Segura P, Beltrán-Montoya J, Espinosa-Maldonado NC et al (2019) Influence of pre-pregnancy body mass index (p-BMI) and gestational weight gain (GWG) on DNA methylation and protein expression of obesogenic genes in umbilical vein. Laganà AS. PLoS ONE 14((12):e0226010
42. Ott R, Stupin JH, Melchior K, Schellong K, Ziska T, Dudenhausen JW et al (2018) Alterations of adiponectin gene expression and DNA methylation in adipose tissues and blood cells are associated with gestational diabetes and neonatal outcome. Clin Epigenet 10(1):131
43. Arbel EJ, Myneni AA, Boccardo JD, Simmonds I, Link H, Hoffman AB et al (2025) Maternal and neonatal outcomes after metabolic and bariatric surgery among women with severe obesity. Surgery Obes Related Diseases:S1550728925000103
44. Bönisch C, Irmler M, Brachthäuser L, Neff F, Bamberger MT, Marschall S et al (2016) Dexamethasone treatment alters insulin, leptin, and adiponectin levels in male mice as observed in DIO but does not lead to alterations of metabolic phenotypes in the offspring. Mamm Genome 27(1–2):17–28
45. Khurana I, Kaspi A, Ziemann M, Block T, Connor T, Spolding B et al (2016) DNA methylation regulates hypothalamic gene expression linking parental diet during pregnancy to the offspring's risk of obesity in Psammomys obesus. Int J Obes 40(7):1079–1088
46. Ornellas F, Carapeto PV, Mandarim-de-Lacerda CA, Aguila MB (2017) Obese fathers lead to an altered metabolism and obesity in their children in adulthood: review of experimental and human studies. Jornal de Pediatria 93(6):551–559
47. Cheng M, Conley D, Kuipers T, Li C, Ryan CP, Taeubert MJ et al (2024) Accelerated biological aging six decades after prenatal famine exposure. Proc Natl Acad Sci USA 121(24):e2319179121
48. Reed SA, Ashley R, Silver G, Splaine C, Jones AK, Pillai SM et al (2022) Maternal nutrient restriction and over-feeding during gestation alter expression of key factors involved in placental development and vascularization. J Anim Sci 100(6):skac155
49. Liu J, Yu C, Doherty TM, Akbari O, Allard P, Rehan VK (2020) Perinatal nicotine exposure-induced transgenerational asthma: effects of reexposure in F1 gestation. FASEB 34(9):11444–11459
50. Meyer KF, Krauss-Etschmann S, Kooistra W, Reinders-Luinge M, Timens W, Kobzik L et al (2017) Prenatal exposure to tobacco smoke sex dependently influences methylation and mRNA levels of the *Igf* axis in lungs of mouse offspring. Amer J Physiol Lung Cellular Molecul Physiol 312(4):L542–L555
51. Deng WQ, Cawte N, Campbell N, Azab SM, De Souza RJ, Lamri A et al (s) Maternal smoking DNA methylation risk score associated with health outcomes in offspring of European and South Asian ancestry. eLife 13:RP93260
52. Castro-Quintas A, Palma-Gudiel H, Eixarch E, San Martín González N, Röh S, Sauer S et al (2025) Placental epigenetic signatures of maternal distress in glucocorticoid-related genes and newborn outcomes: a study of Spanish primiparous women. European Neuropsychopharmacol 90:36–47
53. Barker DJ (1998) In utero programming of chronic disease. Clin Sci (Lond) 95(2):115–128
54. Patisaul HB (2021) [citato 11 febbraio 2025] Endocrine disrupting chemicals (EDCs) and the neuroendocrine system: beyond estrogen, androgen, and thyroid. In: Advances in Pharmacology [Internet]. Elsevier. pp 101–150. Disponibile su: https://linkinghub.elsevier.com/retrieve/pii/S1054358921000211
55. Ragusa A, Svelato A, Santacroce C, Catalano P, Notarstefano V, Carnevali O et al (2021) Plasticenta: first evidence of microplastics in human placenta. Environ Int 146:106274

56. Zhu M, Li X, Lin W, Zeng D, Yang P, Ni W et al (2024) Microplastic Particles Detected in Fetal Cord Blood, Placenta, and Meconium: A Pilot Study of Nine Mother–Infant Pairs in South China. Toxics 12(12):850
57. Braun T, Ehrlich L, Henrich W, Koeppel S, Lomako I, Schwabl P et al (2021) Detection of microplastic in human placenta and meconium in a clinical setting. Pharmaceutics 13(7):921
58. Zhang S, Wu Q, He W, Zhu H, Wang Z, Liang H et al (2025) Bisphenol a alters JUN promoter methylation, impairing steroid metabolism in placental cells and linking to sub-representative phenotypes. Gene 941:149210
59. Ragusa A, Matta M, Cristiano L, Matassa R, Battaglione E, Svelato A et al (2022) Deeply in plasticenta: presence of microplastics in the intracellular compartment of human placentas. IJERPH 19(18):11593
60. Ragusa A, Cristiano L, Di Vinci P, Familiari G, Nottola SA, Macchiarelli G, Svelato A, De Luca C, Rinaldo D, Neri I, Facchinetti F (2025, in press) Artificial plasticenta: how polystyrene nanoplastics affect in-vitro cultured human trophoblast cells. Cell Develop Biol. https://doi.org/10.3389/fcell.2025.1539600
61. Amereh F, Amjadi N, Mohseni-Bandpei A, Isazadeh S, Mehrabi Y, Eslami A et al (2022) Placental plastics in young women from general population correlate with reduced foetal growth in IUGR pregnancies. Environ Pollut 314:120174
62. Oral Concurrent Session 1 (2025) Equity, public health, and policy: Thursday, January 30, 2025 1:30 PM—4:00 PM. Pregnancy 1(S1):e12004
63. Naviaux RK (2020) Perspective: cell danger response biology—the new science that connects environmental health with mitochondria and the rising tide of chronic illness. Mitochondrion 51:40–45
64. Huang YS, Hung PL, Wang LJ, Tsai CM, Tsai CK, Tiao MM et al (2024) Distinct impacts of prenatal and postnatal phthalate exposure on behavioral and emotional development in children aged 1.5 to 3 years. Toxics 12(11):795
65. Ragusa A, Notarstefano V, Svelato A, Belloni A, Gioacchini G, Blondeel C et al (2022) Raman microspectroscopy detection and characterisation of microplastics in human breastmilk. Polymers (Basel) 14(13):2700
66. Hales CN, Barker DJ (2001) The thrifty phenotype hypothesis. Br Med Bull 60:5–20
67. Rothstein MA, Harrell HL, Marchant GE (2017) Transgenerational epigenetics and environmental justice. Environ Epigenetics [Internet] 3(3) Disponibile su: http://academic.oup.com/eep/article/doi/10.1093/eep/dvx011/4061580
68. Ragusa A, Principi G, Matta M (2022) Pregnancy in the era of the environmental crisis: plastic and pollution. Clin Exp Obstet Gynecol 49(10):216

Epigenetics and Cancer

6

Laura Giunti, Marco Tellini, and Iacopo Sardi

Contents

References 94

Cancer is a complex medical condition characterized by the transformation of cells with acquired oncological capabilities and the dysregulation of individual cellular processes or functions that operate synergistically. The main dysregulated cellular processes involved in cancer development include proliferation signals, cell growth, apoptosis, mortality, induction of angiogenesis, avoidance of immune destruction, and activation of invasion and metastasis [1].

It is known that the origins of tumorigenesis are triggered by the constitutive hereditary, sporadic, or somatic activation/inactivation of oncogenes (OCG)/tumor suppressor genes (TSG) that promote or inhibit cancer cell division/growth, respectively, by generating a group of target tumor cells. Recently, several studies report that OCGs and TSGs may have multiple and conflicting functions and, their transition from tumor promoters to tumor suppressors or *vice versa* appears to be controlled at different levels: DNA (variants and gene fusions), RNA (alternative splicing and regulation by non-coding RNAs), or protein (isoforms, mislocalization, post-translational modifications, proteolytic cleavage). Thus, their oversimplified and often inaccurate classification, based on the traditional function of their gene product, could lead to a misunderstanding of the tumorigenic process in different tumor types.

L. Giunti (✉) · M. Tellini · I. Sardi
Neuro-Oncology Unit, Meyer Children's Hospital IRCCS, Florence, Italy
e-mail: laura.giunti@meyer.it; marco.tellini@meyer.it; iacopo.sardi@meyer.it

L. Emmi, V. Fanos (eds.), *Epigenetics and Rare Diseases*, Rare Diseases of the Immune System, https://doi.org/10.1007/978-3-032-02902-7_6

Not all variants in tumor genome contribute to carcinogenesis and its progression. Tumorigenesis is caused by driver variants that confer an advantage to growth by promoting cancer development with also effect on gene product function. Passenger variants, on the other hand, are induced by tumor proliferation and often do not confer advantage or functional consequences. The development of cancer requires 5–8 somatic driver variants that occur in both the coding region of genes (1% of the human genome) and non-coding regions. Non-coding variants can affect RNA-binding proteins (RBPs), the binding of transcription factors (TFs) and micro RNAs (miRNAs) and their effect of transcriptional and post-transcriptional gene regulation has yet to be fully understood and experimentally validated. Additionally, OCGs are defined as driver genes in which driver mutations are activating or result in new functions and TSGs are driver genes in which driver mutations are inactivating. During cancer development, tumor cells can modulate molecular and cellular pathways to evade protective mechanisms that prevent tumors through tumor-specific recurrent genetic pathways. Activation of OCGs and TSGs can occur through different genetic mechanisms such as single-nucleotide variants (SNVs), insertion-deletion (indel), copy number variants (CNVs), polyploidy, balanced/unbalanced chromosome rearrangements (translocations, paracentric and pericentric inversions, tandem duplications, and reciprocal insertions) [2] and complex structural variants generating through a catastrophic event engaging one or multiple chromosomes (chromothripsis and chromoplexy) respectively [3]. Structural variations are the emerging key mutational process with clinical involvement in cancer and it was reported that signatures of structural variations are often highly clustered, with differences across patients in size distribution especially for tandem duplication and deletion [4]. The most common structural variants are deletions, followed by unbalanced translocations and tandem duplications [4] which can have important functional consequences such as the formation of gene fusions and chimeric transcripts. Furthermore, chromosomal structural variants can change the copy number of the coding regions of an OCG or TSG and eliminate the binding of OCGs or TSGs to their regulatory elements. All these genomic consequences can explain the dysregulation of gene expression in structural variants. The genesis of structural variants is not yet fully known, and numerous mechanisms have been reported in the literature [4]. Several evidences suggest that structural variation can lead to a change three-dimensional (3D) conformational composition of chromatin. Genome is organized in a compact three-dimensional (3D) structure, and its principal components are topologically associated domains (TADs), lamin-associated domains, CCCTC-binding factor domains, and cohesin. TADs and the protein folding architecture are the basic structural and functional units of chromatin and are directly related to genome functionality. Indeed, through TADs, target genes contact regulatory genomic regions (enhancers) playing a role in gene and transcriptional regulation [5]. Chromosomal structural variants can cause disruption or formation of TADs, alterations in enhancer-promoter communication, resulting in dysregulation of gene expression with abnormal overexpression of associated nearby genes.

The new knowledge in molecular biology has reinterpreted the traditional scenario of tumor hallmarks, suggesting that genetic mechanisms alone might be insufficient to drive carcinogenesis. Tumorigenesis processes have been enriched by including epigenetic changes of tumor DNAs, which modulate gene expression in cancer cells, dedifferentiation of non-stem cells in cancer stem cells (CSCs), and alterations of the microbiome/neuronal signaling. Epigenetics and genetics lead a main role in the development and progression of cancer. Indeed, cancer cells reprogram their genome and epigenome, re-arrange their nuclear space, and introduce variations in the DNA. To understand cancer development, it is essential to evaluate, holistically, genetic alterations of SNVs and CNVs, epigenetic dysregulations, and chromatin remodeling defects. Epigenetic mechanisms comprise several sophisticated regulatory systems that are not yet fully understood. Through interaction with genetic and physiological environmental factors, they control gene expression and chromatin structure, conditioning a wide range of biological processes. Therefore, it is clear that the study of cancer requires new strategies that contemplate genetic and epigenetic traits. In cancer, this close link between genetics and epigenetics in controlling chromatin structure and modulating gene expression can be inferred from various mechanisms. In particular, genetic alterations have been reported in genes encoding transcription factors (TFs) that bind regulatory elements and recruit cofactors (co-activators or co-repressors), in genes encoding epigenetic effectors (e.g., chromatin remodelers, histone modifiers, DNA methylases) and in DNA regulatory sequences that prevent the binding of TFs and epigenetic modifiers.

The role of epigenetic in tumorigenesis is described over 90 years ago and Holliday refined the significance of epigenetics as "heritable changes in gene expression without DNA sequence alteration with phenotype alteration without genotype alteration". The term "epi" is used to identify the molecular process that operates "above" the gene. In non-cancerous cells, the epigenome is well balanced and is essential for homeostasis and development through the regulation of tissue-specific gene expression in mammals and regulates physiological processes such as genetic imprinting, X-chromosome inactivation in female, and chromatin structure. However, various environmental and lifestyle factors can compromise the stability and the typical epigenetic signature of normal mammalian cells leading to the development of genetic disorders associated with epigenetic defects, including tumor.

As mentioned previously for genetic modifications introduced during tumorigenesis, epigenetic modifications can also be diverse in nature, interact with each other, and lead to a change in the organization of the DNA architecture. They alter the accessibility and compactness of the chromatin structure, which also have direct and indirect effects on gene expression, modulation of DNA replication and repair processes. Epigenetic modifications are mostly stable and heritable changes in gene expression caused by chemical modifications of DNA or related regulatory/structural proteins without altering the primary DNA sequence. The patterns of these modifications, called *epigenome* or *epigenetic landscape*, in specific cell types and/or cell states confer unique properties to the cell by regulating and controlling its identity, cell fate, and gene activity.

Epigenome of differentiated cells exhibits typical epigenetic signatures stable across multiple cell divisions unlike stem cells' (SCs) epigenome which presents more plastic properties crucial for developmental processes.

Epigenetic modifications in eukaryotic cells can be divided into the following categories involving closely related biological processes: methylation and demethylation of DNA cytosines, post-translational modifications of histones (hPTMs), variants in the genes that encode histones, ATP-dependent chromatin remodeling complexes, Polycomb (PcG) and Trithorax (TrxG) complexes, and more recently, non-coding RNAs in particular microRNAs (miRNAs) mediate both the targeting of different genes without altering in DNA sequence and, nuclear dynamics and three-dimensional genome architecture. The epigenetic landscape is governed by the dynamic action of enzymes called *writers* (enzymes that attack groups such as acetyl and methyl), *readers* (enzymes that recognize epigenetic modifications and then control the effects of epigenetics), and *erasers* (enzymes that remove post-translational modifications).

DNA methylation is the most common mechanism to silence gene expression in mammals. The pattern of DNA methylation is established by DNA methyltransferase (DNMT) family (DNMT1, DNMT3A, and DNMT3B). DNMT transfers a methyl group from S-adenyl methionine (SAM) at position C5 (carbon 5) of cytosine within CpG dinucleotide, mainly at the 5′ ends of many human genes, to produce C5-methyl cytosine (5mC). It is known that there are regions preferentially unmethylated and extremely methylated in the genome. Groups of regions called "CpG islands" have been identified in human DNA. They are often found on the promoters of genes and housekeeping genes and are between 500 and 1000 base pairs (bp) long. Most of the CpG islands in gene promoters are unmethylated during normal development or in adult cell renewal pathways and they activate the chromatin expression states of CpG island genes (CGI genes). The repressive phenomena linked to the methylation of the CpG islands in promoter are reported in X-inactivation in female and in imprinting process. There are also regions known as unmethylated "shores" CpG islands, located around two kilobase pairs (2 Kb) from the CpG islands associated with genetic activity, and CpG-poor distal regulatory regions (tissue-specific enhancers). Enhancers modulate the expression of their genes through post-translational histone modifications and nucleosome positions.

Hypermethylation of CpG islands, particularly in gene promoters and distal enhancers, induces chromatin compaction, which locks the gene in an inactive state and silences TSG gene expression, contributing to carcinogenesis. On the other hand, methylation of cytosines in the gene body of active genes has the opposite effect of promoters, favoring their transcriptional elongation and enhancing gene expression.

Silencing of gene expression is reported in the literature with promoter methylation of TSGs such as *RB1*, *MLH1*, *p16INK4a/CDKN2A*, and *GSTP1* (π-class glutathione transferase) in different types of tumors. DNA hypermethylation might appear early in tumorigenesis and can be estimated to identify tumor subgroups with high levels of DNA methylation (CpG island methylator phenotype; CIMP) associated with a worse prognosis. The Cancer Genome Atlas reported CIMP phenotypes and stratification of disease subtypes in glioblastoma and colon tumors.

Global DNA hypomethylation has also been described in interspersed DNA repetitive regions, oncogenes, and retrotransposon elements in cancer. DNA hypomethylation causes decompaction of chromatin with induction of homologous recombination and genomic instability, structural variants (deletions, inversions, translocation, and amplifications) and chromosomal abnormalities such as aneuploidies. Moreover, at the molecular level, DNA hypomethylation can determine activation of transposable element (TE), gene expression of repeat elements, and OCGs activation. During tumorigenesis, the global decrease in methylation of tumor's DNA has been described in colon adenocarcinoma, small cell lung cancer, prostate cancer, and chronic lymphocytic leukemia. Otherwise, CpG-poor enhancer elements and CpG island shores acquire hypermethylation in tumors. The mechanism underlying the global loss of methylation in the genome of cancer cells is not yet known. DNA methylation can be actively removed from DNA. Demethylation is a recurring mechanism during early development of an organism and has subsequently been demonstrated in other cell types. It is a dynamic process during embryogenesis, physiological cell biology, and disease. An imbalance in the DNA methylation/demethylation state results in a disruption of gene expression patterns and genomic stability in the physiological development of an organism, in the development of disease including cancer. It has been hypothesized that genes that regulate methylation homeostasis, such as members of the *ten-eleven translocation (TET) family* or *DNMT* with loss of function, are involved in this imbalance. These involved genes, together with structural changes in chromatin, may lead to functional effects in the vicinity of hypomethylated regions.

The demethylation process is principally driven by the TET family (TET1–3) of enzymes, which are involved in oxidization of 5-methylcytosine (5mC) to 5-hydroxymethylcytosine (5hmC), a key intermediate in the active demethylation pathway. Reduction of 5hmC level has been reported in melanoma, breast, prostate, colon, liver, lung, and pancreatic cancer. Underexpression or reduced activity of hydroxylase TET proteins or mutations/depletion in TET genes may be responsible for reduced 5hmC levels. Alteration of isocitrate dehydrogenases 1 and 2 (IDH1/2) enzymatic activities represents an indirect effector of the TET pathway. IDH1/2 enzymes catalyze the oxidative decarboxylation of isocitrate to 2-oxoglutarate, a cofactor of TET catalytic activity. *IDH1/2* gene mutations determine a reduction of 2-oxoglutarate with aberrant accumulation of oncometabolite (R-)-2-hydroxyglutarate. (R-)-2-hydroxyglutarate represents a competitive inhibitor of TET enzyme activity in glioma, acute myeloid leukemia, and melanoma. Mutations in the *TET* and *IDH* genes are mutually exclusive and bring an increased CpG island hypermethylation.

It has been shown that hypermethylation in tumors alters the binding affinity of the insulator DNA-binding protein CTCF (CCCTC-binding factor) between an enhancer and a gene in the TADs. Reduced binding of CTCF, induced by epigenetic dysregulation, leads to aberrant interactions between enhancers and genes. Constitutive activation of platelet-derived growth factor receptor A (*PDGFRA*), a known glioma oncogene, exemplifies such a mechanism. Loss of CTCF isolation can be transmitted during cell divisions, leading to further activation of other oncogenes.

In epigenetic context, gene expression patterns can also be influenced by genetic variants in epigenetic modifier genes: variants in the *DNMT3a* gene have been reported to be associated with 22% of cases of acute myeloid leukemia (AML) with poor outcome, and in the *TET2* gene with 15% of myeloid neoplasms.

Post-translational modifications of histones and variants in genes encoding histones are another epigenetic mechanism that alter chromatin structure. Chromatin is organized into repetitive units called nucleosomes. A nucleosome is a histone octamer of four histone proteins (H3 and H4 and H2A and H2B) with 147 bp of DNA wrapped around it. The modulation of accessibility and compactness of the chromatin structure depends on histone proteins and their hPTMs, mainly in their charged N-terminal amino acid tails of core histones (especially H3 and H4). hPTMs are introduced by chromatin-modifying enzymes that recruit DNA-bound transcription factors to chromatin altering its structural properties. Chromatin modifying enzymes are classified based on their functions: acetylation by histone acetyltransferases (HATs), deacetylation by histone deacetylases (HDACs), methylation by methyltransferases (HMTs), and demethylation by histone demethylases (HDMs). Acetylation relaxes the structure of the chromatin and allows the access of RNA polymerase and transcription factors, piloting to active gene transcription. In contrast, histone methylation generally leads to condensation of the chromatin structure, which represses gene transcription. HATs add acetyl groups to lysine residues on histones and HDACs carry out histone deacetylation. Similarly, HMTs promote the addition of mono-, di-, or tri-methyl groups to arginine and/or lysine residues on histones, while HDMs are responsible for removing these methyl groups. The majority of hPTMs are reversible and associated with different chromatin states. They could function together or alone to promote chromatin accessibility, activation/repression of chromatin-mediated gene expression including tumor-associated genes [6] and genome integrity. Conversely to DNA methylation, histone modifications not only silence the expression of certain genes, but also promote transcription. Acetylation, methylation, phosphorylation, ubiquitination, ADP ribosylation, and sumoylation are the principal post-translational modifications of histone proteins that determine chromatin remodeling. There are specific amino acid hotspots in histones involved in hPTMs. The histone methylation sites are located at H3K4, H3K9, H3K27, H3K36, H3K79, and H4K20, the acetylation sites are at H3K9, H3K14, H3K18, and H3K23 and H4K5, H4K8, H4K12, and H4K16 and the phosphorylation site is indicated at H3S10. The ubiquitination site is randomly designated in H2A and H2B.

There is a "physiological" balance between specific hPTMs and modifiers that regulates the structure of chromatin, and consequently the genetic expression program. In tumors, this balance is changed and cellular phenotypes can be altered. Several cancers have been linked to abnormalities in histone methylation and acetylation.

During histone acetylation, HATs add acetyl groups to lysine residues on the N-terminal tails of histone nucleosomes, resulting in loss of positive charge, reduced histone-DNA affinity, and subsequent unwinding to allow access to RNA polymerase and transcription factors. The DNA helix undergoes subsequent unwinding which allows access to RNA polymerase and transcription factors. HDACs, on the other hand, remove acetyl groups and act as transcriptional repressors. Histone methylation can promote or inhibit various gene transcriptions, depending on the location and level of methylation of the target amino acid. Methylation of H3K4 and H3K36 can activate gene transcription. Methylation of H3K9, H3K27, H3K79, and H4K20 can repress gene transcription.

HMTs can add one, two, or three methyl groups, and trivalent methylation is known as hypermethylation. EZH2, core subunit of polycomb repressive complex 2 (PRC2) protein, mediates trivalent methylation of lysine 27 on histone H3 (H3K27me3), causing chromatin condensation and repression of TSGs transcription.

Altered histone patterns and dysregulated levels of hPTM are responsible for silencing TSGs and activating OCGs in multiple tumor types. For example, decreased levels of H3K9ac, H3K9me3, and H4K16ac have been identified in non-small cell lung cancer and loss of H3K9me3 in promoter genes has been documented in patients with acute myeloid leukemia. More, increase or reduction levels of specific hPTM signature have been linked to the prognosis of several types of tumors. Dysregulation of the enzymes HMTs and HDMs has been identified in various tumors (breast, prostate, lung, and brain). HTMs and HDMs are mainly involved in the regulation of the methylation status of residues K4, K9, K27, and K36 on histone H3 in multiple tissues.

Over the last twenty years, mutations in histone genes, called oncohistones have been identified in several cancers. Alterations in these genes are responsible for disruption of histone-DNA contacts, destabilization of nucleosomes, and alterations in chromatin-associated remodeling activities. This leads to changes in chromatin compaction and altered epigenetic signaling. Different pathogenetic somatic missense variants in the 4 histones have been identified in different pathologies and in particular in different types of cancer. The most studied are the variants at two hotspots that map to the educated N-terminal tail of *H3F3A* (Lys27 and Gly34) and of *HIST1H3B* (Lys27) genes encoding histone H3.3 and H3.1, respectively. They have been identified mainly in rare pediatric brain and bone tumors. Lys27 hotspot of *H3F3A* compromises the repressive function of the PRC, causing deregulation and changes in hPTM in tumor cells. Diffuse midline gliomas (DMG), a devastating childhood brain tumor, carry mutations in the *HIST1H3B* or *H3F3A* genes (80%). In addition, H3 wild-type (H3WT) DMGs harbor overexpression of the enhancer of *Zeste Homologs Inhibitory Protein* (*EZHIP*), also termed *CXORF67*. It can inhibit PRC2 by interacting with EZH2.

Trimethylation of H3K27 (H3K27me3) is reduced by both H3K27M mutations and overexpression of EZHIP, resulting in dysregulation of gene expression, including upregulation of an RNA-binding protein (RBP), LIN28B. LIN28A/B cooperates with the let-7 family of microRNAs during physiological development and oncogenesis (the LIN28B–let-7–oncogene axis). The epigenetic effect of the H3K27M variant can induce the expression of LIN28B, which decreases the let-7 level, leading to increased expression of the oncogene promoting DMG formation. Dysregulation of the LIN28B-let-7 oncogene axis has also been reported in other types of pediatric brain tumors, including embryonal multilayered rosette tumor, atypical teratoid rhabdoid tumor (ATRT), and medulloblastoma.

Histone mutations have also been identified in the globular and C-terminal domains of the linker histone, also known as histone H1. Such mutations occur in 50% of Hodgkin lymphomas, 30–40% of diffuse large B-cell lymphomas, and 30% of follicular lymphomas.

Another mechanism implicated in epigenetic regulation is the involvement of ATP-dependent chromatin remodeling complexes. Their central component has a helicase domain that requires energy from ATP hydrolysis to induce nucleosome sliding and conformational changes with nucleosome displacement. Different families of ATP-dependent chromatin remodeling complexes are identified: the Chromodomain Helicase DNA (CHD) family which bind directly to methylated lysines at histone H3, the SWI/SNF complex containing over 10 subunits that if mutated lead to rhabdoid tumors and other tumors; the INO80/SWR family and the ISWI family which includes two independent ATPase subunits, SNF2H and SNF2L. As mentioned above, the genes encoded for SWI/SNF subunits are mutated in several types of cancer, but it should also be reported that epigenetic defects can cause dysregulation of SWI/SNF. Rhabdoid brain tumors are characterized by genetic deactivation of the *SMARCB1* gene encoding the NF5 component of SWI/SNF. This complex works together with nucleosomes and hPTMs to occlude or expose DNA regulatory sequences and regulate the exchange of histone variants in specific genomic regions.

The multiprotein complexes Polycomb (PcG) and Trithorax (TrxG) regulate the chromatin accessibility of homeotic genes. They can affect gene expression and genome stability at different levels. In particular, the PcG complex is composed of the Polycomb Repressive Complex 2 (PRC2, the initial complex) and the PRC1 complex (the maintenance complex). PRC2 has two complexes (PRC2.1 and PRC2.2) that incorporate histone H3K27me3 into their target chromatin regions. PcG has been implicated in the development of breast, prostate, and many other cancers.

Non-coding RNA activities are dysregulated in many diseases, including cancer. Non-coding RNAs (ncRNAs) interact with nucleic acids and proteins in cells to modulate post-transcriptional gene expression, control genomic and epigenomic regulation [7], and their mechanisms of action are poorly understood.

Many tumors exhibit specific microRNA (miRNAs) expression patterns, a particular type of ncRNAs. The downregulation or loss of tumor suppressor miRNAs can increase the expression of OCGs and lead to tumorigenesis, and upregulation of oncogenic miRNAs can inhibit TSGs and lead to tumor formation as well. They have been classified as small non-coding RNAs (sncRNAs; < 200 nucleotides in length) and long non-coding RNAs (lncRNAs >200 nucleotides in length). sncRNAs are involved in gene transcription, protein synthesis, and splicing, instead lncRNAs are highly cell type-specific, have regulatory effects at the chromatin structure, contribute to epigenetic regulation, and are implicated in carcinogenesis and disease progression in various types of cancer. Some lncRNAs interact with chromatin-modifying complexes such as Polycomb Repressive Complex 2 (PRC2) and nuclear DNA-binding factors such as YY1 and CTCF. HOTAIR is an example of lncRNAs overexpressed in cancer metastasis that promotes increased chromatin compaction and gene silencing through interaction with the PRC2 complex and LSD1 (lysine-specific histone demethylase 1). It affects the expression of at least 800 genes. Recent data indicate that processes such as DNA methylation and histone modification can be regulated by different miRNAs (let7a, miR9, miR34a, miR21, miR124, miR137, miR148, and miR203). In addition, the expression of important epigenetic regulators such as DNA methyltransferases, histone deacetylases, and polycomb group genes is controlled by a subset of specific miRNAs. This supports the existence of an epigenetic-miRNA regulatory circuit that organizes and controls the entire gene expression profile. CpG island methylation of miR-148/152 family member genes and DMNT1 overexpression in gastric cancer reduce the expression of these miRs with inactivation of their functions. Epigenetic inactivation of the miR-148 family results in increased TGFβ signaling and consequently tumor growth and metastasis. miR-34a, a controller of various cell cycle progression and apoptosis proteins, is inactivated by CpG island methylation upstream of its transcription start site in various tumors. In addition, the expression of miR-34a and miR-34b/c is epigenetically silenced in cholangiocarcinoma cells and in soft tissue sarcomas, respectively. Further miRs downregulated by CpG island methylation are miR-137 in colorectal cancer, melanoma, breast cancer, and miR-124 in high-grade gliomas and non-Hodgkin's lymphoma.

The genesis of the tumor cell from embryonic or adult stem cells (SCs) plays a primary role. SCs are pluripotent, self-renewing cells capable of generating new stem cells and different lineages of daughter cells to replenish the functional cell pool. SCs are highly plastic cells with a marked ability to switch from one phenotypic state to another. SCs contain a more plastic epigenome required for developmental processes compared to the epigenome of the differentiated cell, which remains stable through multiple cell divisions.

Genetic and epigenetic changes in SCs can result in loss of differentiation capacity, increase cell growth and invasion, leading to the formation of cancer stem cells (CSCs). Furthermore, differentiated cancer cells can undergo a process of re-programming and de-differentiation into CSCs. They are a minor subpopulation of malignant cells so called due to their "stem-like" properties. These cells possess an extensive capacity for self-renewal (symmetric and asymmetric) and differentiation, and thus can generate the different cell types that constitute the entire tumor. CSCs can divide asymmetrically into two different cell types or symmetrically into two similar cells to achieve self-renewal, multi-lineage differentiation and infinite proliferation. Cell surface markers expression, gene expression signatures, and subsequent pathway activations identify a CSC. CSCs have a low level of mitotic activity, and are protected from the action of specific therapeutic agents that target highly dividing cells. The intrinsic plasticity of CSCs confers the ability to give rise to all the cells that will make up the entire tumor and is tightly regulated via the complex interplay between the 3D epigenome, the epigenetic machinery, factors produced by stem cells, and structural proteins that modulate gene expression. It has been hypothesized that 3D chromatin architectures are pre-established during CSC reprogramming. The hypothesis is that some of the chromatin interactions (e.g., enhancer-promoter interactions) might be pre-established in pre-cancerous stem cells (pre-CSCs) by transcription factor binding. This interaction drives the transcription of the gene involved in CSC reprogramming, leading to the formation of CSCs [8]. In addition, to their role in cancer initiation, progression, and metastasis, CSCs have other properties, such as upregulation of drug efflux pumps, self-renewal properties, tendency to quiescence, high plasticity, and their tumor protective niche, which can confer resistance to therapy and promote tumor survival of poor responder/recurrent tumors leading to disease relapse. The involvement of CSCs in the tumorigenesis has been demonstrated in various types of tumors including acute myeloid leukemia (AML), brain, breast, prostate, and colon cancers. The tumor microenvironment in which CSCs reside is complex, highly specialized, and heterogeneous, and provides essential insights into the self-renewal and propensity of CSCs. During tumor growth, cancer cells tend to acquire genetic mutations and epigenetic modifications that generate a heterogeneous set of epigenomes/genomes. The tumor heterogeneity enriches the proliferative potential and aggressiveness of the tumor and represents an obstacle to the treatment of malignancies. The two theories that explain the phenomenon of tumor heterogeneity are that of clonal evolution and that of CSCs. The two theories are not mutually exclusive and they can overlap. In 2006, Feinberg proposed the epigenetic progenitor model of tumorigenesis. This model suggests that tumors originate from early, polyclonal epigenetic dysregulation of the progenitor cell population, followed by initial mutations and the acquisition of genetic and epigenetic plasticity to evolve their phenotype in a stable manner. Several direct or indirect epigenetic mechanisms have been implicated in the genesis of CSCs: DNA hypomethylation/methylation mediated by TET and DMNTs proteins, remodeling of nucleosomes and their associated complexes due to mutations in SWI/SNF, alterations in 3D chromatin architecture due to post-translational modification of histones or Polycomb group protein activity and altered expression of non-coding RNAs (high and low levels of Lin28 and low let-7, respectively).

Hypermethylation of genes involved in cell differentiation and TSGs, together with hypomethylation of specific pluripotency loci, is important in the generation of CSCs. In detail, the specific methylation signature of 5mC and 5hmC is much more important than the global methylation pattern in establishing the cellular identity of CSCs. Physiologically, methylation of CpG islands occurs during cell differentiation and demethylation of CpG islands in the induction of pluripotency of both SCs and induced pluripotent stem cells (iPSCs). In tumors, CSCs generation has been shown to be associated with loss-of-function mutations in the gene encoding DMNT3A and loss of methylation of the Nanog promoter through inhibition of DNMT1. However, some aspects remain to be clarified. For example, DMNT1 has been shown to play opposing roles in CSC formation in different tumors. DNMT has also been shown to be important for CSC function in leukemia, breast, lung, and pancreatic ductal adenocarcinoma (PDAC) tumors and is required for colon cancer initiation. In contrast, DNMT1 inhibition promoted CSCs in prostate cancer. It has been shown that certain regions of the genome are exclusively demethylated in CSCs. In particular, in PDAC tumors, the *GATA6*, *SOX9*, and *BMP4* genes, which are involved in the development of CSCs, are specifically demethylated in the invasive cell population with stem cell characteristics. Most importantly, hypomethylation process is associated with CSCs maintenance after tumor formation. For example, the CpG region of the *CD133* gene promoter, a gene that encodes a tumor stem cell marker, has been shown to be hypomethylated in several types of cancer, including ovarian, breast, colorectal, and glioblastoma.

The chromatid structure of the genomes of differentiated cells and SCs differs by approximately one-third. Notably, SCs contain smaller regions with the marks H3K9me3 and H3K27me3 and more acetylated chromatin in comparison to differentiated cells, resulting in an enrichment of transcriptionally active euchromatin and less heterochromatin. Moreover, enzymes involved in histone modification can produce tumor-specific effects. The inhibition of histone methyltransferase EZH2 is associated with the generation of CSCs in some cancers and it has been hyperactivated in PDAC and breast cancer.

Nucleosome repositioning may be important for CSC formation and may occur through aberrant functions of various types of ATP-dependent chromatin remodelers in leukemia, lung cancer, rhabdoid tumors, epithelioid sarcoma, and renal medullary carcinoma. ATP-dependent chromatin remodeling complexes can influence gene expression through the incorporation of histone variants into nucleosomes which directly alter nucleosome structure and stability or through the recruitment of readers of histone modifications to induce local changes of the chromatin. These two aspects generate dynamism and they are responsible for the plasticity of SCs and CSCs.

Many lncRNAs and mi-RNAs are implicated in the phenomenon of pluripotency, cellular plasticity, and in the control of many types of adult SCs. Long intergenic ncRNAs (lincRNAs) are essential for the cellular identity because of their tissue-specific expression and some of them are highly expressed in CSCs. Specifically, H19 is involved in glioblastoma CSCs and LINK-RoR is involved in the conversion of cancer cells to CSCs during tumorigenesis and it is important in a wide range of cancer types. Regarding mi-RNAs it is important to mention Let7 and Lin-28. These two miRNAs create a miRNA axis that controls the genesis of CSCs in cancer and the pluripotency of non-cancerous cells. Let7 miRNA expression is controlled by a negative feedback loop with Lin-28. In fact, in cancer cells Lin-28 binds to let-7 precursors to block their maturation, whereas let-7 targets Lin-28 for degradation in differentiated cells. Nowadays, it is known that in breast cancer Let7 is a negative regulator of CSCs where it is implicated in cell differentiation. Instead, Lin-28 promotes CSC formation in breast and prostate cancer, and induces tumorigenesis in multiple tumors. A further role of mi-RNAs is to establish a rich crosstalk network with the pathways and epigenetic mechanisms involved in CSCs genesis. In breast cancer, miR-22 directly inhibits TET activity, preventing demethylation of the miR-200 promoter, thereby silencing its expression and promoting SCs formation.

Stem cell activity is modulated by a complex network of signaling pathways that can be deregulated because of epigenetic changes occurring during tumor formation. Their aberrant activation can occur through genetic alterations, DNA methylation, aberrant histone modifications, or dysregulation of non-coding RNAs. The result of these epigenetic changes is that tumor cells acquire their own self-renewal capabilities with unlimited proliferation, maintenance of CSCs status, progression, tumor invasion, and drug resistance. The main signaling pathways activated are the Wnt/β-catenin, Hedgehog, and Notch signaling pathways, usually involved in physiological development and maintenance of normal tissues and organs, but are also known to be involved in the tumorigenesis of various cancers. At present, the epigenetic regulation of these signaling pathways and their modifiers may be targets for therapy or be considered as potential mechanisms for targeted therapy against CSCs.

Several bulk methodologies have been developed to study different aspects of epigenetics (DNA methylation, chromatin accessibility, histone modifications, DNA-protein interactions, and 3D chromatin architecture) but, due to the complex cellular heterogeneity of many tumor types, the use of single-cell sequencing technologies has become indispensable. These methodologies can study only one (mono-omic approach) or multiple (multi-omic approach) epigenetic mechanism simultaneously on a single cell. In the near future, single-cell epigenetic technologies will be able to identify complete epigenetic signature associated with each type of tumor, discover new diagnostic/prognostic/predictive markers to develop personalized therapeutic strategies that will avoid the phenomenon of resistance to therapy, and allow the detection of early metastatic events [9].

Table 6.1 shows the main recurrent epigenetic mechanisms in a series of hematological and solid tumors.

A better characterization of mechanisms of chromatin biology and non-coding RNAs (ncRNAs) has allowed the development of epigenetic drugs able to interfere with transcriptional programs implicated in tumorigenesis, progression, and drug resistance mechanisms [27].

Epidrugs aim at targeting various epigenetic factors that introduce, recognize, or remove modifications at the level of DNA or histones [28]. The safety and efficacy of so-called epidrugs have been evaluated in single-agent trials, but also in combinatorial approaches with conventional anti-cancer drugs or natural epigenetic modulatory bioactives (epi-diets) [29].

According to the above-mentioned updated molecular dissection of cancer epigenetics, we can didactically divide potential epidrugs under investigation or approved for clinical use according to protein family as reported in Table 6.2.

Aberrant DNA methylation represents a pivotal target of several classes of epigenetic drugs. Among those, DNMT inhibitors have symbolized a historically forerunner class of epidrugs since the Food and Drug Administration (FDA) approval of azacytidine in 2004 and decitabine in 2006 for the treatment of myelodysplastic syndrome (MDS) and chronic myelomonocytic leukemia (CMML). To date, their use has been extended to other conditions as juvenile myelomonocytic leukemia (JMML) and AML, but, in contrast, the toxicity and bioavailability profile pose primary concerns for their use against solid tumors. With these premises, SGI-110 (Guadecitabine) has been synthesized as an oral bioavailable prodrug of decitabine,

Table 6.1 Epigenetic aberrations by tumor type

Tumor type	Epigenetic mechanisms	References
ALL	*Global hypomethylation; Aberrant DNA methylation: TP73, TET2, CDNK1A (p21), CDKN2B (p15), CDKN2A (p16), sFRP2, DKK3, WIF1, PTEN, SYK, C/EBPA, PAX5, ABCG2; mutations of DNMT3A, IDH1/2, TET1 leading to DNA hypermethylation; Aberrant Histone modification by mutations of PHF6, EZH2, SUZ12, EED, JARID2, BMI1,UTX, MLL1, MLL2, EP300, SETD2; Overexpression of HDACs; Loss of global histone H4 acetylation; Inhibition of HATs:* CREBBP; *Overexpression of Arginine Methyltransferase Family: PRMT5; Upregulation of H4R3sme2; Overexpression of Lysine methyltransferases DOT1L; ATPase remodeling complexes*: ETV6-INO80D fusion; *miRNA deregulation:* miR-128, miR-181a, miR-155, miR-146a, miR-497, other miRNAs reported in [10, 11]	[10, 11]
AML	*Aberrant DNA methylation by DNMT3A, TET2, IDH1/2 mutations; Deregulation of HATs:* CREBBP, KAT6A, EP300; *Deregulation of HDAC2/3; Aberrant Histone methylation by mutations of KDM5A, KDM6A, KMT2A, NSD1, JARID2, SUZ12,ASXL1EZH2; Mutations of ASXL2; miRNAs deregulation*: miRNA-10a, miRNA-21, miR-542, miR-509 has-miR-146a, other ncRNAs reported in [12]	[12]
Breast cancer	*Global hypomethylation; Aberrant DNA methylation: CCND2, CDKN2A, BRCA1, GSTP1, BCL2, DAPK, RASSF1A, RARβ, TWIST, HIN1, HOXA5, CDH1, APC, CTNNB, MINT17, MINT13, ERα, ERβ and THRβ; Hypomethylation* of genes *IL10, MDR1, FEN1, NAT1, CDH3, urokinase, synuclein, JAGGED1 and NOTCH1; Aberrant ncRNA promoter methylation; Histone modifications:* hyperacetylation of H3K4, hypomethylation of H3K9me2 and H3K9me2 by KDM3A/JMJD1A, increase in H3K4me3; Abnormal patterns of H1 phosphorylation (threonine 146); *Overexpression HDAC1, HDAC2* and *HDAC3* genes; *Overexpression of Demethylase* 1 (LSD1); *Inhibition of HATs*: H4K20 acetylation by hMOF; Aberrant H4K20me3 levels; *miRNA deregulation:* Let-7c, miR-769–3p, miR-10b, miR125b, miR-128a, other miRNAs reported in [13, 14]	[13, 14]

(continued)

Table 6.1 (continued)

Tumor type	Epigenetic mechanisms	References
Colorectal cancer	*Aberrant DNA methylation: ALX4, APC, BMP3, CDKN2A, CHFR, EVL, GATA4, HIC1, HLTF, HPP1, IGF2, IGFBP3, ITGA4, LINE-1, MDG1, MGMT, MLH1, NDRG4, NEUROG1, NGFR, NPTX2, OSMR, p14ARF, PCDH10, PDL1, PHACTR3, PRIMA1, QKI, RARB, RASSF1A, RASSF2, RET, SDC2, SEPT9, SFRP1, SFRP2, SHOX2, SPARC, SST, TFAP2E, TFPI2, TMEFF2, UCHL1, VIM, WIF1; Aberrant histone methylation:* H3K9me2, H3K27me3, H4K20me3, H3K4me2, H3K4me3, H3K9me3, H3K20me3, H3K27me2*; Aberrant histone acetylation:* H3K9Ac, H3K56ac, H4K16ac, H3K18ac, H3K27ac, H4K12ac*; miRNA deregulation:* let-7b, let-7c, let-7 g, miR-103, miR-103a-3p, miR-106a, other ncRNAs reported in [15]	[15]
DMG	*Somatic mutation* p.Lys27Met in *H3F3A* or *HIST1H3B*; *EZHIP overexpression* in H3-wt DMGs; *Deregulation of the LIN28B-let-7-oncogene axis*	[16, 17]
GIST	*Global hypomethylation*; *Aberrant DNA methylation: MGMT, CDKN2A, RASSF1A, CDH1, hMLH1, APC, MINT1/2, p73, MINT31, SDHx, PTEN, REC8, PAX3, MBD2* and *SPP1; DNMT and MBD2 Overexpression; Involvement of histone* H2AX; *miRNAs deregulation:* miR-34a and miR-335	[18]
HCC	*Aberrant DNA methylation: CDKN2A, RASSF1, APC and SMAD6; Hypomethylation of IGF2, CCL20, NQO1; Histone methylation: EZH2, EHMT2, SETDB1 and SETD2; Histone acetylation: p300 and MOF; Histone ubiquitination: USP22 and ATXN7L3;Histone phosphorylation: Aurora A, pH3S10, pH3S28, MTA1; miRNAs deregulation:* miR-15a/16–1, miR-122, miR-139 miR-192, miR-199a/b-3p, other ncRNAs reported in [19]	[19]
HGG	*Aberrant DNA methylation*: *p14ARF, CDKN2A, NDRG2, MGMT* and *MLH1; Histone modification*: hyperacetylation of H3; histone lysine methylation by MLL1, MLL2, EZH2 and G9a Histone Methyltransferases; EZH2 overexpression and Gly34Arg/Val in *H3F3A* gene; *Chromatin remodeling*: up-regulation of CR factor (LSH); *miRNAs deregulation:* miR-451, miR-1275, miR-137, miR-30, miR-17-92, other miRNAs reported in [20, 21]	[20, 21]
Melanoma	*Global hypomethylation: IDH-2* downregulation or mutation; *Aberrant DNA methylation*: *PTEN, CDKN2A, p14ARF, RASSF1A, cGAS* and *STING; Chromatin remodeling perturbation:* Histone hypoacetylation or hypermethylation, Chromatin modification Recognition, SWI/SNF complex regulation and NURF complex regulation; *Loss of histone variant* macroH2A *and overexpression of histone variant* H2A.Z.2; *miRNAs deregulation*: miR-200c, miR-149, miR-21, MiR-1908, miR-199a-5p, other ncRNAs reported in [22]	[22]

NSCLC	*DNMTs overexpression; CIMP: RASSF1A, SEMA3B, DAPK, P14, CDKN2A, CDH1, CDH13, BLU, TBX-2, WIF1; Histone modifications:* hyperacetylation of H4K5/H4K8, hypoacetylation of H4K12/H4K16 and loss of H4K20 trimethylation; *Aberrant histone modifications that deregulate signaling pathways* (PI3K/AKT/mTOR and Wnt/-catenin); *Overexpression HDAC1 and HDAC3 genes; miRNAs deregulation:* miR-9-3, miR-34b and miR-126, miR-487b, miR-30a, other miRNAs reported in [23, 24]	[23, 24]
Pancreatic nets	*Aberrant DNA methylation: DAXX* gene and several oncogenic pathways (mTOR, NOTCH, Wnt, CDKN2A); *Increased levels of HDAC5* in metastatic PNETs; *Histone Modification Status:* Mutations in chromatin remodeler *ATRX* and the histone chaperone *DAXX;* Loss of Menin causes H3K4me3 loss and sporadic PNETs; Loss of MEN1 alters the epigenetic landscape of its target genes; Overexpression of Histone demethylase Rbp2	[25]
PROSTATA	*Aberrant DNA methylation: GSTP1, PTEN, CDKN2A* genes; *Overexpression of KMT*: KMT1A, KMT1E, SET, SMYD3, PRMT5; *Overexpression* of EZH2 and NSD2; *Increased activity of HATs*; *Decreased activity of HDACs*; *Deregulation in BRD-containing proteins:* BRD4, BAZ2A, CHD1, ATAD2, BRD8, CREBBP, or KTM2A; *Decreased expression of TET enzymes* (TET1, TET2); *Deregulation Demethylase* 1: LSD1/KDM1A, *H3K4 Demethylase KDM5D, KDM5C, KDM4B, KDM5B, PHF8*	[26]

ALL Acute lymphoblastic leukemia, *AML* Acute myeloid leukemia, *BANCR* BRAF-activated non-coding RNA, *BRD* Bromodomain, *CCAT* Colon cancer-associated transcript 1, *CREBBP* CREB binding protein, *CRNDE* Colorectal neoplasia differentially expressed, *DMG* Diffuse midline glioma DMG, *EZHIP* Enhancer of Zeste homologs inhibitory protein, *GAS5* Growth arrest-specific 5, *GIST* Gastrointestinal stromal tumors, *HCC* Hepatocellular carcinoma, *HGG* High-grade gliomas, *HOTAIR* HOX transcript antisense RNA, *IDH-2* Isocitrate dehydrogenase 2, *KMT* Lysine methyltransferases, *miRNAs* microRNAs, *MTA1* Metastasis-associated 1, *ncRNAs* Non-coding RNAs, *NETs* neuroendocrine tumors, *NSCLC* non-small cell lung cancer, *MALAT1* metastasis-associated lung adenocarcinoma transcript 1, *PRMT5* Arginine methyltransferase family, *PVT1* Plasmacytoma variant translocation 1, *Rbp2* Retinoblastoma binding protein 2, *SMYD3* MYND domain-containing protein 3, *USP22* Ubiquitin-specific protease 22, *XIST* X-inactive specific transcript

Table 6.2 Epigenetic therapies: approved or under investigation drugs

Protein family/target	Approved drug	Conditions	Under investigation
DNMT inhibitors	*Decitabine*	AML, MDS, CMML	CC-486 (oral azacitidine) guadecitabine, ASTX727, CP-4200, 4′ -Thio-20-deoxycytidine and RX-3117
	Azacitidine	AML, MDS, CMML, JMML	
IDH1 and IDH2 inhibitors	*Ivosidenib*	AML, Cholangiocarcinoma	
	Enasidenib	AML	
	Olutasidenib	AML	
	Vorasidenib	LGG	
HDAC inhibitors	*Vorinostat*	NHL	Resminostat (4SC-201), givinostat (ITF2357), pracinostat (SB939), abexinostat (PCI-24781), quisinostat (JNJ-26481585), CHR-3996, entinostat, tacedinaline, mocetinostat, domatinostat, OKI-179, fimepinostat, NBM-BMX
	Romidepsin	NHL	
	Belinostat	NHL	
	Panobinostat	MM	
	Chidamide (tucidinostat)	NHL[a]	
HDAC6 inhibitors			Ricolinostat, Citarinostat, KA2507
EZH2 methyltransferase inhibitors	*Tazemetostat*	NHL, rhabdoid tumor, epithelioid sarcoma	CPI0209, CPI1205, Valemetostat
DOT1L methyltransferase inhibitor			Pinometostat
LSD1 inhibitors			Iadademstat, CC90011, INCB059872
BET inhibitors			Molibresib, Birabresib, ZEN003694, PLX51107
miRNA based-drugs			RX34 (miR-34a mimic), MRG-110 (anti-miR-92), MRG-201 (anti-miR-29), MRG-106 (anti-miR-155), MesomiR 1 (TargomiRs), Anti-miRNA-34a and anti- miRNA-194

Modified from [28, 29]

[a]It has received regulatory approval from China. *AML* Acute myeloid leukemia, *BET* Bromodomain and extraterminal, *CMML* chronic myelomonocytic leukemia, *JMML* juvenile myelomonocytic leukemia, *DNMT* DNA methyltransferase, *HDAC* Histone deacetylase, *IDH* Isocitrate dehydrogenase, *LGG* low-grade glioma, *LSD1* Lysine-specific demethylase 1, *MDS* myelodysplasia, *MM* multiple myeloma, *NHL* non-Hodgkin lymphoma

currently tested in phase I and II trials for hematological malignancies. Further promising nucleoside analog DNMT inhibitors are CP-4200, 4′-Thio-20-deoxycytidine, and RX-3117, now under investigation in trials for AML, MDS, and ovarian and liver cancers. Though non-nucleoside analogs (e.g., laccaic acid, procainamide, SGI-1027, RG-108) have been shown to overcome some concerns of nucleoside counterparts, as genotoxicity, the potency of these drugs seems very limited [28, 29].

Mutations in *TET2*, *IDH1,* and *IDH2* in AML highlight the malignant potential of DNA hypermethylation in several tumor types (e.g., AML, glioma, chondrosarcoma, and intrahepatic cholangiocarcinoma), but the evidence of ineffectiveness of DNMT inhibitors opened the route to other classes of compounds. Hot-spot mutation of *IDH1* and *IDH2* in AML have been targeted effectively by specific IDH inhibitors Ivosidenib and Enasidenib, respectively, now approved for this condition [28]. In solid tumors, Vorasidenib has shown a significant objective response in the phase III trial for grade 2 IDH-mutant glioma [30].

Several histone lysine methyltransferases (KMT) writers have been implicated in oncogenesis, including EZH2, NSD2, SETD2, and DOT1L. Tazemetostat is currently the most advanced EZH2 inhibitor, approved for non-Hodgkin's lymphoma (NHL) follicular lymphoma but its use has been extended to rhabdoid tumor and epithelioid sarcoma with loss-of-function mutations in SWI/SNF family proteins as *SMARCB1* deletion [28]. Since the central role of histone acetylation in epigenetic machinery, HATs and HDACs become a suitable target of specific inhibitors. In preclinical studies, several HAT inhibitors showed submicromolar efficiency in inhibiting HATs; however, selectivity and efficiency in clinical trials are the significant challenges in developing HAT inhibitors.

Moving from writers to erasers, HDAC inhibitors include some of the most paradigmatic epidrugs. These compounds are divided by chemical structure into hydroxamic acids, cyclic peptides, short-chain fatty acids, and amino-benzamides. To the first class belongs vorinostat that has been the first HDAC inhibitor approved by the US-FDA in 2006 for primary cutaneous T-cell lymphoma (CTCL) and now under investigation in several trials for solid tumors. Other relevant hydroxamic acid-based HDAC inhibitors are belinostat, a pan-HDAC inhibitor approved for peripheral T-cell lymphoma (PTCL), and panobinostat approved in combination with bortezomib and dexamethasone for drug-resistant multiple myeloma. Several promising compounds of this class named resminostat (4SC-201), givinostat (ITF2357), pracinostat (SB939), abexinostat (PCI-24781), quisinostat (JNJ-26481585), and CHR-3996 are under various stages of clinical trials for both solid and hematological malignancies. Romidepsin is a cyclic peptide HDAC inhibitor that has been approved for relapsed/refractory CTCL in 2009; while apicidin is another cyclic peptide that however showed limited effectiveness despite preclinical antitumor activities. In the class of short-chain fatty acid HDAC inhibitors, there are sodium butyrate, valproic acid (VPA), sodium phenylbutyrate as promising agents against several solid and hematological malignancies. Lastly, among benzamide HDAC inhibitors, Chidamide has achieved approval by the Chinese regulatory agency for the treatment of recurrent/refractory PTCL while other compounds as entinostat, ricolinostat, tacedinaline, and mocetinostat are undergoing clinical trials for the treatment of several malignancies [28].

To date, there are several interesting compounds under investigation in clinical trials which have not yet been granted regulatory approval. They act on different pathways to restore epigenetic imbalance as DOT1L methyltransferase inhibitor Pinometostat, Lysine-specific demethylase 1 (LSD1) inhibitors (Iadademstat, CC90011, INCB059872), and Bromodomain and extraterminal (BET) inhibitors (Molibresib, Birabresib, ZEN003694, PLX51107) [28].

In the same way, miRNA-based drugs are a promising ongoing strategy for cancer therapy, where drugs act either as miRNA mimics or miRNA antagonists for the treatment of different cancer. In addition, ncRNAs can also be used as next-generation biomarkers with great prognostic and predictive value [28].

A new and growing class of compounds, epigenetic agents, is currently available for standard-of-care treatment, but many challenges need to be overcome, as reported by Bates et al. [28]. In particular, sometimes tumors with different epigenetic patterns show some sensitivity to the same drug. In addition, epigenome restoration by treatment has not always been shown to be the exclusive mechanism of therapeutic action, as evidenced by the ineffectiveness of DNMT inhibitors in the context of TET- and IDH-mutated tumors. Finally, the emergence of resistance during epigenetic treatment remains a controversial issue [28].

Cancer resistant to conventional therapy represents a major unmet clinical need. Further molecular dissection of epigenetic landscape of tumors will guide the combinatorial use of epidrugs with other epigenetic agents or with conventional chemotherapies, targeted therapies, and immune checkpoint inhibitors to broaden response rates.

References

1. Hanahan D, Weinberg RA (2011) Hallmarks of cancer: the next generation. Cell 144(5):646–674. https://doi.org/10.1016/j.cell.2011.02.013
2. Alonso S, Dow LE (2021) Engineering chromosome rearrangements in cancer. Dis Model Mech 14(9):dmm049078. https://doi.org/10.1242/dmm.049078
3. Pellestor F, Gaillard JB, Schneider A, Puechberty J, Gatinois V (2022) Chromoanagenesis, the mechanisms of a genomic chaos. Semin Cell Dev Biol 123:90–99. https://doi.org/10.1016/j.semcdb.2021.01.004
4. Li Y, Roberts ND, Wala JA, Shapira O, Schumacher SE, Kumar K, Khurana E, Waszak S, Korbel JO, Haber JE, Imielinski M, PCAWG Structural Variation Working Group, Weischenfeldt J, Beroukhim R, Campbell PJ, PCAWG Consortium (2020) Patterns of somatic structural variation in human cancer genomes. Nature 578(7793):112–121. https://doi.org/10.1038/s41586-019-1913-9
5. Kempfer R, Pombo A (2020) Methods for mapping 3D chromosome architecture. Nat Rev Genet 21:207–226. https://doi.org/10.1038/s41576-019-0195-2
6. Sun L, Zhang J, Gao P (2022) Metabolic reprogramming and epigenetic modifications on the path to cancer. Protein Cell 13:877–919. https://doi.org/10.1007/s13238-021-00846-7
7. Herman AB, Tsitsipatis D, Gorospe M (2022) Integrated lncRNA function upon genomic and epigenomic regulation. Mol Cell 82:2252–2266. https://doi.org/10.1016/j.molcel.2022.05.027

8. Feng Y, Liu X, Pauklin S (2021) 3D chromatin architecture and epigenetic regulation in cancer stem cells. Protein Cell 12(6):440–454. https://doi.org/10.1007/s13238-020-00819-2
9. Casado-Pelae M, Bueno-Costa A, Esteller M (2022) Single cell cancer epigenetics. Trends Cancer 8(10):820–838. https://doi.org/10.1016/j.trecan.2022.06.005
10. Drożak P, Bryliński Ł, Zawitkowska J (2022) A comprehensive overview of recent advances in epigenetics in pediatric acute lymphoblastic leukemia. Cancer 14(21):5384. https://doi.org/10.3390/cancers14215384
11. Van der Meulen J, Van Roy N, Van Vlierberghe P, Speleman F (2014) The epigenetic landscape of T-cell acute lymphoblastic leukemia. Int J Biochem Cell Biol 53:547–557. https://doi.org/10.1016/j.biocel.2014.04.015
12. Xu H, Wen Y, Jin R, Chen H (2022) Epigenetic modifications and targeted therapy in pediatric acute myeloid leukemia. Front Pediatr 10:975819. https://doi.org/10.3389/fped.2022.975819
13. Pasculli B, Barbano R, Parrella P (2018) Epigenetics of breast cancer: biology and clinical implication in the era of precision medicine. Semin Cancer Biol 51:22–35. https://doi.org/10.1016/j.semcancer.2018.01.007
14. Feng J, Meng X (2022) Histone modification and histone modificationtargeted anti-cancer drugs in breast cancer: fundamentals and beyond. Front Pharmacol 13:946811. https://doi.org/10.3389/fphar.2022.946811
15. Jung G, Hernández-Illán E, Moreira L, Balaguer F, Goel A (2020) Epigenetics of colorectal cancer: biomarker and therapeutic potential. Nat Rev Gastroenterol Hepatol 17(2):111–130. https://doi.org/10.1038/s41575-019-0230-y
16. Cooney TM, Lubanszky E, Prasad R, Hawkins C, Mueller S (2020) Diffuse midline glioma: review of epigenetics. J Neuro-Oncol 150:27–34. https://doi.org/10.1007/s11060-020-03553
17. Knowles T, Huang T, Qi J, An S, Burket N, Cooper S, Nazarian J, Saratsis AM (2023) LIN28B and Let-7 in diffuse midline glioma: a review. Cancer 15:3241. https://doi.org/10.3390/cancers15123241
18. Jasek K, Kasubova I, Holubekova V, Stanclova A, Plank L, Lasabova Z (2018) Epigenetics: an alternative pathway in GISTs tumorigenesis. Neoplasma 65(4):477–493. https://doi.org/10.4149/neo_2018_170726N504
19. Braghini MR, Re OL, Romito I, Fernandez-Barrena MG, Barbaro B, Pomella S, Rota R, Vinciguerra M, Avila MA, Alisi A (2022) Epigenetic remodelling in human hepatocellular carcinoma. J Exp Clin Cancer Res 41:107. https://doi.org/10.1186/s13046-022-02297-2
20. Uddin MS, Al Mamun A, Alghamdi BS, Tewari D, Philippe Jeandet M, Sarwar S, Ashraf GM (2022) Epigenetics of glioblastoma multiforme: from molecular mechanisms to therapeutic approaches. Semin Cancer Biol 83:100–120. https://doi.org/10.1016/j.semcancer.2020.12.015
21. Hasan H, Afzal M, Castresana JS, Shahi MH (2023) A comprehensive review of miRNAs and their epigenetic effects in glioblastoma. Cells 12:1578. https://doi.org/10.3390/cells12121578
22. Giunta EF, Arrichiello G, Curvietto M, Pappalardo A, Bosso D, Rosanova M, Diana A, Giordano P, Petrillo A, Federico P, Fabozzi T, Parola S, Riccio V, Mucci B, Vanella V, Festino L, Daniele B, Ascierto PA, Ottaviano M, On Behalf of Scito Youth (2021) Epigenetic regulation in melanoma: facts and hopes. Cells 10:2048. https://doi.org/10.3390/cells10082048
23. Ansari J, Shackelford RE, El-Osta H (2016) Epigenetics in non-small cell lung cancer: from basics to therapeutics. Transl Lung Cancer Res 5(2):155–171. https://doi.org/10.21037/tlcr.2016.02.02
24. Munteanu R, Tomuleasa C, Iuga C-A, Gulei D, Ciuleanu TE (2023) Exploring therapeutic avenues in lung cancer: the epigenetic perspective. Cancer 15:5394. https://doi.org/10.3390/cancers15225394
25. Saleh Z, Moccia MC, Ladd Z, Joneja U, Li Y, Spitz F, Hong YK, Gao T (2024) Pancreatic neuroendocrine tumors: signaling pathways and epigenetic regulation. Int J Mol Sci 25(2):1331. https://doi.org/10.3390/ijms25021331
26. López J, Añazco-Guenkova AM, Monteagudo-García Ó, Blanco S (2022) Epigenetic and epitranscriptomic control in prostate cancer. Genes 13:378. https://doi.org/10.3390/genes13020378

27. Asano T (2020) Drug resistance in cancer therapy and the role of epigenetics. J Nippon Med Sch 87(5):244–251. https://doi.org/10.1272/jnms.JNMS.2020_87-508
28. Bates SE (2020) Epigenetic therapies for cancer. N Engl J Med 383(7):650–663. https://doi.org/10.1056/NEJMra1805035
29. Mondal P, Natesh J, Penta D, Meeran SM (2022) Progress and promises of epigenetic drugs and epigenetic diets in cancer prevention and therapy: a clinical update. Semin Cancer Biol 83:503–522. https://doi.org/10.1016/j.semcancer.2020.12.006
30. Mellinghoff IK, van den Bent MJ, Blumenthal DT, Touat M, Peters KB, Clarke J, Mendez J, Yust-Katz S, Welsh L, Mason WP, Ducray F, Umemura Y, Nabors B, Holdhoff M, Hottinger AF, Arakawa Y, Sepulveda JM, Wick W, Soffietti R, Perry JR, Giglio P, de la Fuente M, Maher EA, Schoenfeld S, Zhao D, Pandya SS, Steelman L, Hassan I, Wen PY, Cloughesy TF, INDIGO Trial Investigators (2023) Vorasidenib in IDH1- or IDH2-mutant low-grade glioma. N Engl J Med 389(7):589–601. https://doi.org/10.1056/NEJMoa2304194

Epigenetics in Autoimmune Diseases

7

Giacomo Bagni, Edoardo Biancalana, Emanuele Chiara, Iole Costanzo, Danilo Malandrino, Giacomo Emmi, Elena Silvestri, and Lorenzo Emmi

Contents

7.1 Introduction ... 98
7.2 Epigenetic Mechanisms ... 101
7.2.1 DNA Methylation ... 101
7.2.2 Histone Modifications ... 102
7.2.3 Non-coding RNAs ... 103
7.3 Specific Epigenetic Mechanisms in Autoimmune Diseases ... 104
7.3.1 Systemic Lupus Erythematosus (SLE) ... 104
7.3.2 Multiple Sclerosis (MS) ... 104
7.3.3 Rheumatoid Arthritis (RA) ... 105
7.3.4 Type 1 Diabetes Mellitus (T1DM) ... 105
7.4 Conclusions ... 106
7.5 Future Perspectives ... 106
References ... 107

G. Bagni
Department of Experimental and Clinical Medicine, University of Florence, Florence, Italy

Department of Clinical and Biological Sciences, University of Turin, Turin, Italy
e-mail: giacomo.bagni@unifi.it

E. Biancalana · I. Costanzo · D. Malandrino · E. Silvestri
Department of Experimental and Clinical Medicine, University of Florence, Florence, Italy

SOD Interdisciplinary Internal Medicine Unit, Careggi University Hospital, Florence, Italy
e-mail: edoardo.biancalana@unifi.it; iole.costanzo@unifi.it; danilo.malandrino@unifi.it; elena.silvestri@unifi.it

E. Chiara · G. Emmi (✉)
Department of Medical, Surgery and Health Sciences, University of Trieste, and Clinical Medicine and Rheumatology Unit, Cattinara University Hospital, Trieste, Italy
e-mail: emanuele.chiara@phd.units.it; giacomo.emmi@units.it

L. Emmi
Department of Neuroscience, Psychology, Drug Research and Child Health, University of Florence, Florence, Italy

L. Emmi, V. Fanos (eds.), *Epigenetics and Rare Diseases*, Rare Diseases of the Immune System, https://doi.org/10.1007/978-3-032-02902-7_7

7.1 Introduction

Autoimmune diseases are a group of disorders characterized by the immune system's inability to distinguish between self and non-self, leading to an attack on the body's tissues. These diseases can affect virtually any organ system, resulting in a broad spectrum of clinical manifestations and complications. Some of the most prevalent autoimmune conditions include systemic lupus erythematosus (SLE), multiple sclerosis (MS), rheumatoid arthritis (RA), and type 1 diabetes mellitus (T1DM) (Fig. 7.1). The chronic nature of these conditions often leads to a significant burden on healthcare systems, as well as profound impacts on the quality of life for those affected.

The impact of autoimmune diseases extends beyond the individual; it can affect families, communities, and healthcare resources. Patients often experience a wide range of symptoms that can fluctuate in severity, making diagnosis and treatment particularly challenging. Many patients endure a lengthy journey to receive a correct diagnosis, which can take years due to the overlap of symptoms with other health conditions. Moreover, treatment often involves a combination of medications aimed at reducing immune system activity and managing symptoms, which can lead to significant side effects.

Although genetic predisposition plays a crucial role in the development of autoimmune diseases, it is increasingly recognized that environmental factors also significantly influence disease onset and progression. Factors such as infections, dietary habits, exposure to toxins, and even psychological stress can trigger or exacerbate autoimmune responses. The interplay between genetic susceptibility and environmental triggers is complex, and researchers are actively investigating how these factors converge to promote autoimmunity. For instance, certain infections may mimic self-antigens, leading the immune system to mistakenly target the body's own tissues. This phenomenon, known as molecular mimicry, is one of the mechanisms that can initiate autoimmune responses.

In addition to molecular mimicry, other environmental factors such as the gut microbiome have also emerged as influential in autoimmune disease development. The microbiome plays a critical role in shaping immune responses, and imbalances in gut bacteria can contribute to dysregulation of immune function. Research has shown that alterations in gut microbiota composition can exacerbate inflammation and promote autoimmunity, emphasizing the need for further exploration of dietary interventions and probiotics as potential therapeutic strategies.

Recent advancements in the field of epigenetics have provided new insights into the mechanisms underlying autoimmune diseases [1]. Epigenetics refers to heritable changes in gene expression that do not involve alterations to the DNA sequence itself [2, 3]. These changes are mediated by various mechanisms, including DNA methylation, histone modifications, and non-coding RNAs. Epigenetic modifications play a crucial role in regulating gene expression, influencing cellular identity and function in response to environmental stimuli. In autoimmune diseases, dysregulation of these epigenetic mechanisms can lead to aberrant immune responses and the breakdown of tolerance to self-antigens [4–6].

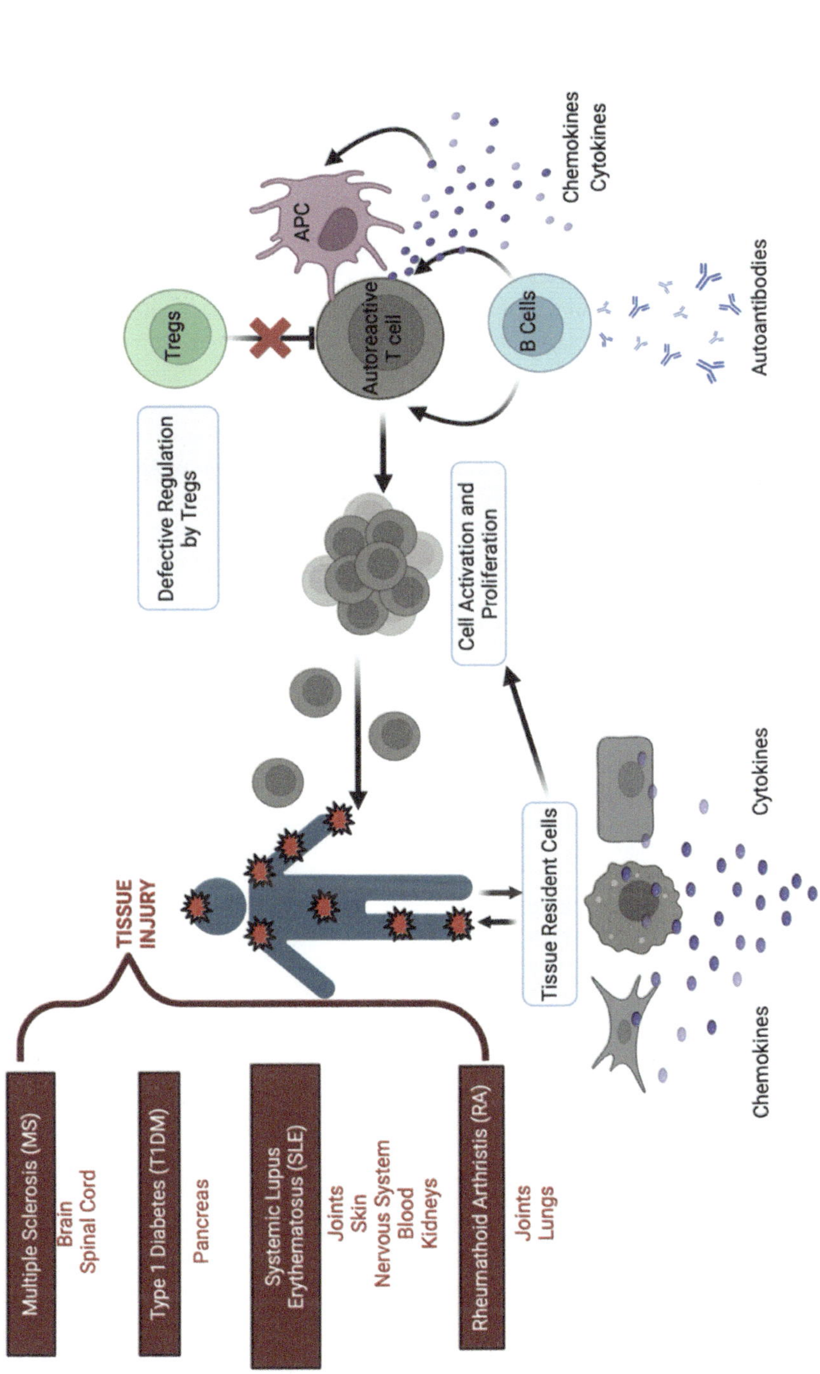

Fig. 7.1 Schematic representation of autoimmune mechanisms causing tissue damage in different types of diseases. *APC* antigen-presenting cell, *Tregs* regulatory T cells

Understanding the role of epigenetics in autoimmune diseases can provide valuable insights into isease pathogenesis and open new avenues for diagnosis and treatment (Tables 7.1 and 7.2). This review will focus on the role of epigenetics in specific autoimmune disorders, emphasizing the mechanisms of DNA methylation, histone modifications, and non-coding RNAs, as well as their potential as therapeutic targets. We will also discuss the implications of these findings for future research and clinical practice. Moreover, investigating how lifestyle interventions, such as dietary changes and stress management, may modulate epigenetic markers could further enhance treatment strategies. Identifying specific epigenetic signatures associated with different autoimmune conditions may lead to the development of personalized medicine approaches, allowing for more effective management of these complex disorders.

In conclusion, a deeper understanding of the interplay between genetics, environment, and epigenetics is essential for advancing our knowledge of autoimmune diseases and improving outcomes for affected individuals. By unraveling the intricate mechanisms that contribute to the development and progression of these diseases, researchers and clinicians can develop more targeted therapies and intervention strategies (Tables 7.1 and 7.2). Continued research efforts in this field

Table 7.1 Future directions in epigenetics research for autoimmunity

Research focus	Description	Potential impact on treatment
Targeting epigenetic modifications	Investigating therapies that restore normal epigenetic patterns	Personalized medicine approaches for autoimmune diseases
Understanding environmental influences	Exploring how environmental factors shape epigenetic changes	Development of preventive strategies targeting epigenetics
Biomarker identification	Identifying epigenetic biomarkers for diagnosis and prognosis	Improved patient stratification and treatment response prediction

Table 7.2 Therapeutic implications of targeting epigenetics in autoimmune diseases

Epigenetic therapy	Mechanism of action	Autoimmune diseases studied	Stage of development
DNA methyltransferase inhibitors	Restore normal gene expression by inhibiting DNA methylation	SLE, RA	Preclinical, phase I/II
Histone deacetylase inhibitors (HDACi)	Increase gene transcription by preventing histone deacetylation	MS, RA	Phase II clinical trials
miRNA modulators	Regulate miRNA levels to control gene expression	T1DM, SLE	Preclinical studies
CRISPR-based epigenetic editing	Precise editing of epigenetic markers to modulate gene expression	T1DM, SLE	Experimental, early-stage

will not only enhance our understanding of autoimmune diseases but also empower patients with more effective treatment options and improved quality of life. Addressing these multifaceted challenges requires a collaborative approach that encompasses genetic, environmental, and lifestyle factors, underscoring the complexity of autoimmune disorders and the need for comprehensive care.

7.2 Epigenetic Mechanisms

Epigenetic mechanisms involve modifications that regulate gene expression without changing the DNA sequence itself. They are essential for maintaining normal cellular function and responding to environmental cues, making them particularly relevant in the context of autoimmune diseases (Table 7.3).

7.2.1 DNA Methylation

DNA methylation is one of the most well-characterized epigenetic modifications. It involves the addition of a methyl group to the cytosine residues of DNA, primarily occurring at CpG dinucleotides. This process is catalyzed by DNA methyltransferases (DNMTs) and generally leads to transcriptional repression by preventing the binding of transcription factors to DNA. Methylation patterns are critical for regulating gene expression and maintaining cellular identity. However, these patterns can be altered by environmental exposures, such as infections, toxins, and dietary components, leading to immune dysregulation [7].

In the context of autoimmune diseases, aberrant DNA methylation patterns have been observed across multiple conditions. For example, in SLE, hypomethylation of specific genes in T cells leads to the overexpression of genes that promote inflammation and autoantibody production. Reduced DNMT1 activity has been implicated in the global hypomethylation observed in lupus T cells, contributing to disease pathogenesis [8]. This overactivity not only perpetuates the autoimmune response but also complicates the management of the disease over time.

Similar mechanisms have been observed in MS, where hypomethylation of genes involved in immune regulation contributes to the breakdown of immune tolerance and the development of autoreactive immune cells. The dysregulation of these

Table 7.3 Epigenetic modifications and their implications in autoimmunity

Epigenetic mechanism	Description	Implications in autoimmunity
DNA methylation	Addition of methyl groups to DNA, primarily at CpG sites	Alters gene expression; hypomethylation linked to inflammation
Histone modifications	Chemical modifications of histones affecting chromatin	Influences gene expression; increased acetylation leads to inflammation
Non-coding RNAs	Regulatory RNA molecules that do not code for proteins	Modulate gene expression; dysregulation linked to disease severity

pathways creates a feedback loop that exacerbates the inflammatory response, leading to increased tissue damage.

In T1DM, hypermethylation of the FOXP3 promoter impairs the function of regulatory T cells (Tregs), disrupting immune tolerance and contributing to the immune-mediated destruction of insulin-producing beta cells in the pancreas. The immune system's failure to distinguish between self and non-self is a hallmark of T1DM, and understanding the epigenetic factors involved could provide new avenues for intervention.

Furthermore, DNA methylation patterns are influenced by environmental factors such as diet, infections, and chemical exposures, which may contribute to the development and progression of autoimmune diseases. For instance, dietary components such as folate and other B vitamins play a critical role in one-carbon metabolism and thus affect DNA methylation status. Additionally, exposure to environmental pollutants has been shown to induce changes in DNA methylation patterns, which may further complicate the etiology of autoimmune diseases.

7.2.2 Histone Modifications

Histone modifications represent another critical aspect of epigenetic regulation. Histones are proteins that package DNA into chromatin, and their post-translational modifications, such as acetylation, methylation, phosphorylation, and ubiquitination, can influence chromatin structure and gene expression. These modifications regulate the accessibility of DNA to transcription factors, thereby controlling gene expression [9].

Histone acetylation, mediated by histone acetyltransferases (HATs), generally promotes gene expression by loosening chromatin structure, while deacetylation by histone deacetylases (HDACs) results in chromatin compaction and gene silencing. In autoimmune diseases, aberrant histone modifications can disrupt immune cell function and promote autoimmunity [10]. For example, increased histone acetylation has been observed in RA, leading to the overexpression of pro-inflammatory cytokines such as tumor necrosis factor-alpha (TNF-α) and interleukin-6 (IL-6). These cytokines are key players in perpetuating the inflammatory process, resulting in joint destruction and functional impairment.

In MS, histone modifications also play a significant role in disease progression. Increased histone H3 citrullination, a process catalyzed by peptidylarginine deiminase (PAD), has been associated with the demyelination observed in MS patients. This modification alters the structure of myelin basic protein (MBP), contributing to the instability of the myelin sheath and leading to neurodegeneration. The alteration of histone marks can create a favorable environment for the activation of autoreactive T cells, driving the disease forward.

Additionally, PAD2 overexpression due to hypomethylation has been implicated in the pathogenesis of MS, suggesting a link between histone modifications and DNA methylation. This intricate relationship emphasizes the need for a comprehensive understanding of epigenetic regulation in autoimmune diseases.

Histone modifications also influence immune cell differentiation and function. In T1DM, altered histone acetylation patterns in pancreatic beta cells and T cells have been associated with immune cell activation and beta-cell destruction. Histone deacetylase inhibitors (HDACi) are being investigated as potential therapeutic agents in autoimmune diseases due to their ability to modulate gene expression and immune responses. The therapeutic potential of targeting histone modifications represents a promising area of research for managing autoimmune diseases [11].

7.2.3 Non-coding RNAs

Non-coding RNAs (ncRNAs) are a diverse class of RNA molecules that do not code for proteins but play critical roles in regulating gene expression. Two major types of ncRNAs involved in autoimmune diseases are microRNAs (miRNAs) and long non-coding RNAs (lncRNAs). These molecules regulate gene expression at both the transcriptional and post-transcriptional levels, influencing various cellular processes, including immune cell differentiation, activation, and function [12, 13].

MiRNAs are short RNA molecules that regulate gene expression by binding to complementary sequences in target messenger RNAs (mRNAs), leading to their degradation or translational repression [12]. MiRNAs are involved in the regulation of immune responses, and their dysregulation has been linked to the pathogenesis of autoimmune diseases [12]. For example, miR-146a and miR-155 have been implicated in the regulation of T cell differentiation and activation, processes that are critical in autoimmune diseases such as MS, RA, and SLE.

In RA, miR-155 has been shown to promote the differentiation of pro-inflammatory Th17 cells, contributing to joint inflammation and tissue damage. In contrast, miR-146a acts as a negative regulator of inflammation by targeting key components of the nuclear factor-kappa B (NF-κB) pathway, which is involved in the production of pro-inflammatory cytokines. Dysregulation of these miRNAs in RA patients is associated with disease severity and progression, and miRNA-based therapies are being explored as potential treatments. These therapies could provide a means of modulating inflammatory pathways in a targeted manner [12].

LncRNAs are longer RNA molecules that regulate gene expression through various mechanisms, including chromatin remodeling, transcriptional regulation, and post-transcriptional control [13]. In autoimmune diseases, lncRNAs have been shown to modulate immune cell function and contribute to disease pathogenesis. For example, the lncRNA NEAT1 has been implicated in the regulation of inflammatory responses in SLE by modulating the expression of genes involved in cytokine production. The potential for lncRNAs as therapeutic targets and biomarkers in autoimmune diseases is an exciting area of research.

Recent studies have identified several lncRNAs associated with autoimmune diseases, including T1DM and MS. For instance, lncRNA HOTAIR has been shown to promote inflammation by modulating the expression of key inflammatory cytokines. Targeting these lncRNAs presents a novel therapeutic approach to restoring immune homeostasis and preventing disease progression [13].

7.3 Specific Epigenetic Mechanisms in Autoimmune Diseases

Understanding the role of epigenetics in specific autoimmune diseases can offer valuable insights into their unique pathogenesis and potential treatment avenues.

7.3.1 Systemic Lupus Erythematosus (SLE)

SLE is a systemic autoimmune disease characterized by chronic inflammation and tissue damage in multiple organ systems. Epigenetic modifications, particularly DNA hypomethylation, play a critical role in the pathogenesis of SLE. Hypomethylation of genes involved in immune activation, such as ITGAL and CD40LG, leads to their overexpression, contributing to the development of autoreactive T and B cells [14].

In SLE, global hypomethylation of T cells is associated with reduced DNMT1 expression, leading to the overactivation of immune cells and the production of autoantibodies. Histone modifications, such as increased acetylation, have also been observed in SLE, contributing to the overexpression of pro-inflammatory genes [15, 16]. Dysregulated miRNAs, including miR-146a and miR-21, further contribute to the breakdown of immune tolerance in SLE by promoting the production of type I interferons, which are central to SLE pathogenesis [17].

Epigenetic therapies, such as HDAC inhibitors, are being investigated as potential treatments for SLE. These therapies aim to restore normal epigenetic patterns and modulate immune responses, offering a novel approach to managing this complex disease [18]. By re-establishing proper epigenetic regulation, it may be possible to enhance the immune system's ability to distinguish between self and non-self, thereby ameliorating disease symptoms.

7.3.2 Multiple Sclerosis (MS)

MS is an autoimmune disease that targets the central nervous system, leading to demyelination and neurodegeneration. The pathogenesis of MS involves a combination of genetic susceptibility, environmental triggers, and epigenetic modifications. DNA hypomethylation of genes involved in immune regulation has been linked to the breakdown of immune tolerance in MS, contributing to the development of autoreactive T cells that attack the myelin sheath [19, 20].

Histone modifications also play a critical role in MS progression. Increased histone citrullination, driven by PAD2 overexpression, destabilizes myelin proteins, leading to the demyelination observed in MS patients [19]. MiRNAs, such as miR-155, miR-146a, and miR-326, regulate the inflammatory response in MS by modulating the activity of immune cells, including T cells and macrophages [21, 22]. These miRNAs are being explored as potential biomarkers for MS diagnosis and as therapeutic targets. The ability to modify miRNA expression could provide a new strategy for regulating inflammation in MS.

Recent studies have also suggested a role for epigenetic factors in MS treatment responses. For example, the use of certain disease-modifying therapies (DMTs) has been associated with changes in DNA methylation patterns and miRNA expression in MS patients [23]. Understanding how these therapies interact with the epigenetic landscape of MS may provide insights into optimizing treatment strategies for individual patients.

7.3.3 Rheumatoid Arthritis (RA)

RA is a chronic autoimmune disease that primarily affects the joints, leading to inflammation, pain, and joint destruction. Epigenetic modifications, including DNA methylation and histone modifications, play a critical role in RA pathogenesis. Hypomethylation of immune-related genes in synovial fibroblasts and immune cells leads to their overactivation, perpetuating inflammation and joint damage [24, 25].

Histone modifications, such as increased histone acetylation, have been observed in RA synovial tissues, contributing to the overexpression of pro-inflammatory cytokines. HDAC inhibitors are being investigated as potential therapies for RA, as they can modulate gene expression and reduce inflammation. MiRNAs, such as miR-155 and miR-146a, play key roles in regulating immune responses in RA, and their dysregulation is associated with disease severity [26]. MiRNA-based therapies are indeed being explored as novel treatment options for RA. The potential to fine-tune immune responses through epigenetic modulation presents a promising frontier in RA management.

In recent years, the identification of epigenetic biomarkers for RA has gained attention. These biomarkers could serve as diagnostic tools or prognostic indicators, aiding in the stratification of patients based on disease severity and treatment response. For instance, specific patterns of DNA methylation have been associated with increased risk of developing RA, providing an opportunity for early intervention [27].

7.3.4 Type 1 Diabetes Mellitus (T1DM)

T1DM is an autoimmune disease characterized by the destruction of insulin-producing beta cells in the pancreas. DNA methylation patterns in immune cells and pancreatic beta cells are altered in T1DM, contributing to immune dysfunction and beta-cell destruction. Hyper-methylation of the FOXP3 promoter impairs the function of Tregs, disrupting immune tolerance and leading to beta-cell destruction [28].

Histone modifications, such as altered acetylation patterns, have been observed in T1DM and are associated with immune cell activation and inflammation. MiRNAs, including miR-21 and miR-29, are dysregulated in T1DM and contribute to beta-cell dysfunction and apoptosis [29]. Epigenetic therapies aimed at restoring normal methylation patterns and miRNA expression are being explored as potential treatments for T1DM. The modulation of epigenetic factors may provide a pathway to preserving beta-cell function and preventing the progression of diabetes.

Research has also highlighted the role of environmental factors in shaping the epigenetic landscape in T1DM. For instance, early exposure to certain infections, dietary factors, and the gut microbiome has been associated with changes in DNA methylation patterns that influence the risk of developing T1DM [30]. Understanding these environmental influences could lead to preventative strategies targeting the epigenetic mechanisms involved in T1DM.

7.4 Conclusions

Epigenetic modifications play a central role in the regulation of immune responses and the development of autoimmune diseases (Fig. 7.1). DNA methylation, histone modifications, and non-coding RNAs are key regulators of gene expression and immune cell function, and their dysregulation contributes to the pathogenesis of diseases such as SLE, MS, RA, and T1DM. Understanding these epigenetic mechanisms provides valuable insights into disease development and offers new opportunities for the diagnosis, prognosis, and treatment of autoimmune diseases (Tables 7.1 and 7.2).

Targeting epigenetic modifications holds promise for the development of novel therapeutic strategies that can modulate immune responses and restore immune tolerance. Future research in this area will continue to advance our understanding of the complex interplay between genetics, environment, and epigenetics in autoimmune diseases, leading to more effective and personalized treatments for patients. The exploration of epigenetic landscapes not only enhances our comprehension of disease mechanisms but also opens doors for innovative treatment approaches that could transform the management of autoimmune disorders. Continued efforts in this field will be crucial in unraveling the complexities of autoimmunity and improving patient outcomes.

7.5 Future Perspectives

The exploration of epigenetic modifications in autoimmune diseases is an exciting and rapidly evolving field. Future research should focus on several key areas:

1. **Personalized Medicine**: Understanding the epigenetic landscape of individual patients may enable the development of personalized treatment strategies. By identifying specific epigenetic alterations associated with disease severity and progression, clinicians can tailor therapeutic interventions to optimize patient outcomes (Table 7.1).
2. **Therapeutic Targeting**: The potential of targeting epigenetic modifications as a therapeutic strategy in autoimmune diseases is promising. Further research is

needed to investigate the efficacy and safety of epigenetic drugs, such as DNMT inhibitors, HDAC inhibitors, and miRNA-based therapies, in clinical trials (Table 7.2).

3. **Environmental Interactions**: Investigating the interactions between environmental factors and epigenetic modifications will deepen our understanding of the etiology of autoimmune diseases. This knowledge can inform preventative strategies and guide public health initiatives aimed at reducing exposure to environmental triggers.
4. **Longitudinal Studies**: Longitudinal studies assessing epigenetic changes over time in patients with autoimmune diseases can provide valuable insights into disease progression and treatment response. This approach may elucidate the dynamic nature of epigenetic modifications and their relevance to disease outcomes.

References

1. Meda F, Folci M, Baccarelli A, Selmi C (2011) The epigenetics of autoimmunity. Cell Mol Immunol [Internet] 8(3):226–236. Available from: https://doi.org/10.1038/cmi.2010.78
2. Bjornsson HT, Fallin MD, Feinberg AP (2004) An integrated epigenetic and genetic approach to common human disease. Trends Genet [Internet] 20(8):350–358. Available from: https://doi.org/10.1016/j.tig.2004.06.009
3. Mazzone R, Zwergel C, Artico M, Taurone S, Ralli M, Greco A, et al (2019) The emerging role of epigenetics in human autoimmune disorders. Clin Epigenetics [Internet] 11(1):34. Available from: https://doi.org/10.1186/s13148-019-0632-2
4. Huang B, Jiang C, Zhang R (2014) Epigenetics: the language of the cell? Epigenomics [Internet] 6(1):73–88. Available from: https://doi.org/10.2217/epi.13.72
5. Quintero-Ronderos P, Montoya-Ortiz G (2012) Epigenetics and autoimmune diseases. Autoimmune Dis [Internet] 2012:593720. Available from: https://doi.org/10.1155/2012/593720
6. Surace AEA, Hedrich CM (2019) The role of epigenetics in autoimmune/inflammatory disease. Front Immunol [Internet] 10:1525. Available from: https://doi.org/10.3389/fimmu.2019.01525
7. Sun B, Hu L, Luo Z-Y, Chen X-P, Zhou H-H, Zhang W (2016) DNA methylation perspectives in the pathogenesis of autoimmune diseases. Clin Immunol [Internet] 164:21–27. Available from: https://doi.org/10.1016/j.clim.2016.01.011
8. Farivar S, Shaabanpour Aghamaleki F (2018) Effects of major epigenetic factors on systemic lupus erythematosus. Iran Biomed J [Internet] 22(5):294–302. Available from: https://doi.org/10.29252/ibj.22.5.294
9. Song H, Shen R, Liu X, Yang X, Xie K, Guo Z, et al (2023) Histone post-translational modification and the DNA damage response. Genes Dis [Internet] 10(4):1429–1444. Available from: https://doi.org/10.1016/j.gendis.2022.04.002
10. Verdone L, Caserta M, Di Mauro E (2005) Role of histone acetylation in the control of gene expression. Biochem Cell Biol [Internet] 83(3):344–353. Available from: https://doi.org/10.1139/o05-041
11. Hamminger P, Rica R, Ellmeier W (2020) Histone deacetylases as targets in autoimmune and autoinflammatory diseases. Adv Immunol [Internet] ;147:1–59. Available from: https://doi.org/10.1016/bs.ai.2020.06.001
12. Rauniyar VK, Wu W, Yang H (2011) MicroRNAs in autoimmune disease. Auto Immun Highlights [Internet] 2(2):59–65. Available from: https://doi.org/10.1007/s13317-011-0022-z

13. Statello L, Guo C-J, Chen L-L, Huarte M (2021) Gene regulation by long non-coding RNAs and its biological functions. Nat Rev Mol Cell Biol [Internet] 22(2):96–118. Available from: https://doi.org/10.1038/s41580-020-00315-9
14. Hedrich CM, Mäbert K, Rauen T, Tsokos GC (2017) DNA methylation in systemic lupus erythematosus. Epigenomics [Internet] ;9(4):505–525. Available from: https://doi.org/10.2217/epi-2016-0096
15. Hu N, Qiu X, Luo Y, Yuan J, Li Y, Lei W et al (2008) Abnormal histone modification patterns in lupus CD4+ T cells. J Rheumatol 35(5):804–810
16. Wang Y, Yang Y, Luo Y, Yin Y, Wang Q, Li Y, et al (2013) Aberrant histone modification in peripheral blood B cells from patients with systemic sclerosis. Clin Immunol [Internet] 149(1):46–54. Available from: https://doi.org/10.1016/j.clim.2013.06.006
17. Dai Y, Huang Y-S, Tang M, Lv T-Y, Hu C-X, Tan Y-H, et al (2007) Microarray analysis of microRNA expression in peripheral blood cells of systemic lupus erythematosus patients. Lupus [Internet] 16(12):939–946. Available from: https://doi.org/10.1177/0961203307084158
18. Reilly CM, Regna N, Mishra N. HDAC inhibition in lupus models. Mol Med [Internet] 2011;17(5–6):417–425. Available from: https://doi.org/10.2119/molmed.2011.00055
19. Moscarello MA, Mastronardi FG, Wood DD (2007) The role of citrullinated proteins suggests a novel mechanism in the pathogenesis of multiple sclerosis. Neurochem Res [Internet] 32(2):251–256. Available from: https://doi.org/10.1007/s11064-006-9144-5
20. Kumagai C, Kalman B, Middleton FA, Vyshkina T, Massa PT (2012) Increased promoter methylation of the immune regulatory gene SHP-1 in leukocytes of multiple sclerosis subjects. J Neuroimmunol [Internet] 246(1–2):51–57. Available from: https://doi.org/10.1016/j.jneuroim.2012.03.003
21. Shademan B, Nourazarian A, Nikanfar M, Biray Avci C, Hasanpour M, Isazadeh A (2020) Investigation of the miRNA146a and miRNA155 gene expression levels in patients with multiple sclerosis. J Clin Neurosci [Internet] 78:189–193. Available from: https://doi.org/10.1016/j.jocn.2020.04.071
22. Junker A, Krumbholz M, Eisele S, Mohan H, Augstein F, Bittner R, et al (2009) MicroRNA profiling of multiple sclerosis lesions identifies modulators of the regulatory protein CD47. Brain [Internet] 132(Pt 12):3342–3352. Available from: https://doi.org/10.1093/brain/awp300
23. Manna I, De Benedittis S, Porro D (2024) A comprehensive examination of the role of epigenetic factors in multiple sclerosis. Int J Mol Sci [Internet] 25(16):8921. Available from: https://doi.org/10.3390/ijms25168921
24. Nemtsova MV, Zaletaev DV, Bure IV, Mikhaylenko DS, Kuznetsova EB, Alekseeva EA, et al (2019) Epigenetic changes in the pathogenesis of rheumatoid arthritis. Front Genet [Internet] 10:570. Available from: https://doi.org/10.3389/fgene.2019.00570
25. Ai R, Laragione T, Hammaker D, Boyle DL, Wildberg A, Maeshima K, et al (2018) Comprehensive epigenetic landscape of rheumatoid arthritis fibroblast-like synoviocytes. Nat Commun [Internet] 9(1):1921. Available from: https://doi.org/10.1038/s41467-018-04310-9
26. Bagheri-Hosseinabadi Z, Mirzaei MR, Hajizadeh MR, Asadi F, Rezaeian M, Abbasifard M (2021) Plasma MicroRNAs (miR-146a, miR-103a, and miR-155) as potential biomarkers for rheumatoid arthritis (RA) and disease activity in Iranian patients. Mediterr J Rheumatol [Internet] 32(4):324–330. Available from: https://doi.org/10.31138/mjr.32.4.324
27. Liu Y, Aryee MJ, Padyukov L, Fallin MD, Hesselberg E, Runarsson A, et al (2013) Epigenome-wide association data implicate DNA methylation as an intermediary of genetic risk in rheumatoid arthritis. Nat Biotechnol [Internet] 31(2):142–147. Available from: https://doi.org/10.1038/nbt.2487
28. Wang Z, Zheng Y, Hou C, Yang L, Li X, Lin J, et al (2013) DNA methylation impairs TLR9 induced Foxp3 expression by attenuating IRF-7 binding activity in fulminant type 1 diabetes. J Autoimmun [Internet] 41:50–59. Available from: https://doi.org/10.1016/j.jaut.2013.01.009

29. Margaritis K, Margioula-Siarkou G, Giza S, Kotanidou EP, Tsinopoulou VR, Christoforidis A, et al (2021) Micro-RNA implications in type-1 diabetes mellitus: a review of literature. Int J Mol Sci [Internet] 22(22):12165. Available from: https://doi.org/10.3390/ijms222212165
30. Houeiss P, Luce S, Boitard C (2022) Environmental triggering of type 1 diabetes autoimmunity. Front Endocrinol (Lausanne) [Internet] 13:933965. Available from: https://doi.org/10.3389/fendo.2022.933965

Epigenetics and Systemic Vasculitis

8

Michelangelo Tesi and Augusto Vaglio

Contents

8.1 Introduction 112
8.2 ANCA-Associated Vasculitis 113
8.2.1 DNA Methylation and Histone Modifications in GPA and MPA 114
8.2.2 miRNAs in GPA and MPA 115
8.3 IgA Vasculitis (Henoch-Schönlein Purpura) 117
8.3.1 Epigenetics of IgA Vasculitis 118
8.4 Kawasaki Disease 120
8.4.1 DNA Methylation in Kawasaki Disease 120
8.4.2 miRNAs in Kawasaki Disease 122
8.4.3 lncRNAs in Kawasaki Disease 125
8.5 Large-Vessel Vasculitis 126
8.5.1 Epigenetics of Large-Vessel Vasculitis 127
8.6 Behçet's Disease 129
8.6.1 DNA Methylation and Chromatin Accessibility in Behçet's Disease 130
8.6.2 miRNAs in Behçet's Disease 132
8.6.3 lncRNAs and circRNAs in Behçet's Disease 135
8.7 Conclusions 136
References 137

M. Tesi · A. Vaglio (✉)
Nephrology and Dialysis Unit, Meyer Children's Hospital IRCCS, Florence, Italy

L. Emmi, V. Fanos (eds.), *Epigenetics and Rare Diseases*, Rare Diseases of the Immune System, https://doi.org/10.1007/978-3-032-02902-7_8

8.1 Introduction

Primary systemic vasculitides are a group of rare autoimmune-autoinflammatory diseases characterized by the destruction and necrosis of blood vessels, which leads to impaired blood flow, tissue ischemia, and organ damage. These diseases affect blood vessels of various sizes and can involve multiple organs. The exact etiology of most vasculitides remains unclear, but they are believed to result from a combination of genetic predisposition, environmental factors, and immune dysregulation. Clinical manifestations of systemic vasculitis vary significantly depending on the vessels and organs involved, but common symptoms include fever, fatigue, weight loss, and specific organ dysfunction. The 2012 Chapel Hill Consensus Conference (CHCC) provides a widely used classification system for vasculitis, based on the size of the vessels predominantly affected: large vessels (aorta and its major branches), medium vessels (main visceral arteries and their branches), and small vessels (intraparenchymal arteries, arterioles, capillaries, and venules). This classification framework also includes variable-vessel vasculitis, where vessels of varying caliber can be involved [1].

A recent operational definition describes epigenetics as "the study of molecules and mechanisms that can perpetuate alternative gene activity states in the context of the same DNA sequence" [2]. This concept encompasses gene expression regulation mechanisms that can be inherited mitotically or transgenerationally, as well as the persistence of gene activity or chromatin states through extended periods of time, even without cell division. Crucially, these mechanisms—though inheritable—alter gene expression without modifying the underlying DNA sequence. Key mechanisms include DNA methylation, histone modifications, and non-coding RNAs. DNA methylation involves the addition of methyl groups to cytosines, typically in CpG dinucleotides near gene promoters, which can lead to stable gene silencing. Histone modifications, such as acetylation and methylation on specific amino acids of histone tails, alter chromatin compaction and accessibility, with acetylation generally associated with gene activation and methylation potentially leading to either activation or repression depending on the context. Thus, DNA methylation and histone modifications regulate gene expression pre-transcriptionally, by either blocking or facilitating transcription of specific genes. On the other hand, non-coding RNAs, including microRNAs (miRNAs) and long non-coding RNAs (lncRNAs), participate in post-transcriptional regulation, guiding protein complexes to specific DNA regions or directly interfering with messenger-RNA (mRNA) stability and translation. Collectively, these mechanisms allow cells to modulate gene expression dynamically in response to environmental cues, bridging genetic and environmental influences in health and disease.

Epigenetic research has advanced the understanding of autoimmune diseases, highlighting the role of DNA methylation, histone modifications, and non-coding RNAs in regulating immune responses and maintaining immune tolerance. Reviews, such as those by Mazzone et al. [3] and Surace and Hedrich [4], illustrate how epigenetic alterations influence the pathogenesis of autoimmune conditions, including systemic lupus erythematosus, rheumatoid arthritis, and psoriasis. These gene expression alterations can affect immune cell function, cytokine production, and inflammatory pathways, potentially serving as biomarkers or therapeutic targets. In systemic vasculitis, exploring these epigenetic mechanisms has yielded novel insights into disease processes and treatment opportunities, bridging genetic susceptibility with environmental triggers to shape disease outcomes. This chapter will examine the role of epigenetic mechanisms in systemic vasculitis. The discussion will start by exploring anti-neutrophil cytoplasmic antibody (ANCA)-associated vasculitis (AAV), including the distinct roles of epigenetic factors in granulomatosis with polyangiitis (GPA) and microscopic polyangiitis (MPA). Next, findings on the epigenetic regulation in IgA vasculitis (IgAV) and Kawasaki disease (KD) will be reviewed. Then, the contributions of epigenetic changes in large-vessel vasculitis (LVV), namely giant cell arteritis (GCA) and Takayasu arteritis (TAK), will be addressed, and finally the epigenetic landscape of Behçet's disease (BD) will be discussed.

8.2 ANCA-Associated Vasculitis

AAV is a group of disorders involving severe small-vessel vasculitis, characterized by the development of autoantibodies to the neutrophil proteins leukocyte proteinase 3 (PR3-ANCA) or myeloperoxidase (MPO-ANCA). AAV includes three distinct subtypes, namely microscopic polyangiitis (MPA), granulomatosis with polyangiitis (GPA), and eosinophilic granulomatosis with polyangiitis (EGPA) [5]. The present discussion will focus solely on GPA and MPA for two reasons: (1) EGPA is substantially different from the other two forms of AAV, both from a clinical point of view (EGPA involving eosinophilia and asthma is one of the macroscopic differences) and from a pathophysiologic and genetic standpoint, with Th2-driven immune responses; (2) to date, no epigenetic studies have yet explored EGPA, possibly due to its lower prevalence as compared to GPA and MPA.

GPA is typically associated with PR3-ANCA and characterized by necrotizing granulomatous inflammation involving the respiratory tract and kidneys, while microscopic polyangiitis (MPA) is associated with MPO-ANCA, marked by necrotizing vasculitis affecting the kidneys and lungs, but without granulomatous inflammation. The pathophysiology of both GPA and MPA involves loss of immunological tolerance to neutrophil proteins PR3 or MPO, leading to the formation of ANCAs, which activate neutrophils and induce endothelial damage and inflammation. Genetic studies reveal distinct yet overlapping associations, such as *HLA-DP* in PR3-ANCA-positive vasculitis and *HLA-DQ* in MPO-ANCA-positive vasculitis, suggesting a stronger genetic link to ANCA specificity than clinical diagnosis, while shared genetic elements like *PTPN22* or *CTLA4* indicate common autoimmune mechanisms [5, 6].

8.2.1 DNA Methylation and Histone Modifications in GPA and MPA

In the study of AAV, DNA methylation and histone alterations have been identified as mechanisms influencing the abnormal expression of PR3 and MPO—two critical autoantigens implicated in the pathogenesis of the disease. Three main aspects are prominent in this field of AAV research: (1) alterations in DNA methylation and histone modifications leading to abnormal PR3 and MPO expression, (2) the presence of anti-sense and alternative transcripts that expand the range of potential autoantigens, and (3) the potential for epigenetic modifications to serve as biomarkers for tracking disease activity and predicting clinical outcomes.

Studies by Yang et al. [7] and Ciavatta et al. [8] point out how disrupted epigenetic regulation in neutrophils and monocytes contributes to the abnormal expression of PR3 and MPO in AAV. Normally, these genes are silenced in mature immune cells, which restricts the availability of these autoantigens. However, in AAV, an aberrant expression of PR3 and MPO occurs, which Yang et al. [7] proposed as part of a "second hit" hypothesis. This hypothesis suggests that a secondary reactivation of *PRTN3* (encoding PR3) and *MPO* (encoding MPO) genes in mature neutrophils and monocytes increases antigen levels, making these cells vulnerable to ANCA binding and immune-mediated damage. Ciavatta et al. [8] extended this understanding by examining the epigenetic basis of this re-expression, revealing that specific histone modifications, particularly the H3K27me3 repressive mark, were depleted at *PRTN3* and *MPO* loci in AAV patients. This depletion was linked to a reduction in the expression of RUNX3, a transcriptional regulator that typically recruits histone modifiers to maintain silencing at these loci. Without sufficient H3K27me3, PR3 and MPO genes escape repression, leading to their expression and thus providing targets for ANCAs. Additionally, Ciavatta et al. found that while DNA methylation of *PRTN3* gene remained unchanged, partial methylation of the *MPO* gene in healthy individuals contrasted with the unmethylated state observed in AAV patients, further implicating DNA methylation as an additional silencing mechanism that fails in AAV.

Further complexity in AAV pathogenesis comes from the production of anti-sense and alternative transcripts of PR3 and MPO, which increase the diversity of potential autoantigenic targets. Ciavatta and Falk [9] in their review hypothesized that abnormal epigenetic regulation in AAV could lead to the generation of anti-sense transcripts for the *PRTN3* and *MPO* genes. These anti-sense transcripts would lead to the formation of complementary proteins to PR3 and MPO, potentially serving as new targets for autoantibodies. This phenomenon might broaden the immune response beyond the classical PR3 and MPO proteins, potentially reinforcing autoimmunity in AAV. In addition to anti-sense transcripts, McInnis et al. [10] identified alternative transcripts of the *PRTN3* gene in the neutrophils of AAV patients, specifically an isoform known as $p24^{PR3/MBN}$. This isoform represents an alternative spliced form of PR3, which may act as a novel autoantigen, further expanding the immune targets in AAV. The existence of these alternative transcripts suggests that dysregulation of transcriptional processes in AAV not only elevates antigen levels but also introduces new forms of these proteins, creating an additional layer of immune dysregulation and potentially contributing to the chronic, relapsing nature of the disease.

Histone modifications and DNA methylation have shown potential as biomarkers for monitoring disease activity and predicting clinical outcomes in AAV. Yang et al. [11] demonstrated that specific histone modifications, including increased H4K16 acetylation at *PRTN3* and *MPO* loci, were more prevalent in active disease states. These activating histone modifications correlated with disease flares, suggesting that histone acetylation and methylation patterns could serve as indicators of disease activity. This finding introduces the possibility of using epigenetic profiles to monitor disease status, providing a dynamic tool to track AAV progression. Similarly, Jones et al. [12] explored DNA methylation patterns at *PRTN3* and *MPO* promoters and their potential as predictors of disease remission and relapse. They found that hypomethylation at these promoters was associated with active disease, while remethylation was linked to stable remission. This longitudinal study revealed that patients with remethylated *PRTN3* loci during remission had a lower likelihood of relapse, offering a prognostic tool based on individual DNA methylation profiles.

Collectively, these studies underscore the role of histone modifications and DNA methylation not only in driving the pathogenesis of AAV by dysregulating PR3 and MPO expression but also in introducing new autoantigenic targets through antisense and alternative transcripts. Furthermore, the dynamic nature of these epigenetic changes provides a promising framework for developing biomarkers that track disease activity and predict outcomes.

8.2.2 miRNAs in GPA and MPA

miRNAs play diverse roles in AAV, impacting immune and inflammatory regulation, vascular and endothelial dysfunction, and providing biomarkers for disease monitoring, prognosis, and assessment of kidney involvement.

In terms of immune and inflammatory regulation, miRNAs play a role in modulating both immune tolerance and pro-inflammatory responses in AAV. Dekkema et al. [13] highlighted how elevated levels of miR-142-3p in regulatory T cells (Tregs) of GPA patients lead to a reduction in the suppressive function of these cells. This occurs through miR-142-3p's inhibition of *ADCY9*, impairing cyclic AMP production, a critical pathway for Treg functionality. This loss of Treg regulation potentially enables unchecked inflammation in GPA, indicating miR-142-3p's significant role in immune dysfunction. Additionally, Bošnjak et al. [14] showed that miRNAs such as the miR-30 family, found in renal tissues of AAV patients, influence macrophage and lymphocyte polarization, thereby contributing to the chronic inflammatory environment typical of AAV. Hu et al. [15] found that microRNA-340-5p plays a protective role in ANCA-mediated glomerulonephritis by inhibiting endothelial apoptosis, inflammation, and pro-coagulation. This effect occurs through miR-340-5p's downregulation of *KDM4C* (a histone demethylase), which enhances B cell activation and reduces inflammatory and apoptotic markers in kidney tissues, suggesting a potential therapeutic target for treating ANCA-induced glomerulonephritis. Of note, Liu et al. [16] investigated the pathogenic role and underlying mechanisms of long noncoding RNAs (lncRNAs) in AAV, and found that LINC02193 acts as a sponge for miR-485-5p (i.e., sequesters it) to relieve the repressive effect of this miRNA on ICAM1, thus promoting ICAM1 expression and resultant neutrophil activation. These findings suggest that miRNAs are involved in immune pathway regulation, affecting both adaptive immunity, as seen in Tregs, and innate immunity, as observed in the renal tissue's immune cell polarization.

A second major aspect of miRNA involvement in AAV is in vascular and endothelial dysfunction, where miRNAs are often associated with extracellular vesicles (EVs) or exosomes that influence endothelial cells, thus driving vascular injury. Bertram et al. [17] examined ADAM17, an enzyme regulated by miR-634, and found it elevated in PR3-ANCA vasculitis, promoting an inflammatory phenotype and correlating with disease activity. Wang et al. [18] highlighted plasma-derived exosomal miRNAs, such as miR-185-3p and miR-125a-3p, which contribute to endothelial apoptosis and inflammation in MPA. Surmiak et al. [19] added to this understanding by showing how neutrophil-derived EVs carrying miR-142-3p and miR-223-3p trigger pro-inflammatory responses in endothelial cells in PR3-ANCA vasculitis. These responses include elevated cytokines like IL-8 and adhesion markers such as VCAM-1, exacerbating endothelial damage. Glémain et al. [20] further emphasized how miRNA-laden neutrophil-derived EVs, particularly those carrying miR-223, miR-142-3p, and miR-451, inhibit angiogenic repair responses and promote endothelial cell apoptosis in both GPA and MPA. Zhu et al. [21] also identified miR-1287-5p in plasma exosomes from MPA patients, which targets the gene encoding CBL, an enzyme that reduces inflammation. By silencing CBL, miR-1287-5p increases neutrophil adhesion and pro-inflammatory markers, aggravating endothelial injury. Together, these studies illustrate the role of miRNAs in vascular and endothelial dysfunction in AAV.

Finally, miRNAs hold potential as biomarkers for disease monitoring, prognosis, and for assessing kidney involvement in AAV. Scullion et al. [22] introduced miR-126 as a marker for vascular health, observing that its reduced levels in PR3-ANCA patients correlate with vascular dysfunction. The study found that miR-126 levels improved with treatment, suggesting that this miRNA could serve as a real-time indicator of vascular health and treatment efficacy. Surmiak et al. [19] expanded the biomarker potential of miRNAs in GPA by examining circulating EV miRNAs, such as miR-223-3p, miR-664a-3p, miR-200b-3p, and miR-769-5p. These miRNAs were differentially expressed in active disease and remission phases, offering potential for monitoring disease activity and predicting relapse risk. Frydlova et al. [23] focused on urinary EVs in AAV patients with renal involvement, finding elevated miRNAs such as miR-30a-5p and miR-182-5p. These miRNAs are associated with pathways essential to cytoskeletal dynamics in glomerular cells, suggesting their role in kidney inflammation and injury in AAV. The study stresses the utility of urinary miRNAs as non-invasive markers for monitoring kidney disease. Similarly, Bošnjak et al. [14] emphasized the role of renal tissue-specific miRNAs, including the miR-30 family, in contributing to AAV's kidney pathogenesis, affecting immune cell function and cytokine production. These miRNAs serve as potential markers for evaluating renal involvement and monitoring chronic inflammatory kidney disease in AAV patients. Additionally, Skoglund et al. [24] demonstrated that circulating miRNA profiles could differentiate AAV patients from healthy controls, identifying seven miRNAs (miR-29a, miR-34a, miR-142-3p, miR-383, miR-20a, miR-92a, and miR-221) with distinct expression patterns.

Overall, miRNAs in AAV contribute to immune regulation by altering immune cell functions and promoting inflammation, particularly within Tregs and renal immune cells. They also participate in vascular and endothelial dysfunction, often by way of EVs or exosomes that exacerbate endothelial cell injury and inflammation. Lastly, specific miRNAs serve as biomarkers, offering non-invasive insights into disease activity, treatment response, and renal involvement, highlighting their significance in AAV prognosis and monitoring.

8.3 IgA Vasculitis (Henoch-Schönlein Purpura)

IgAV, also recognized as Henoch-Schönlein purpura, is a form of small-vessel vasculitis marked by the deposition of IgA1-dominant immune complexes in vessel walls, primarily affecting children but also occurring in adults. The disease manifests through a combination of symptoms, including palpable purpura, commonly on the lower limbs, joint pain, abdominal pain, and, in more severe cases, renal involvement, which can present as hematuria and proteinuria. While often self-limiting in children, IgAV can lead to chronic kidney disease in adults, especially when glomerulonephritis is present. Geographic and seasonal variations suggest environmental triggers, and the condition frequently follows upper respiratory infections, indicating a possible role for mucosal immune stimulation in disease onset [25]. The pathophysiology of IgAV revolves around the presence of galactose-deficient IgA1 (Gd-IgA1), which forms immune complexes due to its altered glycosylation, prompting inflammatory responses when deposited in vessel walls. These immune complexes activate immune cells, particularly through the Fc receptor CD89 on myeloid cells, leading to neutrophil recruitment and further inflammation. Genetic studies reveal that IgAV shares genetic predispositions with IgA nephropathy (IgAN), particularly in genes regulating immune responses and glycosylation pathways. Specific polymorphisms, including *HLA-DRB1*01*, and cytokine-related genes like *CXCL8* (encoding IL-8), contribute to increased susceptibility and severity of IgAV. Additionally, the involvement of mucosal immunity is supported by associations with intestinal immune responses, such as Toll-like receptor (TLR) pathways that may drive Gd-IgA1 production in genetically predisposed individuals [25, 26].

8.3.1 Epigenetics of IgA Vasculitis

The research on epigenetics in IgAV points toward regulatory networks involving DNA methylation, histone modifications, and miRNAs, all of which play roles in disease pathogenesis, progression, and identification of potential biomarkers. The study of miRNAs, in particular, has shown how they can be significant modulators of immune responses and potential biomarkers in IgAV. Further, histone modifications and dysregulated methylation processes add up to the understanding of the inflammatory environment in IgAV and its systemic manifestations.

Research shows that miRNAs play a role in the pathogenesis of IgAV, with multiple studies revealing shifts in miRNA expression that drive inflammatory pathways. Cebi et al. [27] found that plasma levels of miR-33 and miR-34 were elevated in children with active IgAV, while miR-122 and miR-204 were reduced. The increase in miR-33 and miR-34 suggests enhanced inflammatory activity, as miR-33 promotes M1 macrophage polarization, and miR-34 activates immune responses via NF-κB pathways. In contrast, the reduction in miR-122 and miR-204, known for protective roles in endothelial health and immune regulation, may indicate a weakened defense against inflammation. These shifts, especially the decrease in miR-204, highlight the potential use of these miRNAs as biomarkers reflective of IgAV's immune dysregulation.

Yu et al. [28] also explored the miRNA functions in IgAV by examining miR-218-5p and its role in endothelial apoptosis. The downregulation of miR-218-5p was shown to promote apoptosis in human umbilical vein endothelial cells (HUVECs), a commonly used model of vascular endothelium, by affecting the expression of the HMGB1 protein. This study highlights miR-218-5p as a negative regulator of endothelial cell apoptosis and proposes that its modulation could attenuate the observed vascular damage. By targeting HMGB1, miR-218-5p potentially prevents the progression of endothelial injury, indicating it as a potential therapeutic target for reducing IgAV severity.

Research in adult IgAV has also found correlations between altered miRNA expression and histopathological changes, further supporting the role of miRNAs as markers of inflammation. Jurčić et al. [29] identified that miRNAs such as miR-223, miR-17, and miR-155 were significantly elevated in skin lesions characterized by granulocyte infiltration and vessel necrosis, linking these miRNAs to inflammatory amplification in IgAV. Moreover, Hočevar et al. [30] noted that miR-155, miR-223, and let-7b were upregulated in adult IgAV skin lesions with severe purpura, suggesting these miRNAs could serve as indicators of disease severity. Together, these findings provide a basis for the biomarker use of miRNA profiles in IgAV, particularly in evaluating inflammation and predicting disease progression.

Several miRNAs show promise as biomarkers for IgAV for assessing disease severity and renal complications. Cheng et al. [31] analyzed miR-29b expression in renal tissues from Henoch-Schönlein purpura nephritis (HSPN) patients, a severe form of IgAV with renal involvement, and found elevated levels in patients with crescent formation, a marker of renal pathology. This study identifies miR-29b as a pro-inflammatory agent and a potential indicator of renal inflammation, suggesting it could serve as a predictor for adverse renal outcomes in IgAV cases with nephritis.

Histone modifications are also involved in IgAV pathogenesis. Luo et al. [32] identified abnormal histone modifications in peripheral blood mononuclear cells from IgAV patients, particularly enhanced acetylation and methylation at histone H3 at pro-inflammatory loci such as IL-4. This upregulation potentially promotes cytokine transcription, reinforcing the systemic inflammatory characteristics of IgAV. Histone modifications thus might represent a target for therapeutic interventions, as modulating acetylation could help loosen the inflammatory pathways central to IgAV.

DNA methylation, mediated by DNA methyltransferases, also plays a role in IgAV and IgAN. Milillo et al. [33] discovered dysregulation in DNA methyltransferase 1 (DNMT1), as well as poly(ADP-ribose) polymerase 1 (PARP1), a DNA repair enzyme, in patients with both IgAN and IgAV, resulting from defective MAPK/ERK pathway signaling. In patients with IgAV and IgAN, decreased DNMT1 and PARP1 activity was associated with inflammatory processes, especially in renal tissues. This study suggests that abnormal methylation due to DNMT1 deficiency may exacerbate IgAV-related inflammation.

Genetic research has also identified key susceptibility factors for systemic vasculitis, including IgAV. Ortiz-Fernández et al. [34] conducted a cross-phenotype analysis that highlighted the gene *KDM4C* as a susceptibility locus for systemic vasculitides, including IgAV. This gene encodes a histone demethylase involved in transcriptional regulation, and its dysregulation could impact immune response genes, contributing to susceptibility in IgAV and similar inflammatory diseases.

Additional studies have linked lncRNAs to immune and apoptotic regulation in IgAV, particularly in HSPN. Huang et al. [35] identified an immune- and apoptosis-related lncRNA-miRNA-mRNA regulatory network, identifying lncRNAs such as SNHG1 and NEAT1 as potential biomarkers in HSPN. Their findings emphasize that lncRNA dysregulation impacts both immune response and apoptotic pathways in pediatric HSPN, further suggesting these lncRNAs as novel diagnostic and therapeutic targets. Similarly, Jiang et al. [36] showed that lncRNA MEG8 sponges (i.e., sequesters) miR-181a-5p to regulate SHP2 expression, contributing to M1 macrophage polarization in an IgAV rat model. This mechanism suggests that lncRNA MEG8 could play a significant role in mediating macrophage-driven inflammation, stressing its potential in controlling macrophage polarization through miRNA interactions, especially via the JAK2/STAT3 pathway.

In summary, the collective findings on epigenetic mechanisms in IgAV emphasize the roles of miRNAs, as well as histone modifications and DNA methylation, in immune regulation and in promoting inflammatory pathways. Together, these insights suggest that epigenetic mechanisms not only can drive IgAV pathogenesis but also offer biomarker potential, paving the way for more personalized approaches to disease management in IgAV.

8.4 Kawasaki Disease

KD is an acute, febrile vasculitis that primarily affects children under the age of five and is characterized by a systemic inflammation that predominantly targets medium-sized arteries, particularly the coronary arteries. The disease presents clinically with prolonged fever, polymorphous rash, conjunctival injection, changes in extremities, cervical lymphadenopathy, and mucosal involvement, often manifesting as a "strawberry tongue" and cracked lips. KD is the leading cause of acquired heart disease in children in developed countries, as it can lead to coronary artery aneurysms (CAAs), which are observed in about 20–25% of untreated cases. Intravenous immunoglobulins (IVIG) combined with aspirin has become a mainstay in KD treatment, reducing the likelihood of CAAs to around 3–5% in treated children. However, 15–20% of cases are resistant to IVIG, presenting ongoing clinical challenges. The etiology of KD is not yet fully understood, though epidemiological and seasonal patterns suggest an environmental or infectious trigger, acting upon genetically predisposed children. The pathophysiology of KD centers around dysregulated immune responses, involving both innate and adaptive immune mechanisms. The innate immune system likely initiates the inflammatory cascade, with pathogen-associated or damage-associated molecular patterns activating the NLRP3 inflammasome and inducing the release of cytokines such as IL-1β, TNF, and IL-18. These cytokines contribute to vascular endothelial damage, fostering the development of coronary artery lesions. Adaptive immunity also plays a significant role, with increased levels of circulating T cells and infiltration of immune cells, including IgA-producing plasma cells, in affected tissues. Genetic susceptibility appears important in KD, as single-nucleotide polymorphisms in genes involved in immune and inflammatory pathways (such as *ITPKC*, *CASP3*, and *FCGR2A*) are associated with KD risk and severity [37, 38].

8.4.1 DNA Methylation in Kawasaki Disease

Recent studies stress the role of DNA methylation in KD, clarifying how specific epigenetic changes drive immune dysregulation, inflammatory responses, and vascular complications in this disease. Studies on DNA methylation in KD converge on four primary aspects: immune regulation, vascular inflammation and coronary artery complications, methylation changes following IVIG therapy, and the identification of biomarkers. Each aspect adds up to the current understanding of KD's molecular pathology, showing how methylation shifts can regulate immune genes, exacerbate vascular damage, and influence therapy or result from it.

One of the most prominent themes in DNA methylation research on KD involves the modulation of immune and inflammatory responses. These studies emphasize how DNA methylation contributes to the activation and regulation of immune genes that drive the disease's characteristic inflammatory state. In particular, Huang et al. [39] noted that TLR genes, which are fundamental for initiating immune responses to pathogens, are hypomethylated in KD patients, resulting in their upregulation and enhanced inflammatory signaling, which is reversed following IVIG treatment. Ji et al. [40] further corroborated the role of hypomethylation in inflammasome activation by showing that the *ITPKC* gene, involved in calcium signaling, is downregulated in KD due to promoter methylation, which indirectly activates the NLRP3 inflammasome and intensifies inflammation through elevated IL-1β and IL-18 expression. Chang et al. [41] also provide support for this view, identifying key immune pathways, including MAPK, that are subject to differential methylation, linking methylation changes (particularly in *MAPK14* and *PHLPP1*) to immune cell signaling and inflammatory regulation in KD patients. Additionally, Huang et al. [42] found that the S100A gene family, crucial for neutrophil migration, exhibits significant hypomethylation in KD. This hypomethylation leads to increased expression of S100A genes, which enhances neutrophil transendothelial migration—a key factor in KD's immune activation and vascular inflammation. This finding underscores DNA methylation's role in modulating immune responses in KD.

Another theme is the link between DNA methylation and vascular inflammation, specifically concerning the development of coronary artery complications in KD. CAAs have been associated with changes in methylation patterns in genes linked to vascular remodeling and inflammation. Kuo et al. [43] demonstrated that *MMP9*, encoding a homonymous enzyme involved in vascular remodeling, is significantly hypomethylated in KD, leading to increased MMP-9 expression. This change is particularly pronounced in KD patients who develop CAAs, suggesting that *MMP9* hypomethylation may serve as both a biomarker and a contributor to vascular complications. Similarly, Weng et al. [44] explored the methylation status of the gene encoding CTRP1, a protein associated with vascular inflammation, and found that KD patients exhibited significant hypomethylation in the CTRP1 promoter region, which was even more pronounced in patients who developed CAAs. The findings imply that CTRP1 hypomethylation could serve as an early indicator of CAA risk, stressing its potential utility in KD prognosis. Furthermore, the *FCGR2A* gene, which plays a role in immune signaling, was also found to be hypomethylated in patients with IVIG resistance, as shown by Kuo et al. [45]. This study links *FCGR2A* hypomethylation to an increased risk of coronary complications, particularly in cases where patients exhibit resistance to IVIG therapy, thus underscoring the relevance of *FCGR2A* in both vascular complications and treatment response.

Several studies have examined how IVIG affects methylation across different genes, revealing that treatment often leads to increased methylation and subsequent downregulation of inflammation-related genes. Li et al. [46] conducted a comprehensive methylation analysis, finding that IVIG treatment induces hypermethylation in several immune-related genes, which correlates with reduced gene expression and inflammation suppression. Similarly, Yu et al. [47] explored NLRC4 methylation in KD and observed that IVIG treatment reverses NLRC4 hypomethylation, thereby decreasing its expression and reducing inflammasome activation. This methylation change aligns with the therapeutic effects of IVIG, which helps control exaggerated immune activation in KD. Another study by Huang et al. [48] supports this mechanism, demonstrating that genes involved in immune responses, including members of the TLR family and the inflammasome component NLRP12, exhibit hypermethylation and reduced expression following IVIG administration. This shift suggests that IVIG may exert its therapeutic effects by stabilizing the methylation landscape and mitigating inflammatory pathways that are hyperactive during KD. By modulating methylation levels, IVIG potentially attenuates the immune response at an epigenetic level, as shown in the work by Chen et al. [49], where β-catenin and its downstream targets are influenced by IVIG-induced methylation, underscoring the treatment's broader impact on immune modulation in KD.

The search for diagnostic and prognostic biomarkers in KD has driven part of the recent research, with DNA methylation profiles providing a promising avenue for identifying reliable markers of disease severity, treatment response, and risk of complications. Guo et al. [50] identified CD177 hypomethylation as a consistent marker in KD, with its elevated expression associated with increased disease severity. This marker not only aids in KD diagnosis but also offers insights into disease progression, as CD177 methylation status correlates with IVIG resistance and vascular complications. Additionally, hypomethylation of the *HAMP* promoter, as demonstrated by Huang et al. [51], correlates with increased hepcidin expression, suggesting that *HAMP* methylation status and hepcidin levels could serve as additional biomarkers for KD severity and response to treatment. NLRC4 has also shown potential as a prognostic marker: Yu et al. [47] demonstrated that NLRC4 hypomethylation serves as a reliable indicator of KD, with levels returning closer to normal following IVIG treatment, thereby suggesting its utility in assessing treatment response. Chang et al. [41] also contribute to this area by identifying *MAPK14* and *PHLPP1* as hub genes with altered methylation and expression profiles, proposing these genes as markers that could help differentiate KD from other febrile illnesses and provide prognostic value due to their involvement in immune signaling and inflammation pathways. Huang et al. [52] also supports the use of DNA methylation as a diagnostic tool, finding that reduced expression of DNA methyltransferases DNMT1 and DNMT3A is associated with coronary artery lesion formation, suggesting that methylation enzyme levels might help predict vascular risk in KD.

Together, these studies contribute to the current understanding of how DNA methylation influences immune activation, inflammatory regulation, vascular complications, and therapeutic response in KD, offering potential pathways for improved diagnosis, personalized treatment, and risk assessment. The findings across these studies suggest that DNA methylation is not only an important regulatory mechanism in KD but also a promising target for clinical intervention.

8.4.2 miRNAs in Kawasaki Disease

The research on miRNAs in KD provides insights into the disease's pathophysiology, particularly through three main aspects: immunomodulation, vascular and endothelial injury, and genetic susceptibility. Moreover, miRNAs in Kawasaki disease have been shown to be eligible as biomarkers, both in diagnosis and prognosis.

Immunomodulation in KD is a prominent theme in miRNA research, with several studies examining the regulatory roles that miRNAs play in immune responses. miR-155 and miR-21, for example, have been shown to influence the behavior of regulatory T cells (Tregs) in KD. Ni et al. [53] demonstrated that during acute KD, these miRNAs undergo significant downregulation, which correlates with decreased Treg numbers and disrupted immune regulation, contributing to the excessive inflammation seen in KD. Moreover, miR-31 appears to counterbalance these effects, as its upregulation in acute KD antagonizes Treg function, emphasizing how miRNAs orchestrate immune responses at a molecular level. Chu et al. [54] expanded on miR-223's role in immune modulation by illustrating its function as an "endocrine" signal that regulates endothelial inflammation. Their study showed that miR-223, secreted from bone marrow-derived blood cells, contributes to vascular damage in KD by affecting endothelial cells, revealing a pathway where immune signals provoke vascular injury. Luo et al. [55] focused on miR-27a, which is upregulated in KD and interferes with the function of IL-10-producing B10 cells, a subset that normally suppresses inflammation. This miRNA-induced inhibition of B10 cells drives monocyte-mediated inflammation and highlights how miRNAs can tilt the immune response in KD toward persistent inflammation. Yun et al. [56] identified miR-200c and miR-371-5p as significantly elevated in Kawasaki disease, suggesting these miRNAs contribute to disease pathophysiology by targeting key inflammatory pathways, including MAPK and TGF-β, which are involved in immune response and vascular inflammation. Ibáñez-Cabellos et al. [57] provided additional context by examining the role of redox-sensitive miRNAs like miR-223 in autoimmune diseases, emphasizing how oxidative stress and inflammation interact to drive disease progression in KD. Chen et al. [58] focused on several miRNAs, including miR-223-3p, found to be significantly upregulated in acute KD, suggesting miR-223's extensive regulatory role in immune responses and vascular inflammation. Luo et al. [59] also highlighted miR-133a's involvement in immune responses by showing that elevated miR-133a levels can impair endothelial function, potentially contributing to vascular inflammation through its effects on *PPP2CA*, a target related to immune response regulation. Collectively, these studies illustrate how miRNAs intricately regulate immune responses, balancing inflammation and immune tolerance in ways that directly impact KD pathology.

Several studies revealed how miRNAs affect the structural integrity of the vasculature, particularly in coronary arteries, leading to vascular and endothelial injury. A striking finding across multiple studies is the role of miRNAs in promoting endothelial apoptosis and vascular inflammation. Wu et al. [60] identified miR-186 as a key driver of endothelial apoptosis by targeting SMAD6, an inhibitor in the MAPK signaling pathway. Elevated levels of miR-186 in KD serum contribute to endothelial damage by activating pro-apoptotic pathways, which in turn leads to vascular injury. Saito et al. [61], in their study on miR-93, highlighted how this miRNA inversely regulates the expression of vascular endothelial growth factor A (VEGFA) in peripheral blood mononuclear cells. During acute KD, low levels of miR-93 correlate with increased VEGFA mRNA and protein expression, promoting angiogenesis and potentially exacerbating vascular inflammation. Similarly, Rong et al. [62]

found that miR-27b inhibits endothelial proliferation and migration by targeting Smad7, disrupting the TGF-β signaling pathway, which is essential for endothelial repair and homeostasis. Expanding on these findings, Liao et al. [63] revealed that miR-184 and miR-19a-3p contribute to endothelial dysfunction in KD by targeting AGO2, leading to the disruption of the AGO2/PTEN/VEGF axis, a key pathway in endothelial health. Li et al. [64] demonstrated the role of miR-197-3p in KD, showing it modifies endothelial cell behavior by targeting IGF1R and BCL2, thus influencing proliferation, migration, and apoptosis of endothelial cells. Li et al. [65] showed that miR-125a-5p induces apoptosis in endothelial cells by targeting MKK7. Guo et al. [66] found that platelet-derived miR-223 plays a protective role in KD by reducing endothelial inflammation. The study showed that miR-223 limits TNF-α-induced monocyte adhesion to endothelial cells by targeting ICAM-1, thereby decreasing vascular inflammation, suggesting miR-223's potential as a therapeutic target in KD. Zhang et al. [67] further emphasized the role of miR-223 in protecting vascular smooth muscle cells (VSMCs) from dedifferentiation and coronary artery pathology through the miR-223/PDGFRβ axis, showing that low induction of platelet-derived miR-223 in KD patients with severe coronary pathology leads to worsened vascular outcomes. Together, these studies establish miRNAs as important regulators of vascular and endothelial health in KD, with their dysregulation leading to the endothelial injury that characterizes KD's complications.

Some studies investigated the role of miRNA gene polymorphisms in defining KD susceptibility or affecting the disease trajectory. Yao et al. [68] identified a polymorphism in the miR-181c/d gene locus as significantly associated with KD susceptibility in Chinese children, indicating that genetic variations in miRNA regulatory regions may contribute to KD risk. This genetic predisposition perspective is also evident in studies like Fu et al. [69], which links specific miRNA polymorphisms (miR-149, miR-196a2, and miR-499a) with increased risk of coronary artery aneurysms, as well as Che et al. [70], which shows that a polymorphism in the miRNA-137 gene confers a risk of early-onset KD.

The exploration of miRNAs as diagnostic biomarkers has proven promising, as miRNAs can offer a non-invasive means of detecting KD with high specificity. Wang et al. [71] showed that miR-222-3p levels were significantly elevated in KD patients compared to those with other febrile illnesses, indicating miR-222-3p's potential as an early KD biomarker. Kuo et al. [72] advanced diagnostic accuracy by using next-generation sequencing to identify a multi-miRNA panel, including miR-93 and miR-223, to distinguish KD from febrile illnesses. Chen et al. [73] explored a panel of miRNAs (including miR-183-5p and miR-223-3p) correlated with KD symptoms, achieving high sensitivity and specificity in KD detection, especially in differentiating incomplete KD from other conditions. Jia et al. [74] focused on exosomal miRNAs, identifying a specific set that could reliably diagnose KD by distinguishing it from other febrile illnesses. Jone et al. [75] found that circulating miRNAs, particularly miR-210-3p, miR-184, and miR-19a-3p, differentiated KD from other infectious febrile diseases, underscoring their utility as diagnostic biomarkers. Li et al. [76] identified miR-355 and miR-2911 as differentially expressed in KD, with ROC analyses supporting these miRNAs' potential as non-invasive diagnostic markers for KD. The collective findings on diagnostic biomarkers underscore the promise of miRNAs in facilitating early and accurate KD diagnosis.

Indicators of prognosis and treatment response represent another important theme, where miRNAs are investigated for their potential to predict disease severity, IVIG resistance, and coronary artery lesion risk. Jing et al. [77] discovered that miR-221-3p is linked to IVIG resistance in KD, offering a means to identify patients at risk for non-responsiveness to standard treatment. miRNAs such as miR-92a-3p, identified by Rong et al. [78], are also associated with an increased risk of coronary artery lesions, suggesting that miRNA profiles can aid in risk stratification. Fu et al. [69] explored polymorphisms in miR-149, miR-196a2, and miR-499a, finding associations with coronary artery aneurysms, highlighting genetic factors in miRNA expression that may influence disease outcomes. This line of research into prognostic indicators not only aids in anticipating disease progression but also enables clinicians to tailor treatment plans according to individual risk, thereby improving patient outcomes.

In summary, these studies show that miRNAs in KD play a significant role in pathophysiology and susceptibility and may serve as biomarkers. Through immunomodulation, miRNAs mediate the immune dysregulation central to KD. In vascular injury, they drive endothelial damage that leads to coronary complications. Genetic susceptibility studies reveal that miRNA gene polymorphisms contribute to individual risk profiles. Diagnostic biomarkers offer non-invasive tools for early detection, while insights into indicators of prognosis and treatment response potentially enable personalized treatment strategies. Together, this research paints a comprehensive picture of miRNAs as not only molecular drivers of KD but also as tools for improving its diagnosis, prognosis, and treatment.

8.4.3 lncRNAs in Kawasaki Disease

Studies show that lncRNAs take part in processes such as cell proliferation, apoptosis, and inflammation, all involved in KD's pathology. For example, Hao et al. [79] found that the lncRNA Slco4a1 regulates HUVECs through the miR-335-5p/POU5F1 axis. Upregulated in KD patients, Slco4a1 promotes inflammatory factors and apoptosis in these cells, suggesting its importance in the vascular inflammation characteristic of KD. Further insights into KD's inflammatory mechanisms come from Jiang et al. [80], who explored TNF-α-induced endothelial cell apoptosis mediated by the pregnancy-induced noncoding RNA (PINC). This study shows that TNF-α, a crucial inflammatory cytokine in KD, induces apoptosis in vascular endothelial cells by upregulating PINC, contributing to the vascular damage seen in KD. This aligns with the findings of Li et al. [81], who studied the lncRNA THRIL, a regulator of TNF-α expression through its interaction with hnRNPL. Li et al. reported that THRIL levels correlate with TNF-α, underscoring the broader regulatory roles of lncRNAs in the innate immune response.

In the search for potential biomarkers and therapeutic targets, Ko et al. [82] conducted a genome-wide analysis and identified the lncRNA XLOC_006277, which is particularly relevant in KD patients who develop CAAs. This lncRNA not only correlates with disease severity but also influences the expression of enzymes MMP-8 and MMP-9, which are involved in coronary lesions, suggesting that XLOC_006277 may contribute to structural vascular damage in KD.

A different pathway involving the lncRNA HSD11B1-AS1 has been identified by Okabe et al. [83], who linked it to innate immunity regulation in KD via the G0S2 gene. HSD11B1-AS1 is upregulated during acute KD, showing a strong association with inflammation. This finding emphasizes the significant role of innate immune modulation through lncRNAs in KD's inflammatory processes. Additionally, Zhao and Chen [84] investigated the lncRNA SOCS2-AS1 and its interaction with the miR-324-5p/CUEDC2 axis. They found that SOCS2-AS1, which is upregulated in KD, promotes the proliferation and inhibits apoptosis of HUVECs by sponging miR-324-5p and thus increasing CUEDC2 expression. This mechanism suggests that SOCS2-AS1 may play a role in the vascular injury associated with KD and presents a potential target for modulating endothelial function in the disease.

Guo et al. [85] expanded on the genetic predisposition to KD by identifying a single-nucleotide polymorphism in the lncRNA AC008392.1 associated with decreased susceptibility to KD in the Han Chinese population. This single-nucleotide polymorphism, located upstream of the CARD8 gene, highlights a potential genetic mechanism involving inflammasome regulation that could affect KD risk.

Together, these studies illustrate the diverse roles of lncRNAs in KD, influencing inflammation, immune response, and vascular integrity. Their regulatory functions on essential cytokines and cellular pathways reveal new avenues for understanding KD's pathogenesis.

8.5 Large-Vessel Vasculitis

LVV is a type of systemic vasculitis affecting the large arteries, primarily the aorta and its major branches. This group includes two main subtypes: giant cell arteritis (GCA) and Takayasu arteritis (TAK), each with distinct demographic and clinical features. GCA generally occurs in individuals over 50, predominantly affecting branches of the external carotid artery, such as the temporal artery, leading to symptoms like headaches, jaw claudication, and vision loss. Polymyalgia rheumatica, characterized by aching and stiffness in the shoulders and hips, is also frequently associated with GCA. In contrast, TAK primarily affects younger individuals, especially women under 40, and is more common in Asian populations. TAK can involve multiple branches of the aorta, leading to arterial stenosis, aneurysms, and occlusions. Clinically, TAK often presents with symptoms due to decreased blood flow, such as limb claudication, reduced pulses, and blood pressure discrepancies between arms, as well as constitutional symptoms like fatigue and fever. The pathophysiology of LVV is marked by an aberrant immune response involving both innate and adaptive immunity, with a breakdown of immune tolerance in the vessel wall initiating inflammation. In GCA, dendritic cells in the vascular wall become trapped and activated, initiating recruitment of T cells and macrophages that produce inflammatory cytokines, leading to granulomatous inflammation. The immune response is characterized by the differentiation of T cells into Th1 and Th17 cells, which release cytokines, such as IFN-γ and IL-17, that drive vascular remodeling and intimal hyperplasia. In TAK, a similar immune activation occurs but often involves a different composition of T cell subsets, including CD8+ cytotoxic cells. This immune activation leads to similar vascular remodeling processes, including intimal hyperplasia and neovascularization, which contribute to vessel stenosis and potential aneurysm formation. Genetic factors also play a role: GCA is associated with *HLA-DRB104* and genes regulating immune responses, while TAK is linked with *HLA-B52:01*, affecting CD8+ T cell pathways. Shared genetic risk factors, such as *IL12B*, indicate an underlying immune pathway involvement [86, 87].

8.5.1 Epigenetics of Large-Vessel Vasculitis

Epigenetic mechanisms participate to the pathogenesis of GCA and TAK, including DNA methylation, chromatin looping effects, miRNAs, and lncRNAs. DNA methylation has been examined in GCA and TAK by several studies. Coit et al. [88] stressed how aberrant DNA methylation in temporal artery biopsies from GCA patients affects many immune-related genes. Their study revealed that hypomethylated genes are predominantly involved in T cell activation and differentiation, with significant upregulation of Th1 and Th17 pathways. The enhanced calcineurin/NFAT signaling pathway, observed through increased NFAT1 expression, suggests that DNA methylation directly influences the pro-inflammatory response characteristic of GCA. In TAK, Kabeerdoss et al. [89] focused on CD8+ and γδ T cells, finding unique methylation patterns with inflammatory genes like IL-32 hypomethylated in CD8+ T cells, while anti-inflammatory genes (e.g., IL-10) were hypomethylated in γδ T cells. In GCA, Estupiñán-Moreno et al. [90] expanded the epigenetic profile by examining monocytes. Their study showed widespread methylation and transcriptional dysregulation, particularly in active GCA cases, where monocytes exhibited increased pro-inflammatory signaling. Notably, they identified IL-6, IL-1, and IL-11 signaling pathways as being central in active GCA, linking these inflammatory responses directly to hypomethylated genes. They also showed that glucocorticoid treatment in patients in remission leads to downregulation of inflammatory genes, demonstrating an epigenetic effect of glucocorticoid treatment in GCA. Further insights into T cell involvement in GCA came from Estupiñán-Moreno et al. [91], who examined the transcriptome of CD4+ T cells. They discovered altered expression in pathways related to interleukins (e.g., IL-2), which are critical for T cell survival and regulatory functions. Moreover, they identified disrupted apoptotic pathways, which contribute to the persistence of an inflammatory T cell phenotype in GCA. This study also found altered epigenetic interactions, suggesting that methylation might regulate the inflammatory gene expression observed in these CD4+ T cells.

The role of miRNAs in LVV, and specifically in GCA, is another area of significant interest. Bolha et al. [92] identified numerous miRNAs in temporal artery biopsies from GCA patients, focusing on those involved in the modulation of VSMCs and immune cell dynamics. Key findings include the overexpression of miR-146a/b and miR-21, which enhance pro-inflammatory processes by targeting genes associated with the contractile phenotype of VSMCs, thus encouraging VSMCs to adopt a synthetic phenotype that promotes arterial wall remodeling and narrowing. Croci et al. [93] further underscored the role of miR-146a/b in inflammatory regulation within GCA, observing their upregulation in inflamed temporal arteries. This study linked miR-146a/b to cytokines such as IL-1β and IFN-γ, emphasizing how miRNAs amplify inflammatory signaling, possibly by inhibiting anti-inflammatory pathways and reinforcing the chronic inflammation characteristic of GCA. Bonacini et al. [94] added a new dimension by exploring the functional impact of miR-146a and miR-146b on endothelial function, particularly through the regulation of ICAM-1 expression. ICAM-1 is crucial for leukocyte adhesion and transmigration, which are key steps in the recruitment of immune cells to the arterial wall in GCA. This study demonstrated that miR-146a/b overexpression upregulates ICAM-1, suggesting that these miRNAs facilitate leukocyte infiltration into the vascular wall, thus exacerbating inflammation in GCA patients. Suljič et al. [95] took this further by correlating miRNA expression with ultrasound findings in GCA patients. They found that the levels of miR-21-3p and miR-146b-5p were associated with ultrasound markers like the halo sign, which indicates arterial inflammation. These miRNA profiles could serve as potential biomarkers for non-invasive diagnosis of GCA, offering a significant advantage in disease monitoring. Kuret et al. [96] focused on miRNA-mRNA interaction networks in treatment-naïve GCA patients, identifying miRNAs like miR-146a, miR-155, and miR-424 as central players in the molecular pathogenesis of GCA. These miRNAs target genes involved in VSMC proliferation, matrix remodeling, and the inflammatory response, such as YAP1, FGF2, and VEGFA. By linking miRNA expression to the structural changes observed in GCA-affected arteries, this study suggested new potential therapeutic targets aimed at modulating miRNA activity to control disease progression and minimize tissue damage. Bolha et al. [97] contributed further by quantitatively analyzing inflammatory cell composition and immune-related miRNA expression in GCA-affected arteries. They found that miR-132-3p, miR-142-3p, and miR-155-5p are overexpressed, promoting T cell-driven inflammation and macrophage activity, particularly through the CaN/NFAT signaling pathway. This study's findings emphasize that specific immune-related miRNAs are key drivers of the inflammatory cell infiltrates observed in GCA.

In TAK, epigenetic regulation also involves genetic variants impacting chromatin architecture. Kong and Sawalha [98] examined the IL6 locus, revealing that a non-coding variant associated with TAK promotes chromatin looping that represses the anti-inflammatory gene *GPNMB* via recruitment of the MEF2–HDAC complex. This repression decreases *GPNMB* expression, which normally aids in suppressing inflammatory macrophage responses. By reducing *GPNMB*, the TAK-associated allele likely contributes to macrophage-mediated inflammation, suggesting a genetic-epigenetic mechanism by which IL6 variants drive TAK pathogenesis through chromatin structure alterations. This highlights a novel therapeutic potential, targeting chromatin modification pathways to reduce inflammation in TAK.

Lastly, the study by Espinosa-Bautista et al. [99] brings attention to lncRNAs in TAK. This proof-of-concept study identified dysregulation of lncRNAs, including THRIL, HIF1A-AS1, and HOTAIR, in TAK patients. These lncRNAs showed high specificity as diagnostic markers, with THRIL, for instance, known to modulate TNF levels through a feedback loop, possibly dampening inflammation in TAK. Even though the study did not find a correlation with disease activity, their regulatory roles in inflammation and vascular remodeling highlight the importance of lncRNAs in TAK.

Collectively, these studies stress the complex epigenetic landscape in LVV, illustrating that DNA methylation, miRNA dysregulation, polymorphisms related to chromatin architecture, and lncRNAs each play significant roles in disease mechanisms. By orchestrating immune cell behavior and gene expression, these epigenetic factors shape the distinctive inflammatory responses and vascular remodeling seen in GCA and TAK.

8.6 Behçet's Disease

BD, also known as Silk Route disease, is a rare, chronic, multisystem inflammatory variable-vessel vasculitis. It is most prevalent along the Silk Road, with high incidences in countries such as Turkey, Iran, and Japan, though cases have been reported globally. Characterized by a relapsing-remitting course, BD presents with diverse symptoms including painful oral and genital ulcers, skin lesions, ocular inflammation, and vascular, neurological, and gastrointestinal involvement. Its clinical manifestations vary significantly based on geographic and genetic factors, which complicates diagnosis and management. The standard treatment typically involves corticosteroids and immunosuppressive therapies tailored to individual symptoms, though patient outcomes vary due to the syndrome's heterogeneity and complexity. The pathophysiology of BD suggests a multifactorial etiology, combining genetic predisposition, environmental triggers, and immune dysregulation. *HLA-B*51* stands as the most significant genetic marker, increasing the risk fivefold, though other loci, such as those related to IL-10 and IL-23R, also contribute. Environmental factors, particularly microbial agents like *Streptococcus* and *Herpes simplex* virus, may initiate immune responses in genetically susceptible individuals, potentially through molecular mimicry mechanisms. Immunologically, the disorder exhibits both autoimmune and autoinflammatory features, with aberrant activation of neutrophils and T cells, particularly Th1 and Th17 cells, which release inflammatory cytokines like IFN-γ and IL-17 [100, 101].

8.6.1 DNA Methylation and Chromatin Accessibility in Behçet's Disease

The exploration of DNA methylation in cytokine genes provides a view of the inflammatory and immune dysregulation in BD. Zhu et al. [102] investigated methylation patterns in the promoters of GATA3, IL-4, and TGF-β, identifying hypermethylation that corresponded to a marked reduction in expression. Since these genes encode regulators of immune tolerance and anti-inflammatory cytokines, their diminished activity through hypermethylation creates a pro-inflammatory bias in the immune system of BD patients. This dysregulation suggests that specific cytokine gene silencing through methylation directly impacts immune tolerance, potentially exacerbating the disease's chronic inflammatory state. Similarly, Qiu et al. [103] examined the methylation status of IRF8, a transcription factor involved in immune response regulation, in monocyte-derived dendritic cells from patients with ocular BD. The study found hypermethylation of IRF8, which led to decreased gene expression, subsequently influencing the balance of pro-inflammatory and regulatory responses. In BD, the silencing of IRF8, associated with the promotion of Th1/Th17 cell differentiation, reinforces an inflammatory phenotype that perpetuates tissue damage and inflammation. Building upon these findings, Abdi et al. [104] explored SOCS1 methylation and found it to be hypermethylated in BD patients, resulting in diminished expression of SOCS1, a negative regulator of cytokine signaling in the JAK/STAT pathway. This hypermethylation pattern prevents adequate suppression of cytokine production, particularly IL-6 and IL-17, which promote inflammation. The hypermethylation of SOCS1 therefore contributes to excessive and sustained cytokine activity in BD, amplifying immune cell recruitment and cytokine production, which are core drivers of BD pathology. Alipour et al. [105] studied IL-10, an anti-inflammatory cytokine gene, observing hypermethylation at the IL-10 promoter in BD patients. This hypermethylation translated to reduced IL-10 expression, effectively impairing the anti-inflammatory arm of the immune response and tweaking the balance toward an inflammatory state. In their follow-up study in 2020, Alipour and colleagues [106] expanded on this work by examining IL-6 promoter methylation. In contrast to IL-10, they observed hypomethylation of the IL-6 promoter, which correlated with an upregulation of IL-6, a central cytokine in inflammatory processes. This opposing methylation pattern between IL-10 and IL-6 highlights a dual mechanism whereby anti-inflammatory signals are reduced and pro-inflammatory signals are enhanced through epigenetic modifications in BD. Lastly, Aziz et al. [107] examined methylation in both the TNF-α and SOCS3 promoters, finding hypomethylation in the TNF-α promoter that led to increased TNF-α expression, a pro-inflammatory cytokine critical to immune response. Meanwhile, SOCS3 was hypermethylated, though expression levels remained stable. Together, these studies on cytokine gene methylation reveal a consistent pattern of methylation changes that selectively modulate cytokine expression, amplifying inflammation and limiting anti-inflammatory regulation in BD.

Research into DNA methylation in BD extends beyond the genes encoding pro- and anti-inflammatory cytokines and factors. Hughes et al. [108] conducted a genome-wide scan on DNA methylation in monocytes and CD4+ T cells of patients with BD, examining a range of cytoskeletal genes implicated in cellular signaling and immune cell movement. This study identified a pattern of altered DNA methylation that was particularly responsive to treatment, indicating that these methylation changes may not only be involved in disease pathogenesis but also serve as dynamic biomarkers reflecting disease activity and treatment efficacy. The alterations in cytoskeletal gene methylation underscore the broader impact of epigenetic changes on immune cell function and inflammatory response in BD. Building on this approach, Yu et al. [109] also conducted an epigenome-wide association study (EWAS) on BD, with a particular focus on the Han Chinese population. Yu et al. identified several BD-associated methylation loci, such as FKBP5, RUNX2, and NFIL3, which are genes known to regulate immune responses and cellular stress pathways. The results from these two studies suggest that methylation alterations are not isolated to specific genes but rather affect entire networks of immune-related genes, pointing to a systemic epigenetic reprogramming in BD that could have diagnostic and therapeutic implications.

Yüksel et al. [110] examined the methylation status of repetitive DNA sequences, specifically Alu and LINE-1 elements, in BD patients. The study found that these elements were significantly hypomethylated in BD patients compared to controls. Hypomethylation of these repetitive sequences has been linked to genomic instability and has been observed in various inflammatory and autoimmune conditions, suggesting that the destabilization of the genome may contribute to the chronic inflammatory state observed in BD. Asadi et al. [111] investigated the methylation and expression of two apoptosis-regulating genes, Bax and Bcl2, in patients with BD. Bax is a pro-apoptotic gene, while Bcl2 is anti-apoptotic, with both playing key roles in regulating cell survival. Asadi et al. found that the Bax promoter was hypermethylated in BD patients, resulting in reduced expression and diminished apoptosis. This reduction in apoptosis could lead to the persistence of activated immune cells, thereby sustaining the inflammatory environment typical of BD. Farhadi et al. [112] focused on the Foxp3 gene, a critical regulator of Treg development and function. Farhadi et al. observed that the Foxp3 promoter was hypermethylated in BD patients, leading to decreased Foxp3 expression and impaired Treg function. This finding suggests that epigenetic silencing of Foxp3 may contribute to the loss of immune tolerance in BD, exacerbating the inflammatory response.

Kolahi et al. [113] explored the methylation of Toll-like receptor (TLR) genes, specifically TLR2 and TLR4, in BD patients. TLRs are integral to innate immunity as they recognize pathogen-associated molecular patterns and activate downstream signaling pathways that trigger immune responses. Kolahi et al. found that TLR4 was hypomethylated in BD patients, leading to increased expression of TLR4 and heightened immune activation. This overexpression of TLR4 may intensify the inflammatory response by amplifying cytokine production and immune cell recruitment. Önen et al. [114] examined the HLA-B gene, specifically focusing on the HLA-B51 allele, which has been strongly associated with BD susceptibility. This study found that the HLA-B gene, particularly in patients with familial BD, exhibited hypomethylation in the promoter region, resulting in increased expression.

Elevated HLA-B51 expression may enhance the immune system's ability to recognize and respond to antigens, potentially contributing to the heightened immune reactivity seen in BD. The findings from this study underscore the role of HLA-B51 as a genetic and epigenetic factor in BD and point out the potential for targeted approaches to address HLA-related immune dysregulation in the disease. Finally, Shi et al. [115] utilized a multi-omic approach, incorporating single-cell chromatin accessibility and transcriptomic profiling, to analyze immune cells in BD at an unprecedented level of resolution. By mapping regulatory networks and identifying key transcription factors like AP-1 and NF-kB in T cells and monocytes, the study revealed immune cell-specific activation patterns and pro-inflammatory profiles that characterize BD's peripheral blood immune landscape. The findings from Shi et al. broaden the current understanding of immune cell heterogeneity in BD, offering insights into how chromatin accessibility changes may contribute to immune dysregulation and chronic inflammation.

8.6.2 miRNAs in Behçet's Disease

The research into miRNAs in BD has revealed their involvement across various biological processes, including pro-inflammatory cytokine production and the regulation of Th17/Treg cell balance. Many approaches have been used: several studies have tested the involvement of miRNAs in BD by measuring the expression level of specific miRNAs in patients as opposed to controls, while other studies have focused on miRNA gene polymorphisms, by comparing their frequency in patients and controls. Together, these findings portray miRNAs as important modulators of immune function, amplifying or dampening specific pathways implicated in BD's pathogenesis, and indicate specific miRNAs as potential biomarkers.

The production of pro-inflammatory cytokines in BD is influenced by miRNA dysregulation, as shown by several studies. Zhou et al. [116] found that miR-155 expression was significantly decreased in BD patients with active uveitis. The study shows that reduced miR-155 expression in BD promotes inflammation by increasing IL-6 and IL-1β production, while decreasing the anti-inflammatory cytokine IL-10. Additionally, miR-155 directly targets TAB2, a key molecule in the TLR/IL-1 signaling pathway, suggesting that its downregulation may contribute to BD pathogenesis by allowing unchecked inflammatory signaling. Qi et al. [117] examined the role of a functional polymorphism in miRNA-196a2 in increasing susceptibility to BD. This variant was associated with decreased miR-196a2 expression, leading to upregulated Bach1 expression. Bach1, a transcriptional repressor of heme oxygenase-1 (HO-1), influences inflammatory responses through its regulation of HO-1 levels. Qi et al. further demonstrated that this genotype in BD patients results in elevated pro-inflammatory cytokines, including IL-1β and MCP-1, highlighting a pathogenic mechanism where diminished miR-196a2 fails to regulate Bach1, thereby contributing to BD inflammation. Building on this, Zou et al. [118] directly investigated the Bach1/HO-1 pathway in the context of intestinal BD. They found that patients with active intestinal BD exhibited both reduced miR-196a2 and lower

HO-1 expression in PBMCs. This downregulation of HO-1, consistent with increased Bach1 activity, was accompanied by elevated levels of pro-inflammatory cytokines, such as IFN-γ, IL-17, IL-1β, and TNF-α, reflecting an inflammatory profile tied to reduced miR-196a2 activity. Woo et al. [119] found that miR-638, miR-4488, and miR-3591-3p were differentially expressed in BD patients. The study demonstrated that modulating these miRNAs influenced IL-6 mRNA levels in THP-1 cells (a human monocytic cell line commonly used to model immune responses), suggesting that these miRNAs contribute to BD pathogenesis by regulating IL-6 production, a key cytokine in BD's inflammatory processes. Further exploration by Puccetti et al. [120] identified specific miRNA expression signatures, consisting of 47 deregulated miRNAs, linked to active disease. Pathway enrichment analysis showed that these miRNAs target pathways such as TNF, IFN-γ, and VEGF signaling, as well as inflammatory and immune responses. Liang et al. [121] found that miR-155 expression was decreased in BD, which led to defective autophagy control in DCs. This autophagic impairment resulted in increased production of pro-inflammatory cytokines TNF-α, IL-6, and IL-1β. By upregulating miR-155 or downregulating its target TAB2, autophagy was restored, reducing cytokine levels. The study points out that miR-155 modulates inflammation in BD by controlling autophagy through the Akt/mTOR pathway, suggesting potential therapeutic targets for BD management. Saad et al. [122] found that miR-9 levels were significantly upregulated in BD patients compared to controls. They propose that miR-9 acts as a negative feedback regulator of the NF-κB pathway, which controls inflammation. In BS, however, this feedback may fail, allowing persistent NF-κB activation and sustained inflammation.

In parallel with pro-inflammatory signaling, miRNAs play an essential role in maintaining immune cell homeostasis, especially in regulating the Th17/Treg balance, which is critical in BD. Zhou et al. [123] found that individuals with a risk polymorphism associated with higher susceptibility to BD had elevated miR-146a expression levels. This increase in miR-146a expression correlated with higher levels of Th17-related cytokines, including IL-17 and TNF-α, promoting an inflammatory response. Yu et al. [124] identified a miR-182 polymorphism associated with increased miR-182 expression, which downregulates FoxO1, a transcription factor crucial for Treg stability. Reduced FoxO1 levels disrupt Treg function, promoting a shift toward Th17 cell dominance and heightened inflammatory responses. These insights align with findings by Erre et al. [125], who measured differential expression of miR-199a-3p and miR-720 in BD. These miRNAs might be involved in regulating immune responses, particularly within the Toll-like receptor (TLR) and T cell receptor (TCR) signaling pathways, which are critical for innate immunity and T cell function. Na et al. [126] showed that miR-155 is upregulated in BD patients, leading to the suppression of Ets-1, a negative regulator of Th17 differentiation. This suppression of Ets-1 promotes Th17 cell activity and IL-17 production, contributing to BD's inflammatory profile. Kolahi et al. [127] reported a positive correlation between miR-155 levels and TNF-α expression and a negative correlation with CTLA4 expression in BD patients. This suggests that upregulated miR-155 promotes inflammation by enhancing TNF-α production while concurrently inhibiting CTLA4, reducing immune regulation, and contributing to BD's inflammatory profile. Expanding on these findings, Ahmadi et al. [128] observed an elevated Th17/Treg ratio in BD patients, with increased Th17-associated cytokines (IL-17 and IL-23) and decreased Treg-associated factors (IL-10 and TGF-β). The study highlights the role of specific miRNAs in this imbalance, with miR-326 upregulated (supporting Th17 differentiation) and miR-146a and miR-155 downregulated (weakening Treg stability). Jiang et al. [129] found that elevated miR-19b-3p in plasma-derived exosomes from patients with BD uveitis decreases CD46 expression, promoting Th17 cell differentiation and reducing Treg cell differentiation. These studies collectively show that miRNAs significantly influence the Th17/Treg axis in BD, pointing to potential therapeutic targets that could restore immune balance.

In the vascular subtype of BD, where vascular inflammation and thrombosis are prominent, miRNAs also contribute to pathophysiology and might aid in diagnosis. Emmi et al. [130] identified a unique circulating microRNA (ci-miRNA) profile in patients with BD, highlighting miRNAs related to thrombo-inflammation. They found that miR-224-5p, miR-206, and miR-653-5p are upregulated in BD patients, distinguishing them from controls and other inflammatory conditions like lupus and giant cell arteritis. These ci-miRNAs were linked to pathways involving blood coagulation, oxidative stress, and cell-matrix interactions, underlying BD's unique thrombo-inflammatory nature, suggesting potential diagnostic and therapeutic implications. Kaymaz et al. [131] highlighted that miR-195 level is significantly elevated in vascular BD, suggesting that this miRNA could serve as a marker for vascular complications. Additionally, Mekawy et al. [132] examined the roles of vitamin D and miR-155 in BD, finding that miR-155 is significantly upregulated in active BD cases and associated with vascular manifestations, while vitamin D levels are markedly reduced, suggesting that miR-155 and vitamin D may play complementary roles in modulating inflammation and disease activity in BD.

Several studies have investigated miRNAs in terms of gene polymorphisms and have explored their biomarker potential in BD. Oner et al. [133] examined polymorphisms in miR-146a and miR-499, finding that the miR-146a CC genotype was protective, while the TT genotype in miR-499 was linked to increased BD susceptibility. Similarly, Shaker et al. [134] and Ibrahim et al. [135] found that the GG genotype of a miR-146a polymorphism was associated with higher BD risk and correlated with disease activity. Yenmis et al. [136] further confirmed the protective role of the CC genotype in miR-146a against BD. Studies by El Boghdady and Shaker [137] and Jadideslam et al. [138] explored miRNAs as biomarkers, identifying miR-181b and miR-326 as indicators of BD activity, particularly for severe ocular involvement. Kamal et al. [139] linked miR-146a and UTS2 gene polymorphisms with BD susceptibility, suggesting an association between genetic variations and BD manifestations. Abdelaleem et al. [140] specifically linked a miR-146a polymorphism to neurological and ocular complications in BD, emphasizing the role of miRNAs in disease manifestations. Hou et al. [141] highlighted exosomal miRNA profiles in intestinal BD, identifying specific miRNAs (e.g., miR-141-3p and miR-224-5p) that could serve as biomarkers of disease activity. Shan et al. [142], through meta-analysis, reinforced the link between the miR-146a polymorphisms and BD susceptibility, especially in Caucasian populations. Together, these studies underscore the significance of miRNA polymorphisms and expression profiles as both genetic risk factors and biomarkers for BD, suggesting potential pathways for diagnosis and targeted interventions.

In summary, miRNAs in BD contribute to several biological processes including inflammatory responses, immune cell balance, vascular pathology, and genetic predisposition, pointing to their impact on BD's pathogenesis. These insights not only advance the understanding of BD mechanisms but also suggest that miRNAs hold promise as both diagnostic biomarkers and therapeutic targets, providing new tools for tailored interventions in BD management.

8.6.3 lncRNAs and circRNAs in Behçet's Disease

Non-coding RNAs other than miRNAs in the pathogenesis of BD have gained attention recently, emerging as significant players in immune regulation, genetic susceptibility, autophagy, and potential biomarkers. The following studies illustrate how specific lncRNAs and circRNAs interact with immune pathways, influence cellular processes such as autophagy, and provide genetic insights that may be exploited for diagnostic purposes.

Abobakr et al. [143] point out the role of lncRNAs MEG3 and MAFG-AS1 in modulating miR-147b, which is involved in inflammatory responses mediated through TLRs on macrophages. MEG3 is downregulated in BD, leading to an imbalance in pro-inflammatory and anti-inflammatory cytokine production. This downregulation of MEG3 in BD patients, coupled with upregulation of miR-147b, suggests that MEG3 and MAFG-AS1 may function as biomarkers for BD activity, particularly in monitoring ocular involvement. ElMonier et al. [144] further explore immune regulation through the roles of MIAT and PVT1. These lncRNAs sponge miR-93-5p and miR-124-3p, thus modulating immune responses via TNF-α levels, a cytokine involved in BD's inflammatory processes. The overexpression of PVT1 in BD indicates its role in fostering inflammatory cytokine release, whereas MIAT appears to mitigate inflammation, suggesting a regulatory balance that, when disrupted, contributes to BD pathology. Mehana et al. [145] complement this by showing that MEG3 modulates the Th17/Treg balance in BD. Upregulation of MEG3 in BD correlates with increased expression of Th17-related markers and reduced Treg function, promoting an inflammatory state. This regulation of Th17/Treg homeostasis by MEG3 emphasizes its role in the pathogenesis of BD and indicates this lncRNA as a potential therapeutic target.

El-Din et al. [146] address how circRNA-0067835 impacts autophagy in BD through its interaction with miR-155. The study identifies that circRNA-0067835 functions as a sponge to miR-155, which itself is downregulated in BD, thus impairing autophagy. This impairment is linked to enhanced inflammatory cytokine release due to increased TAB2 and mTOR signaling pathways, highlighting the role of autophagy in controlling BD-associated inflammation. This interaction presents circRNA-0067835 as a novel therapeutic target in BD, focusing on restoring autophagic processes to mitigate inflammation.

The genetic susceptibility associated with lncRNA polymorphisms is another aspect with these regulatory molecules in BD. Wang et al. [147] reveal a significant association between a specific polymorphism in the lncRNA C1RL-AS1 and increased susceptibility to ocular BD in Han Chinese males. This gender-specific association suggests that C1RL-AS1 polymorphisms may play a role in BD's clinical variability. Similarly, Zhang et al. [148] identify a particular polymorphism in lncRNA NONHSAT159216.1 as a risk factor for BD. This SNP impacts miR-6778-5p and the downstream gene RPS6KA4, a regulatory network implicated in inflammation. These findings emphasize the importance of lncRNA polymorphisms in BD, suggesting that specific SNPs could be leveraged for risk stratification and understanding genetic predisposition.

Lastly, the diagnostic and prognostic value of lncRNAs in BD is underscored by Mohammed et al. [149], who focus on lncRNAs NEAT1 and lnc-DC (a specific lncRNA predominantly expressed in dendritic cells, regulating the differentiation and function of these cells). These lncRNAs show differential expression in BD, with NEAT1 upregulated and lnc-DC downregulated, both correlating with BD's clinical severity. The expression patterns of these lncRNAs allow for high specificity in distinguishing BD patients from healthy individuals, particularly in cases involving neurological and vascular manifestations. NEAT1 and lnc-DC thus present as accessible biomarkers that could facilitate early diagnosis and monitor disease progression in BD patients.

In summary, these studies collectively underscore the multifaceted roles of lncRNAs and circRNAs in BD, ranging from immune modulation to genetic predisposition, autophagic regulation, and diagnostic biomarker potential.

8.7 Conclusions

The diverse landscape of epigenetic modifications across different forms of systemic vasculitis points out the heterogeneous role of DNA methylation, histone modifications, and non-coding RNA regulation in these diseases. Epigenetic dysregulation of key inflammatory and immune-modulating pathways—such as those governing cytokine production, immune cell activation, and endothelial cell survival—is observed across multiple vasculitides. In AAV, particularly GPA and MPA, the aberrant DNA methylation and histone alterations contribute to the re-expression of PR3 and MPO autoantigens, a critical factor in the immune dysregulation characteristic of this condition. Specific miRNAs influence immune cell function, inflammation regulation, and vascular damage, often by way of EVs, shaping disease progression and symptoms. In IgA vasculitis (IgAV), several miRNAs were studied, driving immune cell recruitment and endothelial apoptosis, and correlating with histopathological changes and disease severity. Histone modifications, DNA methylation, and lncRNAs play a role as well, further affecting immune response and inflammation. In KD, DNA methylation patterns and specific miRNAs are linked to immune dysregulation, vascular inflammation, and IVIG treatment response. These findings suggest potential biomarkers and therapeutic targets that could mitigate coronary artery complications, a severe KD outcome. In large-vessel vasculitis (LVV), epigenetic modifications in genes regulating immune pathways contribute to the vascular remodeling seen in giant cell arteritis (GCA) and Takayasu arteritis (TAK). miRNAs involved in inflammatory signaling further exacerbate endothelial damage and leukocyte recruitment, highlighting the potential for non-invasive biomarkers to aid in diagnosis and monitoring. Epigenetic research on BD also reveals the impact of cytokine gene methylation on immune dysregulation, while miRNA-mediated modulation of the Th17/Treg balance perpetuates BD's characteristic chronic inflammation. These findings underscore the role of miRNAs and methylation patterns as potential diagnostic and prognostic markers, advancing our understanding of BD's pathogenesis.

Future research should incorporate multi-omic approaches, integrating genetic, transcriptomic, and epigenomic data to achieve a more holistic understanding of these mechanisms in systemic vasculitis. Such comprehensive models can better identify shared and distinct pathways across subtypes, guiding biomarker discovery and the development of targeted therapies. Epigenetic markers hold promise in advancing personalized medicine for systemic vasculitis, with potential applications as diagnostic, monitoring, and prognostic biomarkers, and in predicting treatment responses tailored to the unique epigenetic profiles of each subtype. While these insights pave the way for targeted epigenetic therapies, challenges remain in translating these findings into clinical practice. Therapies such as histone deacetylase inhibitors and miRNA-based interventions have potential for restoring immune balance and preventing relapse, yet the complexity of epigenetic regulation in systemic diseases requires further validation. Additionally, the translation of these findings into clinical settings demands careful consideration of individual variability and potential long-term impacts, particularly in developing safe and effective modulation strategies for immune-related epigenetic targets.

References

1. Jennette JC et al (2013) 2012 revised International Chapel Hill Consensus conference nomenclature of vasculitides. Arthritis Rheum 65:1–11
2. Cavalli G, Heard E (2019) Advances in epigenetics link genetics to the environment and disease. Nature 571:489–499
3. Mazzone R et al (2019) The emerging role of epigenetics in human autoimmune disorders. Clin Epigenetics 11:34
4. Surace AEA, Hedrich CM (2019) The role of epigenetics in autoimmune/inflammatory disease. Front Immunol 10:1525
5. Kitching AR et al (2020) ANCA-associated vasculitis. Nat Rev Dis Primers 6:1–27
6. Kronbichler A et al (2020) Immunopathogenesis of ANCA-associated vasculitis. Int J Mol Sci 21:7319
7. Yang JJ et al (2004) Circumvention of normal constraints on granule protein gene expression in peripheral blood neutrophils and monocytes of patients with antineutrophil cytoplasmic autoantibody–associated glomerulonephritis. J Am Soc Nephrol 15:2103

8. Ciavatta DJ et al (2010) Epigenetic basis for aberrant upregulation of autoantigen genes in humans with ANCA vasculitis. J Clin Invest 120:3209–3219
9. Ciavatta D, Falk RJ (2011) Epigenetics and complementary proteins. Clin Exp Immunol 164:17–19
10. McInnis EA et al (2015) Dysregulation of autoantigen genes in ANCA-associated vasculitis involves alternative transcripts and new protein synthesis. J Am Soc Nephrol 26:390
11. Yang J et al (2016) Histone modification signature at myeloperoxidase and proteinase 3 in patients with anti-neutrophil cytoplasmic autoantibody-associated vasculitis. Clin Epigenetics 8:85
12. Jones BE et al (2017) Gene-specific DNA methylation changes predict remission in patients with ANCA-associated vasculitis. J Am Soc Nephrol 28:1175–1187
13. Dekkema GJ et al (2019) Increased miR-142-3p expression might explain reduced regulatory T cell function in granulomatosis with polyangiitis. Front Immunol 10
14. Bošnjak M, Večerić-Haler Ž, Boštjančič E, Kojc N (2022) Renal tissue miRNA expression profiles in ANCA-associated vasculitis – a comparative analysis. Int J Mol Sci 23:105
15. Hu J, Wu W, Yu M, Xia Z, Gao C (2021) MicroRNA-340-5p inhibits endothelial apoptosis, inflammatory response, and pro-coagulation by targeting KDM4C in anti-neutrophil cytoplasmic antibody (ANCA)-mediated glomerulonephritis through activation of B cells. Autoimmunity 54:343–352
16. Liu Y et al (2024) A microRNA sponge, LINC02193, promotes neutrophil activation by upregulating ICAM1 and is correlated with ANCA-associated vasculitis. Rheumatology 63:2295–2306
17. Bertram A et al (2015) Circulating ADAM17 level reflects disease activity in proteinase-3 ANCA-associated vasculitis. J Am Soc Nephrol 26:2860
18. Wang Y et al (2020) Plasma exosomal miRNAs involved in endothelial injury in microscopic polyangiitis patients. FASEB J 34:6215–6228
19. Surmiak M, Wawrzycka-Adamczyk K, Kosałka-Węgiel J, Polański S, Sanak M (2022) Profile of circulating extracellular vesicles microRNA correlates with the disease activity in granulomatosis with polyangiitis. Clin Exp Immunol 208:103–113
20. Glémain A et al (2022) Neutrophil-derived extracellular vesicles induce endothelial inflammation and damage through the transfer of miRNAs. J Autoimmun 129:102826
21. Zhu Y et al (2023) Microscopic polyangiitis plasma-derived exosomal miR-1287-5p induces endothelial inflammatory injury and neutrophil adhesion by targeting CBL. PeerJ 11:e14579
22. Scullion KM et al (2021) Circulating argonaute-bound microRNA-126 reports vascular dysfunction and treatment response in acute and chronic kidney disease. iScience 24
23. Frydlova J et al (2022) Analysis of microRNAs in small urinary extracellular vesicles and their potential roles in pathogenesis of renal ANCA-associated vasculitis. Int J Mol Sci 23:4344
24. Skoglund C et al (2015) Circulating microRNA expression pattern separates patients with anti-neutrophil cytoplasmic antibody associated vasculitis from healthy controls. Clin Exp Rheumatol 33:S-64-71
25. Pillebout E, Sunderkötter C (2021) IgA vasculitis. Semin Immunopathol 43:729–738
26. Song Y et al (2021) Pathogenesis of IgA vasculitis: an up-to-date review. Front Immunol 12
27. Cebi AH, Demir F, Ikbal M, Kalyoncu M (2021) Plasma microRNA levels in childhood IgA vasculitis. Clin Rheumatol 40:1975–1981
28. Yu S-F et al (2018) Down-regulation of miR-218-5p promotes apoptosis of human umbilical vein endothelial cells through regulating high-mobility group box-1 in henoch-schonlein purpura. Am J Med Sci 356:64–71
29. Jurčić V et al (2019) Association between histopathological changes and expression of selected microRNAs in skin of adult patients with IgA vasculitis. Histopathology 75:683–693
30. Hočevar A et al (2019) MicroRNA expression in the affected skin of adult patients with IgA vasculitis. Clin Rheumatol 38:339–345
31. Cheng S, Zhu C-H, Zhang A-H, Huang S-M (2020) MiR-29b expression is altered in crescent formation of HSPN and accelerates Ang II-induced mesangial cell activation. World J Pediatr 16:201–212

32. Luo S, Liang G, Zhang P, Zhao M, Lu Q (2013) Aberrant histone modifications in peripheral blood mononuclear cells from patients with Henoch–Schönlein purpura. Clin Immunol 146:165–175
33. Milillo A et al (2018) Defective activation of the MAPK/ERK pathway, leading to PARP1 and DNMT1 dysregulation, is a common defect in IgA nephropathy and Henoch-Schönlein purpura. J Nephrol 31:731–741
34. Ortiz-Fernández L et al (2018) Cross-phenotype analysis of Immunochip data identifies KDM4C as a relevant locus for the development of systemic vasculitis. Ann Rheum Dis 77:589–595
35. Huang L et al (2022) Integrated analysis of immune- and apoptosis-related lncRNA-miRNA-mRNA regulatory network in children with Henoch Schönlein purpura nephritis. Transl Pediatr 11:1682696–1681696
36. Jiang M et al (2021) LncRNA MEG8 sponging miR-181a-5p contributes to M1 macrophage polarization by regulating SHP2 expression in Henoch-Schonlein purpura rats. Ann Med 53:1576–1588
37. Noval Rivas M, Arditi M (2020) Kawasaki disease: pathophysiology and insights from mouse models. Nat Rev Rheumatol 16:391–405
38. Rife E, Gedalia A (2020) Kawasaki Disease: an update. Curr Rheumatol Rep 22:75
39. Huang Y-H et al (2017) Identifying genetic hypomethylation and upregulation of toll-like receptors in Kawasaki disease. Oncotarget 8:11249–11258
40. Ji M-L et al (2023) Inositol-triphosphate 3-kinase C and DNA methylation involvement in NLRP3 inflammasome activation in Kawasaki disease. Indian J Pediatr 90:22–28
41. Chang D, Qian C, Li H, Feng H (2019) Comprehensive analyses of DNA methylation and gene expression profiles of Kawasaki disease. J Cell Biochem 120:13001–13011
42. Huang L-H et al (2018) Multiomics analyses identified epigenetic modulation of the S100A gene family in Kawasaki disease and their significant involvement in neutrophil transendothelial migration. Clin Epigenetics 10:135
43. Kuo H-C, Li S-C, Huang L-H, Huang Y-H (2017) Epigenetic hypomethylation and upregulation of matrix metalloproteinase 9 in Kawasaki disease. Oncotarget 8:60875–60891
44. Weng H et al (2020) Hypomethylation of C1q/tumor necrosis factor-related protein-1 promoter region in whole blood and risks for coronary artery aneurysms in Kawasaki disease. Int J Cardiol 307:159–163
45. Kuo H-C et al (2015) Identification of an association between genomic hypomethylation of FCGR2A and susceptibility to Kawasaki disease and intravenous immunoglobulin resistance by DNA methylation array. Arthritis Rheumatol 67:828–836
46. Li S-C et al (2016) Major methylation alterations on the CpG markers of inflammatory immune associated genes after IVIG treatment in Kawasaki disease. BMC Med Genet 9:37
47. Yu B et al (2024) NLRC4 methylation and its response to intravenous immunoglobulin therapy in Kawasaki disease: a case control study. BMC Pediatr 24:190
48. Huang Y-H, Lo M-H, Cai X-Y, Kuo H-C (2018) Epigenetic hypomethylation and upregulation of NLRC4 and NLRP12 in Kawasaki disease. Oncotarget 9:18939–18948
49. Chen K-D et al (2018) The human blood DNA methylome identifies crucial role of β-catenin in the pathogenesis of Kawasaki disease. Oncotarget 9:28337–28350
50. Guo MM-H, Chang L-S, Huang Y-H, Wang F-S, Kuo H-C (2020) Epigenetic regulation of macrophage marker expression profiles in Kawasaki disease. Front Pediatr 8:129
51. Huang Y-H et al (2018) HAMP promoter hypomethylation and increased hepcidin levels as biomarkers for Kawasaki disease. J Mol Cell Cardiol 117:82–87
52. Huang Y-H et al (2019) Decreased DNA methyltransferases expression is associated with coronary artery lesion formation in Kawasaki disease. Int J Med Sci 16:576–582
53. Ni F-F et al (2014) Regulatory T cell microRNA expression changes in children with acute Kawasaki disease. Clin Exp Immunol 178:384–393
54. Chu M et al (2017) Bone marrow-derived MicroRNA-223 works as an endocrine genetic signal in vascular endothelial cells and participates in vascular injury from Kawasaki disease. J Am Heart Assoc 6:e004878

55. Luo Y et al (2020) Up-regulation of miR-27a promotes monocyte-mediated inflammatory responses in Kawasaki disease by inhibiting function of B10 cells. J Leukoc Biol 107:133–144
56. Yun KW, Lee JY, Yun SW, Lim IS, Choi ES (2014) Elevated serum level of microRNA (miRNA)-200c and miRNA-371-5p in children with Kawasaki disease. Pediatr Cardiol 35:745–752
57. Ibáñez-Cabellos JS, Pallardó FV, García-Giménez JL, Seco-Cervera M (2023) Oxidative stress and epigenetics: miRNA Involvement in rare autoimmune diseases. Antioxidants (Basel) 12:800
58. Chen Y et al (2018) Identification of differentially expressed microRNAs in acute Kawasaki disease. Mol Med Rep 17:932–938
59. Luo Y et al (2022) The expression and role of microRNA-133a in plasma of patients with Kawasaki disease. Immunol Investig 51:826–838
60. Wu R et al (2018) miR-186, a serum microRNA, induces endothelial cell apoptosis by targeting SMAD6 in Kawasaki disease. Int J Mol Med 41:1899–1908
61. Saito K et al (2016) MicroRNA-93 may control vascular endothelial growth factor A in circulating peripheral blood mononuclear cells in acute Kawasaki disease. Pediatr Res 80:425–432
62. Rong X et al (2018) miR-27b suppresses endothelial cell proliferation and migration by targeting Smad7 in Kawasaki disease. Cell Physiol Biochem 48:1804–1814
63. Liao J, Guo X, Fan X, Zhang X, Xu M (2023) Upregulation of miR-184 and miR-19a-3p induces endothelial dysfunction by targeting AGO2 in Kawasaki disease. Cardiol Young 33:1962–1966
64. Li Y, Wu X, Gao F, Wang X (2019) MiR-197-3p regulates endothelial cell proliferation and migration by targeting IGF1R and BCL2 in Kawasaki disease. Int J Clin Exp Pathol 12:4181–4192
65. Li Z et al (2017) A plasma mir-125a-5p as a novel biomarker for Kawasaki disease and induces apoptosis in HUVECs. PLoS One 12:e0175407
66. Guo M et al (2022) Platelet-derived microRNA-223 attenuates TNF-α induced monocytes adhesion to arterial endothelium by targeting ICAM-1 in Kawasaki disease. Front Immunol 13:922868
67. Zhang Y et al (2020) Reduced platelet miR-223 induction in kawasaki disease leads to severe coronary artery pathology through a miR-223/PDGFRβ vascular smooth muscle cell axis. Circ Res 127:855–873
68. Yao M et al (2022) Association of miR-181c/d gene locus rs8108402 C/T polymorphism with susceptibility to Kawasaki disease in Chinese children. Front Pediatr 10:899779
69. Fu L et al (2022) Association study of miR-149, miR-196a2, and miR-499a polymorphisms with coronary artery aneurysm of Kawasaki disease in southern Chinese population. J Gene Med 24:e3405
70. Che D et al (2018) The rs1625579 T>G polymorphism in the miRNA-13 gene confers a risk of early- onset Kawasaki disease in a southern Chinese population. Infect Drug Resist 11:1055–1060
71. Wang B et al (2019) MiR-222-3p in platelets serves as a distinguishing marker for early recognition of Kawasaki disease. Front Pediatr 7:237
72. Kuo H-C et al (2016) Next-generation sequencing identifies micro-RNA-based biomarker panel for Kawasaki disease. J Allergy Clin Immunol 138:1227–1230
73. Chen C-C et al (2023) Symptom-correlated MiRNA signature as a potential biomarker for Kawasaki disease. Biomed J 47:100684
74. Jia H-L et al (2017) Sets of serum exosomal microRNAs as candidate diagnostic biomarkers for Kawasaki disease. Sci Rep 7:44706
75. Jone P-N et al (2020) Circulating microRNAs differentiate Kawasaki disease from infectious febrile illnesses in childhood. J Mol Cell Cardiol 146:12–18
76. Li X, Sun Y, Cai Y, Zhang X, Zhang X (2022) Identification of miRNA profile in the peripheral blood and clinical significance of miR-355 and miR-2911 expression in children with Kawasaki disease. Am J Transl Res 14:7820–7830

77. Jing F et al (2022) Association between serum miR-221-3p and intravenous immunoglobulin resistance in children with Kawasaki disease. Clin Exp Med 22:621–628
78. Rong X et al (2017) Serum miR-92a-3p as a new potential biomarker for diagnosis of Kawasaki disease with coronary artery lesions. J Cardiovasc Transl Res 10:1–8
79. Hao J et al (2022) Kawasaki disease: lncRNA Slco4a1 regulates the progression of human umbilical vein endothelial cells by targeting the miR-335-5p/POU5F1 axis. Transl Pediatr 11:183–193
80. Jiang C et al (2016) TNF-α induces vascular endothelial cells apoptosis through overexpressing pregnancy induced noncoding RNA in Kawasaki disease model. Int J Biochem Cell Biol 72:118–124
81. Li Z et al (2014) The long noncoding RNA THRIL regulates TNFα expression through its interaction with hnRNPL. Proc Natl Acad Sci USA 111:1002–1007
82. Ko T-M et al (2019) Genome-wide transcriptome analysis to further understand neutrophil activation and lncRNA transcript profiles in Kawasaki disease. Sci Rep 9:328
83. Okabe M et al (2022) G0S2 regulates innate immunity in Kawasaki disease via lncRNA HSD11B1-AS1. Pediatr Res 92:378–387
84. Zhao J, Chen D (2022) Kawasaki disease: SOCS2-AS1/miR-324-5p/CUEDC2 axis regulates the progression of human umbilical vein endothelial cells. Pediatr Res 92:388–395
85. Guo K et al (2021) Single-nucleotide polymorphism LncRNA AC008392.1/rs7248320 in CARD8 is associated with Kawasaki disease susceptibility in the Han Chinese population. J Inflamm Res 14:4809–4816
86. Pugh D et al (2022) Large-vessel vasculitis. Nat Rev Dis Primers 7:93
87. Harky A, Fok M, Balmforth D, Bashir M (2019) Pathogenesis of large vessel vasculitis: implications for disease classification and future therapies. Vasc Med 24:79–88
88. Coit P, De Lott LB, Nan B, Elner VM, Sawalha AH (2016) DNA methylation analysis of the temporal artery microenvironment in giant cell arteritis. Ann Rheum Dis 75:1196–1202
89. Kabeerdoss J et al (2022) Genome-wide DNA methylation profiling in CD8 T-cells and gamma delta T-cells of Asian Indian patients With Takayasu arteritis. Front Cell Dev Biol 10:843413
90. Estupiñán-Moreno E et al (2022) Methylome and transcriptome profiling of giant cell arteritis monocytes reveals novel pathways involved in disease pathogenesis and molecular response to glucocorticoids. Ann Rheum Dis 81:1290–1300
91. Estupiñán-Moreno E et al (2024) Decoding CD4+ T cell transcriptome in giant cell arteritis: novel pathways and altered cross-talk with monocytes. J Autoimmun 146:103240
92. Bolha L et al (2020) Identification of microRNAs and their target gene networks implicated in arterial wall remodelling in giant cell arteritis. Rheumatology (Oxford) 59:3540–3552
93. Croci S et al (2016) MicroRNA markers of inflammation and remodelling in temporal arteries from patients with giant cell arteritis. Ann Rheum Dis 75:1527–1533
94. Bonacini M et al (2024) miR-146a and miR-146b regulate the expression of ICAM-1 in giant cell arteritis. J Autoimmun 144:103186
95. Suljič A, Hočevar A, Jurčić V, Bolha L (2023) Evaluation of arterial histopathology and microRNA expression that underlie ultrasonography findings in temporal arteries of patients with giant cell arteritis. Int J Mol Sci 24:1572
96. Kuret T et al (2021) Dysregulated expression of arterial MicroRNAs and their target gene networks in temporal arteries of treatment-naïve patients with giant cell arteritis. Int J Mol Sci 22:6520
97. Bolha L, Hočevar A, Suljič A, Jurčić V (2021) Inflammatory cell composition and immune-related microRNA signature of temporal artery biopsies from patients with giant cell arteritis. Front Immunol 12:791099
98. Kong X, Sawalha AH (2019) Takayasu arteritis risk locus in IL6 represses the anti-inflammatory gene GPNMB through chromatin looping and recruiting MEF2-HDAC complex. Ann Rheum Dis 78:1388–1397
99. Espinosa-Bautista F et al (2024) Dysregulation of long non-coding RNAs in Takayasu arteritis: a proof-of-concept study. Clin Rheumatol 43:1253–1259

100. Emmi G, Bettiol A, Hatemi G, Prisco D (2024) Behçet's syndrome. Lancet 403:1093–1108
101. Tong B, Liu X, Xiao J, Su G (2019) Immunopathogenesis of Behcet's disease. Front Immunol 10:665
102. Zhu Y et al (2017) Aberrant DNA methylation of GATA binding protein 3 (GATA3), interleukin-4 (IL-4), and transforming growth factor-β (TGF-β) promoters in Behcet's disease. Oncotarget 8:64263–64272
103. Qiu Y et al (2017) Ocular Behcet's disease is associated with aberrant methylation of interferon regulatory factor 8 (IRF8) in monocyte-derived dendritic cells. Oncotarget 8:51277–51287
104. Abdi A et al (2018) Evaluation of SOCS1 methylation in patients with Behcet's disease. Immunol Lett 203:15–20
105. Alipour S et al (2018) Hypermethylation of IL-10 gene is responsible for its low mRNA expression in Behçet's disease. J Cell Biochem 119:6614–6622
106. Alipour S et al (2020) Methylation status of interleukin-6 gene promoter in patients with Behçet's disease. Reumatol Clin (Engl Ed) 16:229–234
107. Aziz SG-G, Aziz SG-G, Khabbazi A, Alipour S (2020) The methylation status of TNF-α and SOCS3 promoters and the regulation of these gene expressions in patients with Behçet's disease. Biomarkers 25:384–390
108. Hughes T et al (2014) Epigenome-wide scan identifies a treatment-responsive pattern of altered DNA methylation among cytoskeletal remodeling genes in monocytes and CD4+ T cells from patients with Behçet's disease. Arthritis Rheumatol 66:1648–1658
109. Yu H et al (2019) Epigenome-wide association study identifies Behçet's disease-associated methylation loci in Han Chinese. Rheumatology (Oxford) 58:1574–1584
110. Yüksel Ş et al (2016) Methylation status of Alu and LINE-1 interspersed repetitive sequences in Behcet's disease patients. Biomed Res Int 2016:1393089
111. Asadi S et al (2020) Promoter methylation of Bax and Bcl2 genes and their expression in patients with Behcet's disease. Int J Immunogenet 47:309–317
112. Farhadi J et al (2019) Analysis of methylation and expression profile of Foxp3 gene in patients with Behçet's syndrome. Iran J Allergy Asthma Immunol 19:1–8
113. Kolahi S et al (2020) Evaluation of DNA methylation status of toll-like receptors 2 and 4 promoters in Behcet's disease. J Gene Med 22:e3234
114. Özkılınç Önen M et al (2024) HLA-B gene methylation and expression in Behçet's syndrome: a potential role of epigenetics in the pathogenesis. Clin Exp Rheumatol 42:2014–2020
115. Shi W et al (2023) Single-cell chromatin accessibility and transcriptomic characterization of Behcet's disease. Commun Biol 6:1048
116. Zhou Q et al (2012) Decreased microRNA-155 expression in ocular Behcet's disease but not in Vogt Koyanagi Harada syndrome. Invest Ophthalmol Vis Sci 53:5665–5674
117. Qi J et al (2013) A functional variant of pre-miRNA-196a2 confers risk for Behcet's disease but not for Vogt-Koyanagi-Harada syndrome or AAU in ankylosing spondylitis. Hum Genet 132:1395–1404
118. Zou J, Ji D-N, Shen Y, Guan J-L, Zheng S-B (2016) Association of reduced heme oxygenase-1 with decreased MicroRNA-196a2 expression in peripheral blood mononuclear cells of patients with intestinal Behcet's disease. Ann Clin Lab Sci 46:675–679
119. Woo M-Y et al (2016) MicroRNAs differentially expressed in Behçet disease are involved in interleukin-6 production. J Inflamm (Lond) 13:22
120. Puccetti A et al (2018) MicroRNA expression profiling in Behçet's disease. J Immunol Res 2018:2405150
121. Liang L, Zhou Q, Feng L (2021) Decreased microRNA-155 in Behcet's disease leads to defective control of autophagy thereby stimulating excessive proinflammatory cytokine production. Arthritis Res Ther 23:135
122. Saad MA, El Gendy HI, Behiry ME, Shaker O, Laymouna AH (2024) Microribonucleic acid-9 (miRNA-9) as a potential diagnostic marker in Behçet's syndrome patients. Egypt Rheumatol 46:190–193

123. Zhou Q et al (2014) MicroRNA-146a and Ets-1 gene polymorphisms in ocular Behcet's disease and Vogt-Koyanagi-Harada syndrome. Ann Rheum Dis 73:170–176
124. Yu H, Liu Y, Bai L, Kijlstra A, Yang P (2014) Predisposition to Behçet's disease and VKH syndrome by genetic variants of miR-182. J Mol Med (Berl) 92:961–967
125. Erre GL et al (2015) Global microRNA profiling of peripheral blood mononuclear cells in patients with Behçet's disease. Clin Exp Rheumatol 33:S72–S79
126. Na SY, Park M-J, Park S, Lee E-S (2016) MicroRNA-155 regulates the Th17 immune response by targeting Ets-1 in Behçet's disease. Clin Exp Rheumatol 34:S56–S63
127. Kolahi S et al (2018) Determination of mir-155 and mir-146a expression rates and its association with expression level of TNF-α and CTLA4 genes in patients with Behcet's disease. Immunol Lett 204:55–59
128. Ahmadi M et al (2019) Disturbed Th17/Treg balance, cytokines, and miRNAs in peripheral blood of patients with Behcet's disease. J Cell Physiol 234:3985–3994
129. Jiang Q et al (2023) Effects of plasma-derived exosomal miRNA-19b-3p on Treg/T helper 17 cell imbalance in Behçet's Uveitis. Invest Ophthalmol Vis Sci 64:28
130. Emmi G et al (2022) A unique circulating miRNA profile highlights thrombo-inflammation in Behçet's syndrome. Ann Rheum Dis 81:386–397
131. Kaymaz S, Aydın D, Uğur K, Çobankara V, Tan S (2024) Expression levels and clinical values of miR-195, miR-424, miR-10b, miR-103a-3p, and miR-542-3p in Vasculo-Behçet's disease. Mediterr J Rheumatol 35:255–262
132. Mohamed Mekawy D et al (2023) Vitamin D and miRNA-155 in Behçet's disease: possible association with the disease and disease activity. Rep Biochem Mol Biol 12:251–258
133. Oner T et al (2015) Association of Pre-miRNA-499 rs3746444 and Pre-miRNA-146a rs2910164 polymorphisms and susceptibility to Behcet's disease. Genet Test Mol Biomarkers 19:424–430
134. Shaker OG et al (2022) MiR-146a and miR-155 polymorphisms in Egyptian patients with Behcet's disease. Arch Med Sci 18:1467–1474
135. Ibrahim W, Sakr BR, Obaya E, Ghonem H (2019) MicroRNA-146a expression and microRNA-146a rs2910164 polymorphism in Behcet's disease patients. Clin Rheumatol 38:397–402
136. Yenmis G et al (2021) Fas and microRNAs variations as a possible risk for Behçet disease. J Clin Rheumatol 27:306–310
137. El Boghdady NA, Shaker OG (2019) Role of serum miR-181b, proinflammatory cytokine, and adhesion molecules in Behçet's disease. J Interf Cytokine Res 39:347–354
138. Jadideslam G et al (2019) Expression levels of miR-21, miR-146b and miR-326 as potential biomarkers in Behcet's disease. Biomark Med 13:1339–1348
139. Kamal A, Elgengehy FT, Elawady Z, Fawzy NA, El Sisi O (2022) Role of miR-146a rs2910164 and UTS2 rs228648 genetic variants in Behçet's disease. Immunol Investig 51:899–908
140. Abdelaleem OO et al (2021) Association of miR-146a rs57095329 with Behçet's disease and its complications. Br J Biomed Sci 78:63–66
141. Hou C-C et al (2022) Expression of miRNAs derived from plasma exosomes in patients with intestinal Behçet's syndrome. Clin Exp Rheumatol 40:1480–1490
142. Shan J et al (2022) Association of miRNA-146a gene polymorphism Rs2910164 with Behcet's disease: a meta-analysis. Ocul Immunol Inflamm 30:1883–1889
143. Abobakr S, Shaker O, Hegazy MT, Hany AM (2024) A possible role of lncRNA MEG3 and lncRNA MAFG-AS1 on miRNA 147-b in the pathogenesis of Behcet's disease. Immunogenetics 76:233–241
144. ElMonier AA, Shaker OG, Ali SO (2024) Regulatory role of the lncRNAs MIAT and PVT1 in Behçet's disease through targeting miR-93-5p and miR-124-3p. Mol Med 30:157
145. Mehana NA et al (2022) LncRNA MEG3 regulates the interplay between Th17 and Treg cells in Behçet's disease and systemic lupus erythematosus. Life Sci 309:120965

146. Saad El-Din S et al (2023) Regulatory role of circRNA-0067835 in Behcet disease through targeting Micro RNA-155: implication of ATG1, AKT and MTOR. Rep Biochem Mol Biol 12:195–204
147. Wang X et al (2024) Association of long non-coding RNA C1RL-AS1 and PTPN6 gene polymorphisms with ocular Behcet's disease in Han Chinese. Ocul Immunol Inflamm 32:336–341
148. Zhang J et al (2022) SNP rs7130280 in lncRNA NONHSAT159216.1 confers susceptibility to Behçet's disease uveitis in a Chinese Han population. Rheumatology (Oxford) 62:384–396
149. Mohammed SR et al (2021) Expression of lncRNAs NEAT1 and lnc-DC in serum from patients with Behçet's disease can be used as predictors of disease. Front Mol Biosci 8:797689

Epigenetics and Allergic Diseases

9

Beatrice Andrenacci, Amelia Licari,
and Gian Luigi Marseglia

Contents

9.1 The Genetics of Allergies: From Candidate Genes to Genome-Wide Association Studies, to Epigenomics 146
9.1.1 "Omics" in Allergy: Pathogenesis and Natural History 147
9.1.2 "Omics" in Allergy: Biomarkers and the "Epigenomic Aging" 149
9.1.3 "Omics" in Allergy: Pharmacogenomics 150
9.2 Epigenetics among Different Pediatric Allergic Diseases: The Atopic March 151
9.2.1 Epigenetics and Respiratory Allergies: From Allergic Rhinitis to Asthma 153
9.2.2 Epigenetics and Eczema 155
9.2.3 Epigenetics and Food Allergy 156
9.3 The Future of Epigenetics in Allergy 157
9.4 Conclusions 158
References 159

Key Points

- "Omics" sciences play a pivotal role in allergy: in the future, their application may shed light on better understanding the pathogenesis of allergic diseases,

B. Andrenacci
S.C. Pediatria Pneumoinfettivologia, Fondazione IRCCS Ca' Granda Ospedale Maggiore Policlinico, Milan, Italy
e-mail: beatrice.andrenacci@policlinico.mi.it

A. Licari · G. L. Marseglia (✉)
Pediatric Unit, Department of Clinical, Surgical, Diagnostic and Pediatric Science, University of Pavia, Pavia, Italy

Pediatric Clinic, Fondazione IRCCS Policlinico San Matteo, Pavia, Italy
e-mail: amelia.licari@unipv.it; gl.marseglia@smatteo.pv.it

L. Emmi, V. Fanos (eds.), *Epigenetics and Rare Diseases*, Rare Diseases of the Immune System, https://doi.org/10.1007/978-3-032-02902-7_9

early identification of new biomarkers, and the design of innovative targeted therapies.
- Epigenomic signatures common to different allergic diseases could be used to predict "atopic trajectories," paving the way for timely preventive strategies.
- Future multicenter, pediatric-focused studies are needed to provide more consistent results.

9.1 The Genetics of Allergies: From Candidate Genes to Genome-Wide Association Studies, to Epigenomics

In the last decades, allergy has been increasingly widespread in developed and developing countries, known as "the allergy epidemic" [1]. Currently, about 20% of the global population is supposed to be atopic [2]. This is particularly relevant in the pediatric setting since allergic diseases are known to be the third most common chronic disease of childhood [3]. Allergic rhinitis (AR), asthma, atopic dermatitis (AD), and food allergy (FA) are complex multifactorial diseases whose clinical expression profoundly varies in terms of phenotypes and severity, according to the different combination of genetic background and lifestyle factors (pre-natal, peri-natal, and post-natal endogenous or exogenous exposures, globally referred as "exposome" [4]) on a geographic basis. Exploring the contribution of genetics to allergy, many different studies were conducted in the last decades to identify genes related to atopic sensitizations, allergy onset and progression, disease severity, and response to treatments [5–7]. Two main approaches were used, as summarized in Table 9.1.

The preferred method was candidate gene studies, focusing on the a-priori evaluation of single nucleotide polymorphisms (SNPs). Such studies identified causative genes involved in immune response regulation, exposome response modulation, and organ-specific response to inflammation [7]. Those studies are penalized by type 1

Table 9.1 Pros and cons of different studies to identify genes involved in allergy susceptibility and severity

Type of study	Pros	Cons
Candidate gene studies	Biological plausibility Functional consequences and molecular pathways are known in advance	Linkage disequilibrium Type I error Does not take into account the effect of exposomes or disease-modifying genes
Genome-wide association studies	Hypothesis-free approach not limited by the current knowledge in the field Identification of genes with even small effect on the disease Identification of organ-specific or endotype-specific susceptibility genes	Genes with overlap among different allergic diseases Missing heritability

statistical errors due to poor study design or inter-individual differences in ethnicity or rural/urban environment. Moreover, candidate gene studies may be affected by linkage disequilibrium. A specific SNP statistically associated with a particular disease phenotype may only represent a proxy marker for another genetic variant nearby, therefore lacking a direct pathogenetic effect [5–7].

Those limits led to hypothesis-free, genome-wide association studies (GWAS), which made possible the identification of many new genes and novel molecular pathways not explored before. Interestingly, GWAS studies proved that genetic factors involved in atopy expression differ from those influencing disease severity [3, 7]. However, although many genes and polymorphisms have been implicated in allergic diseases [8–13], according to GWAS, the molecular underpinnings of disease severity and heterogeneity are still poorly understood [14], and early biomarkers are lacking. Moreover, combining GWAS results into disease-specific polygenetic risk scores still poorly captures the individual's susceptibility to developing allergy [15]. This "missing heritability" is thought to be caused by rare variants with high genetic effect, common copy number variants (CNVs), or trans-generational epigenetic inheritance [16], which cannot be detected in GWAS. This evidence paves the way for new epigenomic studies in the field of allergy, which will focus on epigenome-wide association study (EWAS) of methylation profiles [14].

Although currently limited to research fields, "omics" sciences have broad applicability in the allergy field. Currently, the three main fields of application of "omics" sciences in allergology include:

- Prediction of disease susceptibility and onset through a better definition of the pathogenesis and natural history.
- Earlier diagnosis and definition of disease endotypes and severity, thanks to the identification of new biomarkers.
- Since the early identification of unique phenotypes with specific responses to target therapies [3], pharmacogenomics could help apply future preventative treatments only to those who will benefit.

9.1.1 "Omics" in Allergy: Pathogenesis and Natural History

Exploring the pathogenesis of allergy, epigenomic studies have deepened our understanding of the interaction between genes and the environment, which plays a crucial role in developing allergic diseases. While the polymorphism of a single gene (SNP) minimally contributes to the overall disease phenotype, ongoing research on epigenetic influences, including CpG island methylation and histone modifications, is gradually elucidating these interactions, especially regarding the sensing of "exposome" and the mediation of the subsequent effects. As a matter of fact, "exposome" leads to epigenomic rearrangements, which shape the immune system toward the allergic response. Interestingly, this immune commitment is believed to begin early, even during fetal life [17], offering opportunities for preventive strategies during pregnancy [18]. Despite challenges in predicting susceptibility using multiple genetic loci, the evolving understanding of allergy epigenetics enhances our understanding of disease mechanisms by identifying specific genes associated with various phenotypes, possibly defining methylation risk scores.

Focusing on the early onset of allergic pathogenesis, a recent study of 367 subjects undergoing whole-blood EWAS revealed a correlation between birth season, DNA methylation levels, and allergy risk. Subjects born in autumn had an increased risk of eczema compared to those born in spring due to postnatal exposure to UV radiation, vitamin D levels, pollen, and viruses [17]. Additionally, seasonal variations in DNA methylation (especially within a CpG island in the Zinc finger RNA-binding gene, implicated in RNA transport, localization, apoptosis, and mitosis) were associated with bronchial remodeling and keratinocyte death [17]. Other studies showing a proinflammatory transcriptomic profile [19] and higher levels of IgE at birth [20] in children born during autumn and winter suggest that the season of birth may also influence the development and commitment of the neonatal immunity system. Those studies suggest that pre-natal and peri-natal stimuli may have a long-term impact on allergic disease onset [21], paving the way for in-utero prediction of allergy phenotypes. However, these effects may vary geographically.

Intriguingly, a recent study nested within the Assessment of Lifestyle and Allergic Disease During Infancy (ALADDIN) birth cohort proved the potential for placental prediction of the development of allergic sensitizations in children from 6 months to 5 years. The study evaluated the histone acetylation profile of 173 at-term placentas in the promoter regions of allergy-related immune regulatory genes, proving a decreased risk of sensitization to food allergy for H3 acetylation at the IFNG and SH2B3 genes and a sex-specific, decreased risk of sensitization to both food and inhaled allergens in daughters of mothers with H5 acetylation at HDAC4 placental genes [22]. Another study on the same birth cohort focused on whether longitudinal DNA methylation changes were associated with the early development of IgE sensitization to airborne or food allergens [23]. DNA methylation was periodically re-assessed in peripheral blood mononuclear cells (PBMC) from 74 mother/child pairs in the ALADDIN birth cohort, analyzed during pregnancy, in cord blood, and then at 2 and 5 years of age, thus identifying 34 differentially methylated regions (DMRs) associated with IgE sensitization to airborne allergens and 38 DMRs associated with sensitization to food allergens among the children of 5 years of age. These DMRs were linked to genes involved in various pathways, such as endocytosis, focal adhesion, bacterial invasion of epithelial cells, and leukocyte migration. Additionally, 25 DMRs in maternal PBMCs were associated with IgE sensitization to airborne allergens in their children at five years, functionally annotated to the mTOR signaling pathway, suggesting that maternal DNA methylation levels may influence the child's predisposition to IgE sensitization, supporting early in utero effects on atopy development (Table 9.2).

Moreover, epigenetic alterations induced by exposome may explain the known transgenerational effects of environmental exposures on allergic disease, as supported by the evidence that grandmaternal smoking increases childhood asthma risk in grandchildren regardless of smoking behavior in the mother [24].

Table 9.2 Most relevant DMRs associated with IgE sensitization, according to the results of Acevedo et al., 2021 [23]

Gene	Functional role	Epigenetic variations	Outcome
RUFY1	Endocytosis	Hyper-methylated in mothers and children at 5 years	Sensitization to airborne allergens
TUBGCP5	Microtubule function cell division and intracellular transport	Hyper-methylated at 5 years old	Sensitization to airborne allergens
AIF1	Inflammation and macrophage activity	Hyper-methylated at 5 years old	Sensitization to airborne allergens
DUSP22	MAP kinase signaling pathway	Hypo-methylated at 5 years old	Sensitization to airborne allergens
CLM-1	Immune homeostasis	Hypo-methylated at 5 years old	Sensitization to both airborne and food allergens
ALOX12P2	Metabolism of arachidonic acid and inflammation	Hyper-methylated in mothers	Sensitization to airborne allergens
S100A13	Cell growth, differentiation, and inflammation	Hyper-methylated in mothers	Sensitization to airborne allergens

AIF1 Allograft Inflammatory Factor 1, *ALOX12P2* Arachidonate 12-Lipoxygenase, Peroxide 2 pseudogene, *CLM-1* CMRF35-like molecule 1, *DMR* different methylated region, *DUSP22* Dual Specificity Phosphatase 22, *IgE* immunoglobulin E, *RUFY1* RUN and FYVE domain-containing protein 1, S100A13: S100 calcium-binding protein A13; TUBGCP5: Tubulin Gamma Complex-Associated Protein 5

Finally, as regards the natural history of allergies, the epigenome represents a fascinating strategy to discriminate asymptomatic subjects with allergy sensitization from subjects with actual symptomatic allergic diseases [25] and to characterize the transition from one to the other, an issue that has always been critical in the allergy field [14].

9.1.2 "Omics" in Allergy: Biomarkers and the "Epigenomic Aging"

In terms of biomarkers, applying "omics" on various biological samples may offer cost-effective, specific, and noninvasive tools for the early definition of disease diagnosis and severity and short- and long-term response to therapies. Additionally, these biomarkers could play a crucial role in the early identification of clinical worsening and upcoming exacerbations and in the definition of specific endotypes that may benefit from additional treatments. Epigenomic studies may involve many different tissues and biological samples. Skin, blood, urine, and upper and lower airway epithelium are excellent sources of pathology-specific and tissue-specific biomarkers [26, 27]. Especially in pediatrics, saliva is emerging as another promising source of "omics", being safe and easy to collect, noninvasive, and rich in genetic material and proteins [28]. Its variable composition limits its routine use as a diagnostic medium for allergic diseases [28].

Notably, along with tissue-specificity, some epigenetic variations appear to be age-specific, too: as a matter of fact. At the same time, some loci are associated with both childhood- and adult-onset asthma endotypes [10]; there is evidence of some unique genomic regions [29, 30]. Identifying age-specific methylation profiles adds a crucial dimension to pediatric epigenetic studies, leading to the intriguing concept of "epigenetic aging," which could be accelerated or delayed compared to chronological aging [31]. Some diseases can accelerate "epigenetic aging," influencing certain illnesses' prevalence, onset, and phenotypic variability. In that regard, some authors proved a higher risk of sensitization to airborne or food allergens for each one-year increase in epigenetic age [32]. In contrast, others highlighted sex-specific differentially methylated regions in nasal epithelial samples [33], suggesting that individuals with greater epigenetic age acceleration (EAA) may exhibit more severe asthma. Other studies have also proved a correlation between "epigenetic aging" and classic allergic biomarkers such as total IgE or FeNO [34], so epigenetic aging profiles could be used for early diagnosis and defining disease endotypes.

Epigenetics can also be applied longitudinally to assess diseases and lung function trajectories. In that regard, a study examining DNA methylation changes at specific sites in the Isle of Wight birth cohort reveals a gender-dependent association with lung function development during adolescence. For females, 42 CpGs were identified with significant associations, highlighting the potential for these CpGs to serve as candidate epigenetic markers for lung function changes in females during adolescence. This underscores the potential gender-specific nature of some epigenetic mechanisms in lung function development [35].

9.1.3 "Omics" in Allergy: Pharmacogenomics

Regarding pharmacogenomics, recent improvements in understanding the pathogenesis of severe, therapy-resistant allergic diseases (namely, asthma) have resulted in introducing the new concept of "theratypes" [36]. Theratypes are pragmatically defined according to the patient's clinical response to specific therapies. Using "omics" to identify theratypes in managing pediatric allergic diseases may help identify non-responders early to specific rescue or chronic treatments [37], allowing timely changes in the therapeutic schedule and ultimately reducing healthcare costs and side effects [38].

In that regard, a recent study applied a multi-omics approach to evaluate the association between blood methylome and bronchodilator drug response in pediatric asthma [39]. The research involved 414 African American and Latino subjects aged 4–21 and identified six differentially methylated regions and CpG sites significantly associated with bronchodilator drug response (Table 9.3) [39]. Furthermore, the study conducted machine learning to develop a panel of 70 CpGs as a potential classifier for individuals' treatment response to albuterol therapy [39]. The findings contribute valuable insights into the interplay of genetic and environmental factors

Table 9.3 DMRs and CpG sites significantly associated with bronchodilator drug response and their functional role, according to the findings of Perez-Garcia et al. [39]

Gene	Functional role
FGL2	Immune modulation (suppression of TH1 and TH17 cells)
CEBPD	Interleukin expression, air smooth muscle cell proliferation, and epithelial growth
CDKN1C	Cell cycle regulation
GUSB	Carbohydrates metabolism
ZNF205	Oxidative stress and mitochondrial dysfunction
DNASE2	Lysosomal nuclease

CDKN1C cyclin-dependent kinase inhibitor 1C, *CEBPD* CCAAT/enhancer-binding protein delta, *DNASE2* deoxyribonuclease II, *FGL2* fibrinogen-like protein 2, *GUSB* glucuronidase beta, *ZNF205* Zinc Finger Protein 205

influencing drug responses among minority populations with a high asthma burden, suggesting the clinical applicability of pharmaco-epigenetics in precision medicine for respiratory diseases.

9.2 Epigenetics among Different Pediatric Allergic Diseases: The Atopic March

Epigenetic signatures may be used to diagnose and discriminate specific allergic diseases early. However, some epigenetic signatures are common to different allergic diseases, in accordance with the concept of atopic march.

Atopic march is a term used to define the subsequent development of multiple allergies over time in the same subject within a predefined sequence, going from AD to AR to asthma [40]. Although this paradigm has recently been questioned [41], favoring newer perspectives and the definition of a "new" atopic march, there is no doubt that different allergies coexist and influence each other in the same individual. Heritability explains only part of the disease prevalence in the atopic march, and epigenetic mechanisms like DNA methylation, histone modification, non-coding RNA, and micro-RNA (miRNAs) are critical in post-transcriptionally influencing pro-inflammatory mechanisms. In that regard, a 2020 study applied next-generation sequencing to analyze the eosinophil transcriptome (primary miRNAs) of 130 adults suffering from atopy, AD, AR, or asthma, compared with 15 unaffected ones [42]. By doing so, 18 differential eosinophil miRNAs (13 newly discovered) were identified, associated with atopy, asthma, AR, Ig, and E levels, and response to the methacholine challenge test (Table 9.4) [42]. After clustering individuals based on miRNA counts using Ward's method on Euclidean distances, three main clusters resulted, with significant differences in asthma diagnosis, familiar history of respiratory diseases, AR, and serum neutrophils [42]. Intriguingly, since eosinophils are known to carry miRNAs via exosomes, the authors suggest that these cells can act not only as pro-inflammatory actors but also as miRNA vectors [43], directly regulating gene expression in allergy.

Similarly, a comprehensive meta-analysis conducted within the European Mechanisms of the Development of ALLergy (MeDALL) consortium sought to unveil DNA methylation profiles linked to childhood allergy, encompassing asthma, rhinitis, and eczema [44]. In this exploration, 80 differentially methylated CpG sites (CpGs) were identified, with 21 demonstrating a strong association with childhood allergic diseases after meta-analysis, including 5 newly discovered CpGs [44]. Notably, children with allergies exhibited lower DNA methylation levels at specific CpGs at the ages of 4 and 8. These patterns persisted in independent cohorts until adolescence. Crucially, the 21 replicated CpGs were linked to allergy defined by asthma, rhinitis, or eczema in combination with specific IgE, suggesting shared blood DNA methylation signatures [44]. The most significant DNA methylation pattern was associated with genes involved in T cell balance, while another critical pattern was linked to a gene connected to inflammation and prostaglandin production [44]. Surprisingly, DNA methylation patterns of the same CpGs in cord blood did not serve as a predictive marker for early childhood allergy, highlighting the importance of postnatal changes in DNA methylation [44]. By clustering the 21 allergy-associated CpGs with immune-cell-specific gene expression signatures, the study revealed associations with eosinophil function and activated immune cell subsets, suggesting potential influences from past events or exposures on allergy development [44]. Further studies are required, comparing DNA methylation patterns among different biological samples.

A description of recent epigenomic findings for the most typical pediatric-specific allergic diseases follows.

Table 9.4 Examples of eosinophil micro-RNAs identified in specific allergic phenotypes, according to the results of Bélanger et al., 2020 [42]

Phenotype	Findings
Asthma	miR-1276, miR-29B2, miR-3175, miR-33B, miR-4308, miR-4523, miR-4673, miR-4785, miR-590, miR-638
Atopic dermatitis	miR-614
IgE levels	miR-142, miR-3064, miR-4434
Methacholine challenge response	miR-1304, miR-2355, miR-26A2, miR-645

IgE immunoglobulin E, *miR* micro-RNAs

9.2.1 Epigenetics and Respiratory Allergies: From Allergic Rhinitis to Asthma

Allergic rhinitis affects a significant proportion of children worldwide. The prevalence rates range from 10% to 30% and are constantly increasing [3], although they are currently profoundly underestimated due to the non-life-threatening nature of the disorder [45]. AR pathogenesis involves an immune response to inhaled allergens, primarily mediated by IgE antibodies. Upon exposure to allergens, mast cells in the nasal mucosa release inflammatory mediators, leading to symptoms such as nasal congestion, sneezing, and itching, which can profoundly affect the quality of life of the affected children and their families [45, 46].

Asthma is a chronic respiratory condition characterized by recurrent wheezing, breathlessness, chest tightness, and cough. It affects millions of children worldwide, with a global prevalence estimated at around 10% in pediatric populations [47]. Pathogenesis involves airway inflammation, remodeling, and bronchoconstriction, triggered by various factors, including allergens, infections, and pollutants [47]. Very recently, the conventional understanding of asthma pathogenesis has been challenged by the "epithelial barrier hypothesis" [48]. The current perspective suggests that genetic predisposition, airway dysbiosis, and environmental factors negatively affect the airway epithelium, compromising its ability to repair and heightening vulnerability to harmful or allergenic stimuli. In this way, airway epithelial cells may undergo a targeted pre-commitment, initiating either a type 2-high or type 2-low local response, leading to distinct asthma endotypes [38, 48]. Although most asthma cases are mild or moderate, the impact of severe asthma (defined as asthma whose control requires maximal therapies or is lost as soon as those therapies are stepped down) is substantial. Severe asthma is still a relevant cause of pediatric mortality and morbidity and imposes noteworthy healthcare costs, although affecting less than 2% of asthmatic children worldwide [47]. Therefore, timely identification of subjects suffering from severe-therapy-resistant asthma (STRA) is of the utmost importance.

AR and asthma often co-exist and are mutually influenced. This association primarily arises from the close anatomical, functional, and immunological interconnection between the upper and lower airways, often called the "united airway disease" concept [49–51].

According to this notion, most researchers use nasal epithelium as a less invasive surrogate for lower respiratory tract studies [26, 52], since bronchial epithelial cells (BECs) are more technically challenging to collect. Looking at epigenomics, limited, conflicting data currently exist on nasal epithelial cell methylation profiles and their association with AR, with mixed findings among different cohorts [26, 52]. Moreover, variability in nasal methylome studies poses challenges in distinguishing subgroups of asthma and AR since nasal methylation profiles could be influenced by demographic and clinical aspects, environmental exposures, viral infections, and even local bacterial colonization [26, 52]. Finally, proper definitions of cell-type composition and RNA collection are complex to perform in nasal samples.

Given these limits, Table 9.5 synthesizes the most common differentially methylated loci described in subjects with asthma, detected in blood samples or nasal samples [26]. Recent studies proved differential nasal methylome associated with FeNO, allergic asthma, allergen sensitization, and atopy [26, 34]. Moreover, different studies proved that differential methylation in nasal epithelial cells was associated with asthma severity [26, 53], involving genes like TMEM51, WDR25, HIPK3, and KLF11. Finally, acute systemic steroid treatment seems to modify nasal methylome in asthmatic good responders, with specific genes like VNN1 suggested as potential biomarkers of treatment response [54].

Few studies on adult populations regarding lower airway samples are available in the literature. A recent study focusing on different methylation profiles in BECs cultured with interleukin (IL) 13 identified 6522 differentially methylated CpG sites in asthmatic subjects compared to controls, involving genes like chemokine ligand 26 (CCL26), periostin (POSTH), and serpin family B member 2 (SERPINB2) [55]. Even more intriguingly, the subsequent weighted gene co-expression network analysis identified two clusters of highly correlated genes related to asthma severity, lung function, or eosinophilia [55]. Other studies comparing asthmatics with healthy controls highlighted different methylation profiles in more than 100 genes involved in cellular remodeling, signaling and growth, ionic transport, T cell function, and epithelial barrier integrity [56]. However, other studies did not demonstrate

Table 9.5 Most common differentially methylated loci described in subjects with asthma, detected in blood samples or nasal samples. From Legaki et al. [26]

Gene	Chromosomal position	Functional role
ACOT7	1p36.11	Regulation of lipid metabolism
EPX	17q21.1	Eosinophil function
KCNH2	7q36.1	Regulation of membrane potential
SIGLEC8	19q13.41	Eosinophil function
TNIK	3q29	Intracellular signaling
FOXP1	3p13	Transcriptional regulation
ATPAF2	12q13.3	Mitochondrial function
ZNF862	19p13.2	Transcriptional regulation
ADORA3	1p13.2	Modulation of adenosine signaling
ARID3A	19p13.3	Transcriptional regulation
IL5RA	3q13.2	Eosinophil activation
METRNL	7q22.1	Modulation of metabolic and inflammatory pathways
ZPM1	6p21.1	Transcriptional regulation

ACOT7 acyl-CoA thioesterase, *ADORA3* adenosine A3 receptor, *ARID3A* AT-rich interactive domain-containing protein 3A, *ATPAF2* ATP synthase mitochondrial F1 complex assembly factor 2, *EPX* eosinophil peroxidase, *FOXP1* forkhead box protein P1, *IL5RA* interleukin-5 receptor subunit alpha, *KCNH2* potassium voltage-gated channel subfamily H member 2, *METRNL* meteorin-like protein, *SIGLEC8* sialic acid-binding Ig-like lectin 8, *TNIK* Traf2- and Nck-interacting kinase, *ZNF862* zinc finger protein 862, *ZPM1* zinc finger protein MIZ-1

methylation differences between atopic asthmatics, non-atopic asthmatics, and healthy controls [57]. Future multicenter studies are crucial for the consistency of the results, to provide a better definition of AR and asthma severity and endotypes, discover new biomarkers, and proceed toward personalized medicine.

9.2.2 Epigenetics and Eczema

Atopic dermatitis (AD) is the most common chronic inflammatory skin disease in pediatric populations, with an intermittent-recurrent course. Recent epidemiological studies estimate a prevalence of approximately 16.5% within the Italian pediatric population [58] and 30% worldwide [59]. Being the *primum movens* of the so-called atopic march, AD significantly impacts the quality of life of affected children and their families, and poses relevant socio-economic burdens [60]. AD's pathogenesis involves intermittent-recurrent flares of local inflammation with progressive chronicity. Genetic predisposition, environmental factors, microbiome, and immune dysregulation contribute to its development. In that regard, advancements in genetics, transcriptomics, epigenomics, proteomics, and metagenomics have improved the understanding of AD's pathophysiology [61]. Genetic studies indicate a 70–80% heritability, with critical genes like filaggrin (FLG) identified [61]. These initially disrupt the epithelial barrier integrity and function, promoting allergic sensitizations and developing other allergic diseases.

In-depth studies on epigenomic modifications, specifically DNA methylation, identified significant hypomethylation in childhood AD [62]. Among the globally hypomethylated loci, epigenomic alterations at the golli-mbp locus within the myelin basic protein (MBP) gene correlated with AD severity [62]. Therefore, hypomethylated golli-MBP was proposed as a potential biomarker of AD severity.

Other studies on CD4+ T cell function in patients with moderate-to-severe AD revealed disease-specific alterations in chromatin accessibility, NFKB1 binding, and gene expression [63]. This suggests unique responses in AD, resulting in genotype-dependent epigenetic changes, particularly emphasizing NF-κB-mediated gene regulation [63].

Although most epigenomic studies in atopic dermatitis are conducted on peripheral blood samples, some authors tried to define differential skin DNA methylation profiles in individuals with eczema compared with healthy subjects [64]. Distinct DNA methylation patterns were noted in genes associated with various AD-related processes, such as immune response control, lymphocyte activity, cell proliferation, apoptosis, and epidermis differentiation [64].

Overall, addressing the impact of AD requires enhanced education, diagnostic, and management approaches. The compensation of costs for essential therapy aimed at restoring the normal epidermal barrier function is crucial. Research needs include long-term studies to understand the disease's natural history, interdependency relationships, and advancements in therapeutic approaches. Multi-omics approaches may offer detailed insights into the molecular basis of AD), aiding in personalized diagnostic and therapeutic strategies [61].

9.2.3 Epigenetics and Food Allergy

Food allergy is an immune-mediated, specific, and reproducible adverse reaction after ingesting a particular food. Based on the underlying immunological mechanism, food allergies are divided into IgE-mediated and non-IgE-mediated forms [65]. A rising prevalence marks both of them and profoundly impacts children's quality of life, resulting in substantial economic burdens, growth concerns related to allergen avoidance [25], and risk of nearly-lethal reactions. The diagnostic process for food allergy is complex, requiring a positive oral food challenge against a placebo followed by a negative, open food challenge in a facility equipped to handle emergencies [66]. Dietary management plans are essential, but challenges include cross-reactivity, the need for dieticians' advice, and the potential need for lifelong management [66].

Regarding pathogenesis, epigenomic factors, particularly microbial metabolites like short-chain fatty acids (SCFAs), are thought to profoundly influence the immune response, regulating the activity of regulatory T cells through histone acetylation [67]. In favor of this perspective, abnormal SCFA production and gut microbiota composition have been observed in non-IgE cow milk allergy [25]. Moreover, microbial metabolites, including staphylococcal enterotoxin B and lipopolysaccharides (LPSs), may act as adjuvants in food allergies, particularly in the context of Staphylococcus aureus colonization associated with concomitant atopic dermatitis [25].

Understanding the epigenomic landscape is pivotal for identifying molecular biomarkers that predict food allergy development at different stages, enabling accurate risk assessments and avoiding unnecessary food restrictions. Although food avoidance and oral immune therapy (OIT) are still cornerstones of food allergy management, promoting early introduction of allergenic foods and diet diversity since the first year of life are alternative strategies currently discussed in the literature [68]. Notably, two significant randomized controlled trials, The Learning Early About Peanut Allergy (LEAP) and Enquiring About Tolerance (EAT), provide supportive data. The LEAP study demonstrated a reduced peanut allergy prevalence in children consuming peanuts regularly between 4 and 11 months [69], while the EAT trial suggested that early food introduction from 3 months might decrease the risk of food allergy development, particularly for peanuts and eggs [70]. This shift in expert opinion is reflected in updated guidelines, such as those from the American National Institute of Allergy and Infectious Diseases, recommending peanut-containing food introduction between 4 and 6 months for certain infants [71].

In that regard, whether epigenomic studies may help to schedule the early and safe introduction of allergenic foods in selected children is still to be defined. An integrated transcriptomic and epigenomic study on peanut-allergic children revealed 318 genes and 203 CpG sites associated with reaction severity, providing insights into potential new biomarkers and therapeutic targets [72], such as the NFKBIA, ARG1, PHACTR1, and ZNF121 genes.

Additionally, a randomized controlled trial investigating dietary impacts on epigenetic mechanisms in children with cow's Milk Allergy (EPICMA) revealed that hydrolyzed casein formula with Lactobacillus rhamnosus GG led to accelerated FoxP3 demethylation [73] and distinct modulation of cytokine genes. These findings suggest early dietary interventions can impact epigenetics, influencing the acquisition of immune tolerance.

Finally, a machine learning approach using DNA methylation data aimed to diagnose food allergy within a dataset of 71 subjects with egg or peanut allergies, sensitized individuals, and non-allergic controls aged 11 to 15 months [74]. DNA methylation classifications are highly generalizable since they are data-driven and do not need a priori biological information. This study identified a 13-gene signature associated with clinical reactivity in food allergies [74], with 18 unbiased, highly discriminating CpGs sufficient to accurately differentiate the studied groups [74]. Intriguingly, these data pave the way for better and safer ways to diagnose food allergy in the future, avoiding risky oral food challenges in distinguishing between food sensitization and true food allergy.

Altogether, these studies collectively contribute to a deeper understanding of food allergies, from their molecular basis to potential diagnostic and therapeutic avenues and dietary interventions.

9.3 The Future of Epigenetics in Allergy

As stated before, epigenomics can be applied in allergology for many different purposes, such as the early identification of susceptible subjects or the prediction of the evolution of the atopic march.

A recent multicenter EWAS study involving 739 children from two birth cohorts aimed at correlating DNA methylation profile in mid-childhood peripheral blood with mid-childhood atopic sensitization, mid-childhood aeroallergens sensitization, and mid-childhood food allergen sensitization [14]. The study proved an existing correlation for 705, 1411, and 45 different methylation sites, respectively, involving genes implicated in innate immunity (DICER1), atopic march (AP5B1), asthma (EPX, IL4, IL5RA, PRG2, SIGLEC8, CLU), and eosinophilic esophagitis (SIGLEC8) [14]. Interestingly, most of the methylation marks identified occurred at "putative transcription factor binding motifs". These putative transcription factor binding motifs are DNA sequences thought to be regulatory elements since they are computationally predicted to be recognized and bound by transcription factors. Anyway, experimental validation is required to confirm their functional role. Intriguingly, the same study proved a longitudinal time trend in methylation profiles since the analysis of cord blood DNA methylation profiles was associated with mid-childhood outcomes [14]. Globally, these findings suggest the existence of potential early-life epigenetic markers for later sensitization, highlighting the role of epigenetics in shaping susceptibility to allergies.

EpiMix has recently been proposed as a potential integrative tool for epigenomic subtyping [75]. EpiMix is a sophisticated tool for studying abnormal DNA methylation ("DNAm") and gene expression patterns in populations using a model-based computational approach. Covering DNAm at diverse genomic levels (from cis-regulatory elements to distal enhancers to non-coding RNAs), EpiMix could be particularly useful for discovering new biomarkers and personalized treatment targets within small patient groups. In food allergy, EpiMix helped identify a global gain of DNA methylation at enhancers involved in crucial steps of T cell activation and lipid metabolism [75].

The evidence that some epigenomic signatures are common to one or more allergic diseases paves the way for allergy prevention. In recent years, primary prevention strategies have challenged the conventional wisdom of allergen avoidance, aiming to eliminate or reduce exposure to potentially sensitizing agents before sensitization develops [76]. In that respect, the World Allergy Organization (WAO) promotes primary and secondary prevention [3], emphasizing the urgency of proactive measures. Similarly, the Finnish Allergy Programme, launched in 2008 and shifting from secondary to primary prevention, advocates for a holistic approach [77]. The program aims to decrease the burden and costs of allergic diseases, improve treatment for severe allergies, and endorse health and tolerance. Living on a farm, adhering to an anthroposophical lifestyle, using probiotics, consuming microbe-rich products, spending time in nature, and engaging in outdoor activities positively affect innate immunity [77]. In contrast, excessive avoidance of foods in infancy may be harmful and is not recommended. As regards "omics" sciences, commissioning epigenomic signatures to improve immune tolerance stands as a critical frontier, offering hope for a future with reduced allergic disease prevalence and improved overall well-being.

9.4 Conclusions

In pediatric allergic diseases, epigenomics quietly unfolds as a potential game-changer, filling the current gaps from genetic predispositions to allergy development. Nonetheless, current epigenomic studies face challenges, as they are heterogeneous and lack adequate sample size, age stratification, exclusion of confounding factors, consistent methodologies, and conclusive outcomes. Moreover, since most studies are cross-sectional, they cannot elucidate if different methylation profiles precede or follow allergy development. Furthermore, most studies focus on blood or nasal samples, whereas studies focusing on the methylation profile of bronchial epithelial cells or salivary samples still need to be included.

Despite these limitations, epigenomic and transcriptomic profiling, coupled with genomic data, are emerging as a promising avenue for predictive models and personalized care, especially in pediatrics. In that regard, methylation risk scores [78] present both challenges and opportunities. The intricate genetic architecture, the complexity of gene-environment interactions, the influence of ethnicity, and current methodological constraints still need to improve the generalizability of the results. However, further multicenter, longitudinal studies, including environmental data and comprehensive functional studies, may improve the accuracy and applicability of methylation risk scores, which could be used to predict the severity of future allergic reactions.

Currently, epigenomics is considered a promising source of new biomarkers. Recognizing the limitations of relying on a single biomarker to pinpoint specific allergic endotypes and outcomes, further investigations may help define new multiple omics panels to be evaluated simultaneously in one or more biological samples.

In addition, the emerging concept of "epigenetic aging" may contribute to a better understanding and prediction of the longitudinal trajectory of allergic diseases from birth to adulthood and to better define targeted methylation loci as potential biomarkers. The present findings suggest incorporating epigenetic age into allergy care algorithms to enhance risk stratification and treatment planning.

In conclusion, the synergistic application of different omics sciences represents a promising avenue for personalized care and early diagnosis of allergic diseases, filling diagnostic and therapeutic needs that are still unmet. In the future, comprehensive birth cohorts, comparisons between epigenetic aging and chronological aging, longitudinal EWAS with larger datasets, and detailed cohort phenotyping will help fill current research gaps, moving towards targeted preventive and therapeutic strategies.

References

1. Platts-Mills TAE (2015) The allergy epidemics: 1870–2010. J Allergy Clin Immunol 136(1):3–13
2. Bantz SK, Zhu Z, Zheng T (2014) The atopic march: progression from atopic dermatitis to allergic rhinitis and asthma. J Clin Cell Immunol 5(2):202
3. Pawankar R et al (2013) World allergy organization (WAO) white book on allergy. World Allergy Organization, Milwaukee
4. Celebi Sozener Z et al (2022) Epithelial barrier hypothesis: effect of the external exposome on the microbiome and epithelial barriers in allergic disease. Allergy 77(5):1418–1449
5. Ortiz RA, Barnes KC (2015) Genetics of allergic diseases. Immunol Allergy Clin N Am 35(1):19–44
6. Ober C, Yao T (2011) The genetics of asthma and allergic disease: a 21st century perspective. Immunol Rev 242(1):10–30
7. Holloway JW, Yang IA, Holgate ST (2010) Genetics of allergic disease. J Allergy Clin Immunol 125(2):S81
8. Paternoster L, Waage J, Standl M (2015) Multi-ancestry genome-wide association study of 21,000 cases and 95,000 controls identifies new risk loci for atopic dermatitis. Nat Genet 47(12):1449–1456
9. Torgenson D, Ampleford E, Chiu G (2011) Meta-analysis of genome-wide association studies of asthma in ethnically diverse North American populations. Nat Genet 43(9):887–892
10. Hirota T et al (2011) Genome-wide association study identifies three new susceptibility loci for adult asthma in the Japanese population. Nat Genet 43(9):893–896
11. Van Eerdewegh P et al (2002) Association of the ADAM33 gene with asthma and bronchial hyperresponsiveness. Nature 418(6896):426–430
12. Zhang Y et al (2003) Positional cloning of a quantitative trait locus on chromosome 13q14 that influences immunoglobulin e levels and asthma. Nat Genet 34(2):181–186
13. Allen M et al (2003) Positional cloning of a novel gene influencing asthma from chromosome 2q14. Nat Genet 35(3):258–263
14. Peng C et al (2019) Epigenome-wide association study reveals methylation pathways associated with childhood allergic sensitization. Epigenetics 14(5):445–466

15. Lewis CM, Vassos E (2020) Polygenic risk scores: from research tools to clinical instruments. Genome Med 12(1):44
16. Frazer KA, Murray SS, Schork NJ, Topol EJ (2009) Human genetic variation and its contribution to complex traits. Nat Rev Genet 10:241–251
17. Lockett GA, Soto-Ramírez N, Ray MA (2016) Association of season of birth with DNA methylation and allergic disease. Allergy 71(9):1314–1324
18. Kim HB, Kang MJ, Lee SY, Shin YJ, Hong SJ (2023) Prenatal maternal anxiety promotes atopic dermatitis in offspring via placental DNA methylation changes. Asian Pac J Allergy Immunol 41(1):60–66
19. Thysen AH, Rasmussen MA, Kreiner-Møller E (2016) Season of birth shapes neonatal immune function. J Allergy Clin Immunol 137(4):1238–1246 13
20. Bjerke T, Hedegaard M, Henriksen TB, Nielsen BW, Schiøtz PO (1994) Several genetic and environmental factors influence cord blood IgE concentration. Pediatr Allergy Immunol 5(2):88–94
21. Lockett GA, Huoman J, Holloway JW (2015) Does allergy begin in utero? Pediatr Allergy Immunol 26(5):394–402
22. Harb H et al (2019) Epigenetic modifications in placenta are associated with the child's sensitization to allergens. Biomed Res Int 17:1315257
23. Acevedo N et al (2021) DNA methylation levels in mononuclear leukocytes from the mother and her child are associated with IgE sensitization to allergens in early life. Int J Mol Sci 22(2):801
24. Wheatley LM et al (2022) The role of epigenetics in multi-generational transmission of asthma: an NIAID workshop report-based narrative review. Clin Exp Allergy [Internet] 52(11):1264–1275
25. Childs CE, Munblit D, Ulfman L (2022) Potential biomarkers, risk factors, and their associations with IgE-mediated food allergy in early life: a narrative review. Adv Nutr 13(2):633–651
26. Legaki E, Arsenis C, Taka S, Papadopoulos NG (2022) DNA methylation biomarkers in asthma and rhinitis: are we there yet? Clin Transl. Allergy 12
27. Holgate ST, Yang Y, Haitchi HM (2006) The genetics of asthma: ADAM33 as an example of a susceptibility gene. Proc Am Thorac Soc 3(5):440–443
28. Zainal NHM, Abas R, Mohamad Asri SF (2021) Childhood allergy disease, early diagnosis, and the potential of salivary protein biomarkers. Mediat Inflamm 8:9198249
29. Schauberger EM, Ewart SL, Arshad SH, Huebner M, Karmaus W, Holloway JW (2011) Identification of ATPAF1 as a novel candidate gene for asthma in children. J Allergy Clin Immunol 128:753–760
30. Moffatt MF, Gut IG, Demenais F, Strachan DP, Bouzigon E, Heath S (2010) A large-scale, consortium-based genomewide association study of asthma. N Engl J Med 363:1211–1221
31. Horvath S, Raj K (2018) DNA methylation-based biomarkers and the epigenetic clock theory of ageing. Nat Rev Genet [Internet] 19:371–384
32. Peng C, Cardenas A, Rifas-Shiman SL, Hivert MF, Gold DR, Platts-Mills TA. Epigenetic age acceleration is associated with allergy and asthma in children in project viva J Allergy Clin Immunol 2019;143(6):2263–2270 14
33. Nwanaji-Enwerem JC, Osborne A, Cardenas A (2022) Nasal epigenetic age and systemic steroid response in pediatric emergency department asthma patients. Allergy 77(1):307–309
34. Cardenas A, Sordillo JE, Rifas-Shiman SL, Chung W, Liang L, Coull BA (2019) The nasal methylome as a biomarker of asthma and airway inflammation in children. Nat Commun 10(1):3095
35. Sunny SK, Zhang H, Rezwan FI (2020) Changes of DNA methylation are associated with changes in lung function during adolescence. Respir Res 21:80
36. Ozdemir C, Kucuksezer UC, Akdis M (2018) The concepts of asthma endotypes and phenotypes 775 to guide current and novel treatment strategies. Expert Rev Respir Med 12:733–743
37. Cazzola M, Rogliani P, Calzetta L, Matera MG (2020) Pharmacogenomic response of inhaled corticosteroids for the treatment of asthma: considerations for therapy. Pharmgenomics Pers Med [Internet] 13:261–271. Available from: S231471

38. Andrenacci B et al (2023) Severe pediatric asthma endotypes: current limits and future perspectives. Expert Rev Respir Med 17(8):675–690
39. Perez-Garcia J, Herrera-Luis E, Li A (2023) Multi-omic approach associates blood methylome with bronchodilator drug response in pediatric asthma. J Allergy Clin Immunol 151(6):1503–1512
40. Leung DY, Guttman-Yassky E (2014) Deciphering the complexities of atopic dermatitis: shifting paradigms in treatment approaches. J Allergy Clin Immunol 134(4):769–779
41. Belgrave DC, Simpson A, Buchan IE, Custovic A (2015) Atopic dermatitis and respiratory allergy: what is the link. Curr Dermatol Rep 4(4):221–227
42. Bélanger É, Madore AM, Boucher-Lafleur AM (2020) Eosinophil microRNAs play a regulatory role in allergic diseases included in the atopic March. Int J Mol Sci 21(23):9011
43. Cañas JA, Sastre B, Rodrigo-Muñoz JM, Pozo V (2019) Exosomes: a new approach to asthma pathology. Clin Chim Acta 495:139–147
44. Xu CJ et al (2021) Shared DNA methylation signatures in childhood allergy: the MeDALL study. J Allergy Clin Immunol 147(3):1031–1040
45. Licari A, Andrenacci B, Cicco ME, Leone M, Marseglia GL, Tosca M (2023) Respiratory comorbidities in severe asthma: focus on the pediatric age. Expert Rev Respir Med 17(1):1–13
46. Benazzo M, Leonardi S, Corsico A (2020) Cetirizine modifies quality of life and symptoms in children with seasonal allergic rhinitis: a pilot study. Acta Biomed 92(1)
47. Global strategy for asthma management and prevention (2021) [Internet]. Available from: https://ginasthma.org
48. Gon Y, Hashimoto S (2018) Role of airway epithelial barrier dysfunction in pathogenesis of asthma. Allergol Int 67(1):12–17
49. Licari A, Castagnoli R, Denicolò CF (2017) The nose and the lung: united airway disease? Front Pediatr 5:44
50. Licari A, Brambilla I, Filippo M (2017) The role of upper airway pathology as a co-morbidity in severe asthma. Expert Rev Respir Med 11(11):855–865
51. Brozek JL, Bousquet J, Agache I (2017) Allergic rhinitis and its impact on asthma (ARIA) guidelines—2016 revision. J Allergy Clin Immunol 140(4):950–958
52. Solazzo G, Ferrante G, Grutta S (2020) DNA methylation in nasal epithelium: strengths and limitations of an emergent biomarker for childhood asthma. Front Pediatr 8:256
53. Zhu T, Zhang X, Chen X (2021) Nasal DNA methylation differentiates severe from non-severe asthma in African-American children. Allergy 76(6):1836–1845
54. Zhang X et al (2017) Nasal DNA methylation differentiates corticosteroid treatment response in pediatric asthma: a pilot study. PLoS One 12
55. Marone G, Granata F, Pucino V (2019) The intriguing role of interleukin 13 in the pathophysiology of asthma. Front Pharmacol 10:1387
56. Wawrzyniak P, Krawczyk K, Acharya S (2021) Inhibition of CpG methylation improves the barrier integrity of bronchial epithelial cells in asthma. Allergy 76(6):1864–1868
57. Stefanowicz D, Hackett TL, Garmaroudi FS (2012) DNA methylation profiles of airway epithelial cells and PBMCs from healthy, atopic and asthmatic children. PLoS One 7(9)
58. Cantarutti A, Donà D, Visentin F, Borgia E, Scamarcia A, Cantarutti L (2015) Epidemiology of frequently occurring skin diseases in italian children from 2006 to 2012: a retrospective, population-based study. Pediatr Dermatol 32:668–678
59. Weidinger S, Novak N (2016) Atopic dermatitis. Lancet 387(10023):1109–1122
60. Ali F, Vyas J, Finlay AY (2020) Counting the burden: atopic dermatitis and health-related quality of life. Acta Derm Venereol 100(12)
61. Bin L, Leung DY (2016) Genetic and epigenetic studies of atopic dermatitis. Allergy Asthma Clin Immunol 12:52
62. Chen KD, Huang YH, Guo MM (2022) DNA methylation Array identifies Golli-MBP as a biomarker for disease severity in childhood atopic dermatitis. J Invest Dermatol 142(1):104–113
63. Eapen AA et al (2022) Epigenetic and transcriptional dysregulation in CD4+ T cells in patients with atopic dermatitis. PLoS Genet 18(5)

64. Olisova OY, Kochergin NG, Kayumova LN, Zavarykina TM, Dmitriev AA, Asanov AY (2020) Skin DNA methylation profile in atopic dermatitis patients: a case-control study. Exp Dermatol 29(2):184–189
65. Iweala OI, Choudhary SK, Commins SP (2018) Food allergy, vol 20
66. Abrams EM, Sicherer SH (2016) Diagnosis and management of food allergy. CMAJ 188(15):1087–1093
67. Kim CH (2023) Complex regulatory effects of gut microbial short-chain fatty acids on immune tolerance and autoimmunity. Cell Mol Immunol 20(4):341–350
68. Nuyttens L, Vlieger L, Diels M, Schrijvers R, Bullens DMA (2023) The clinical and immunological basis of early food introduction in food allergy prevention. Front Allergy 4:1111687
69. Du Toit G et al (2015) Randomized trial of peanut consumption in infants at risk for peanut allergy. N Engl J Med 372(9):803–813
70. Perkin MR et al (2016) Randomized trial of introduction of allergenic foods in breast-fed infants. N Engl J Med 374(18):1733–1743
71. Togias A et al (2017) Addendum guidelines for the prevention of peanut allergy in the report of the national institute of allergy and infectious diseases–sponsored expert panel. J Allergy Clin Immunol 139(1):29–44
72. Do AN et al (2020) Dual transcriptomic and epigenomic study of reaction severity in peanut-allergic children. J Allergy Clin Immunol 145(4):1219–1230
73. Paparo L, Nocerino R, Bruno C (2019) Randomized controlled trial on the influence of dietary intervention on epigenetic mechanisms in children with cow's milk allergy: the EPICMA study. Sci Rep 9(1):9504
74. Alag A (2019) Machine learning approach yields epigenetic biomarkers of food allergy: a novel 13-gene signature to diagnose clinical reactivity. PLoS One 14
75. Zheng Y, Jun J, Brennan K, Gevaert O (2023) EpiMix is an integrative tool for epigenomic subtyping using DNA methylation. Cell Rep Methods 3(7):100515
76. Fleischer DM, Chan ES, Venter C (2021) A consensus approach to the primary prevention of food allergy through nutrition: guidance from the American Academy of allergy, asthma, and immunology; American College of Allergy, asthma, and immunology. Canadian Soc Aller Clin Immunol J Allergy Clin Immunol Pract 9(1):22–43 4
77. Haahtela T, Valovirta E, Saarinen K. The Finnish allergy program 2008-2018: society-wide proactive program for change of management to mitigate allergy burden. J Allergy Clin Immunol 2021;148(2):319–326 4
78. Kilanowski A, Chen J, Everson T (2022) Methylation risk scores for childhood aeroallergen sensitization: results from the LISA birth cohort. Allergy 77(9):2803–2817

The Role of Epigenetics in Autoinflammatory Diseases

10

Maria Vincenza Mastrolia, Sarah Abu-Rumeileh, Valerio Maniscalco, Ilaria Maccora, Edoardo Marrani, Ilaria Pagnini, and Gabriele Simonini

Contents

10.1 Introduction ... 164
10.2 Familial Mediterranean Fever (FMF) ... 164
10.3 Cryopyrin-Associated Periodic Syndromes (CAPS) ... 167
10.4 Tumor Necrosis Factor Receptor-Associated Periodic Syndrome (TRAPS) ... 169
10.5 Systemic Juvenile Idiopathic Arthritis (sJIA)/Still's disease ... 170
10.6 Chronic Non-bacterial Osteomyelitis (CNO) ... 172
10.7 Behçet's Syndrome (BS) ... 173
References ... 174

Authors Maria Vincenza Mastrolia, Sarah Abu-Rumeileh, and Valerio Maniscalco have equally contributed to this chapter.

M. V. Mastrolia (✉) · G. Simonini
Rheumatology Unit, ERN ReCONNET Center, Meyer Children's Hospital IRCCS, Firenze, Italy

NEUROFARBA Department, University of Florence, Firenze, Italy
e-mail: maria.mastrolia@unifi.it

S. Abu-Rumeileh
Department of Clinical and Experimental Medicine, Section of Pediatrics, University of Pisa, Pisa, Italy

V. Maniscalco
Rheumatology Unit, ERN ReCONNET Center, Meyer Children's Hospital IRCCS, Firenze, Italy

Department of Pediatrics, Santo Stefano Hospital, Prato, Italy

I. Maccora · E. Marrani · I. Pagnini
Rheumatology Unit, ERN ReCONNET Center, Meyer Children's Hospital IRCCS, Firenze, Italy

L. Emmi, V. Fanos (eds.), *Epigenetics and Rare Diseases*, Rare Diseases of the Immune System, https://doi.org/10.1007/978-3-032-02902-7_10

10.1 Introduction

Autoinflammatory diseases (AIDs) represent a diverse group of disorders caused by defects or dysregulation of the innate immune system, marked by recurrent or continuous inflammation and the absence of a primary pathogenic role for the adaptive immune system (autoreactive T cells or autoantibody production) [1].The understanding of autoinflammation has undergone significant development in the past 25 years, starting with the identification of mutations in the Mediterranean Fever (MEFV) gene as causative factors in Familial Mediterranean Fever (FMF) [2]. Initially, the term "autoinflammatory diseases" was coined in connection with the innate immunity and was confined to classic hereditary monogenic periodic fever syndromes, such as cryopyrin-associated periodic syndromes (CAPS) and FMF [2]. However, advancements in technologies such as next-generation sequencing have led to an expanded classification, encompassing a growing number of recently identified monogenic diseases as well as non-monogenic disorders [1]. This broader classification now extends to include also multifactorial and complex conditions like systemic juvenile idiopathic arthritis (sJIA), chronic non-bacterial osteomyelitis (CNO), and Behçet's syndrome (BS) [1, 2]. Innate immune cells play a pivotal role in driving systemic inflammation in AIDs. Dysregulation in signaling pathways that regulate innate immune defense, such as the hyperactivation of the NLRP3 inflammasome complex, is central to the AIDs pathogenesis [2, 3]. A lack of regulation of these pathways results in excessive production of inflammatory cytokines (e.g., interleukin (IL)-1β and IL-18), leading to the occurrence of systemic manifestations [2, 3]. Recent advancements in understanding the pathogenesis of AIDs have increasingly implicated the role of epigenetics [4–6]. Epigenetic factors can modify gene expression and cellular functions without altering the genomic sequence. These modifications occur through different mechanisms and may vary in monogenic and non-monogenic AIDs [4]. Altered DNA methylation, histone modification, and dysregulation of non-coding RNAs (see Chap. 2 for details) have emerged as contributing pathogenetic factors in these disorders [4–6]. The exact role of epigenetic modifications in the development of AIDs is not yet fully elucidated. However, it is believed that, while disease-causing single gene mutations are responsible for the onset of AIDs, the influence of epigenetic factors could modify the disease manifestations and outcome [5]. Similarly, epigenetic factors may influence the phenotypic expression and disease course of multifactorial AIDs [5]. In light of this, a comprehensive understanding of epigenetic contributions to AIDs may hold the potential to predict individual disease outcomes and pave the way for the development of novel, effective, and targeted therapies. In this chapter, we review the currently available data that elucidate the role of epigenetic factors in AIDs.

10.2 Familial Mediterranean Fever (FMF)

Familial Mediterranean fever (FMF) is a hereditary autoinflammatory disorder characterized by recurrent episodes of fever and inflammation affecting the serosal membranes, particularly those of the abdomen, chest, and joints. It primarily

Table 10.1 Major genetic variants associated with familial Mediterranean fever

	Gene	Pathogen variants	Variants of unknown significance	Variants without clinical significance
FMF	MEFV	M694V, M694I, M680I, V726A, R761H, A744S, I692del, E167D, T267I	E148Q, P369S, F479L, I591T, K695R	R202Q, Q408Q

affects individuals of Mediterranean descent, including Sephardic Jews, Armenians, Turks, and Arabs. The genetic basis of FMF lies in mutations of the MEFV gene, located on chromosome 16p13.3, which encodes a 95 kDa protein named pyrin or marenostrin [7].

The pyrin protein plays a crucial role in regulating inflammation and immune responses. It is primarily expressed in cells of the innate immune system, including granulocytes, monocytes, and dendritic cells. Pyrin consists of five domains: the pyrin domain (PYD), bZIP transcription factor domain, B-box zinc finger domain, α-helical coiled-coil domain, and B30.2 domain [7].

Pyrin, along with pro-caspase-1 and apoptosis-associated specklike protein (ASC), is part of a multiprotein complex called inflammasome. One of the key functions of pyrin is its involvement in the formation of inflammasomes, multiprotein complexes responsible for initiating the inflammatory cascade. In FMF patients, mutations in the MEFV gene lead to dysregulated pyrin function, resulting in the assembly of the pyrin inflammasome and subsequent activation of caspase 1. Activated caspase-1 cleaves pro-IL-1β into its active form, leading to increased secretion of the pro-inflammatory cytokine IL-1β [8]. This dysregulated cytokine production contributes to the systemic inflammation observed in FMF patients, even during asymptomatic periods [9] (Table 10.1).

While FMF is primarily considered a monogenic disorder, its clinical expression is influenced by various genetic and epigenetic factors. DNA methylation, which can regulate gene expression patterns without altering the underlying DNA sequence, represents an important epigenetic factor in FMF patients. The MEFV gene possesses a CpG island spanning 998 base pairs, covering part of the first intron and the entire exon 2. Methylation of this gene was initially thought to influence phenotypic variability in FMF, potentially resulting in a more severe form of the disease [10]. However, available studies have not confirmed this hypothesis regarding the correlation between MEFV methylation and disease severity. Research by Ulum et al. found no correlation between MEFV methylation and clinical symptoms in Turkish children with FMF. They observed lower MEFV gene expression in patients homozygous for the M694V mutation, hinting at other epigenetic factors' involvement in FMF's clinical manifestation [11]. Contrasting results were observed in studies conducted by Kirectepe et al., indicating that methylation of the MEFV gene leads to a reduction in its expression. Methylation of exon 2 was associated with abnormal localization of the pyrin protein in the nucleus instead of the cytoplasm, potentially contributing to FMF pathogenesis [12]. Additionally, NRLP13 inflammasome is epigenetically regulated in FMF patients. In fact, the presence of amyloidosis was associated with hypermethylation of the NRLP13 gene, hence resulting in its silencing and subsequently in the inhibition of inflammasome assembly. NLRP3 inflammasome demethylation was observed in FMF patients, associated with increased IL-1β expression [13, 14].

DNA methylation might also be involved in patients' response to colchicine treatment. A 2023 study examined epigenetic modifications in FMF patients responding differently to this treatment. While non-responders exhibited higher MEFV exon 2 methylation levels, the differences were considered nonsignificant. These findings suggest a complex interplay between DNA methylation, gene expression, inflammasome regulation, and colchicine responsiveness in FMF pathology [10].

The role of microRNAs (miRNAs) in familial Mediterranean fever (FMF) has become a subject of increasing interest, shedding light on various mechanisms underlying FMF pathogenesis, including apoptosis, inflammation, and autophagy. Several studies have explored the expression levels of specific miRNAs in FMF patients compared to healthy individuals, providing insights into potential diagnostic and therapeutic implications [15, 16]. Amarilyo et al. evaluated miRNA expression levels in the peripheral blood mononuclear cells (PBMCs) of FMF patients carrying the M694V mutation under colchicine treatment. Results indicated differential expression of specific miRNAs involved in the immune response, with some upregulated (miR-144-3p, miR-21-5p, miR-4454, and miR-451) and others downregulated (miR-107, let-7d-5p, and miR-148b-3p) compared to healthy controls [17]. However, conflicting results were reported by Wada et al. [18], emphasizing the variability influenced by sample characteristics and quantification methods [18].

Kahraman et al. [19] identified specific miRNAs associated with immune functions in FMF patients, including overexpression of miR-34a-5p, miR-142-3p, miR-216a-5p, miR-340-5p, miR-429, and miR-582-5p, and downregulation of miR-107, miR-569, and miR-1304-5p. Notably, miR-107 exhibited lower expression in patients with the M694V mutation, suggesting its potential as a diagnostic and therapeutic tool [19].

Other investigations highlighted the role of miRNAs in inflammation modulation. Koga et al. [20] and Demir et al. [21] observed downregulation of miR-204-3p during attacks in FMF children in comparison to controls and to children during remission periods. miR-204-3p was shown to target the phosphoinositide 3-kinase gamma (PI3Kγ) pathway, which usually promotes the release of inflammatory cytokines, including IL-12p40 and IL-6. Hence, miR-204-3p could serve as a potential therapeutic tool to inhibit the PI3Kγ pathway in FMF patients and reduce inflammation, targeting the PI3Kγ pathway involved in inflammatory cytokine release [20, 21]. Similarly, Latsoudis et al. [22] identified elevated miR-4520a expression in FMF patients, implicating autophagy deregulation in FMF pathogenesis [22].

Moreover, miRNA profiles differed among FMF patients with different MEFV mutations. Akkaya-Ulum et al. [15] reported differential expression of miRNAs involved in inflammatory pathways, with potential implications for treatment response. Additionally, miR-197-3p emerged as a candidate for targeted therapy in FMF, with proposed applications in colchicine-refractory cases [23].

Furthermore, the influence of colchicine treatment on miRNA expression patterns was explored, revealing alterations in miRNA levels associated with inflammatory pathways. miRNAs such as miR-155, miR-223, and miR-181a were implicated in anti-inflammatory mechanisms, suggesting their potential as therapeutic targets [24].

The involvement of miRNAs in apoptotic pathways was investigated by Karpuzoglu et al. [25], identifying dysregulated miRNAs in FMF children. Notably, miR-181a and miR-125a showed promise as biomarkers of inflammation, exhibiting associations with proinflammatory cytokines such as IL-17 and IFN-γ [19].

In summary, miRNAs play a multifaceted role in FMF pathogenesis, influencing inflammation, apoptosis, and autophagy processes. Differential expression patterns of specific miRNAs in FMF patients underscore their potential diagnostic and therapeutic utility, offering insights into disease mechanisms and treatment response. Further research into miRNA dysregulation in FMF holds promise for advancing personalized medicine approaches and improving patient outcomes.

10.3 Cryopyrin-Associated Periodic Syndromes (CAPS)

The cryopyrin-associated periodic syndromes (CAPS) constitute a clinical continuum of monogenic autoinflammatory disorders caused by gain-of-function mutations in the NLRP3 gene, which encodes the cryopyrin protein [26, 27]. NLRP3 mutations lead to the constitutive activation of the inflammasome and subsequent overproduction of IL-1β, resulting in a multisystem inflammatory disease that affects various organs (including the skin, eyes, ears, musculoskeletal, and central nervous system), with cold-induced urticaria as a hallmark feature [26, 27]. The CAPS clinical spectrum encompasses different disorders with varying degrees of severity, including, in increasing order of severity, familial cold autoinflammatory syndrome (FCAS), Muckle–Wells syndrome, and neonatal-onset multisystem inflammatory disease (NOMID), also known as chronic infantile neurologic cutaneous and articular (CINCA) syndrome [26, 27] (Table 10.2). The presence of different degree of disease severity arising from identical mutation has suggested that additional factors, particularly epigenetic modifications, could modify the disease phenotype and be involved in the variability in the clinical spectrum [6]. A study conducted by Vento-Tormo et al. [28] compared the DNA methylation status of inflammasome genes (i.e., IL1B, NLRC5, PYCARD, AIM2, and CASP1) in monocytes from patients with CAPS and healthy controls. Following stimulation with IL-1β, monocytes from CAPS patients exhibited increased DNA demethylation of inflammasome-related genes in comparison to control subjects. Notably, this difference was reverted by treatment with IL-1 inhibitors, suggesting that elevated IL-1β levels may induce DNA demethylation, potentially exacerbating inflammation [28]. A second investigation by Aubert et al. [29] examined cutaneous biopsies from patients with NOMID, comparing skin lesions and non-lesional skin specimens from NOMID patients with skin samples obtained from healthy controls. Both lesional and non-lesional skin biopsies from NOMID patients showed

Table 10.2 Clinical manifestations and characteristics of cryopyrin-associated periodic syndromes (CAPS)

	Mild		Moderate		Severe	
	CAPS-FCAS		CAPS-MWS		CINCA-NOMID	
Age at onset	Infancy-adult		Early childhood-adult		Perinatal	
Typical attack duration	30 min-72 h		1–2 days or continuous with flares		Continuous with flares	
	Acute inflammation	Chronic damage	Acute inflammation	Chronic damage	Acute inflammation	Chronic damage
Systemic inflammation	Fever, increased acute phase reactants	Amyloidosis uncommon	Fever, increased acute phase reactants	Amyloidosis in up to 25% of cases	Fever, increased acute phase reactants, lymphadenopathy, hepato-splenomegaly	Amyloidosis
Dermatological manifestations	Cold-induced neutrophilic urticaria	None	Neutrophilic urticaria	None	Neutrophilic urticaria	None
Ocular manifestations	Conjunctivitis	None	Conjunctivitis, episcleritis, optic disk edema/ papilledema	Corneal opacification	Conjunctivitis, uveitis, optic disk edema/ papilledema	Chronic papilledema with amaurosis, corneal opacification
Inner ear manifestations	None	None	Cochlear edema	Progressive sensorineural loss	Cochlear edema	Progressive sensorineural loss
Central nervous system	Headache	None	Headache, intermittent meningitis	None	Headache, chronic aseptic meningitis	Increased intracranial pressure resulting in encephalomegaly, brain atrophy, cognitive impairment
Musculoskeletal	Myalgia, arthralgia	None	Myalgia, arthralgia, oligoarthritis	None	Myalgia, arthralgia, arthritis	Chronic arthritis, epiphyseal overgrowth, limb-length discrepancies and contractures

FCAS familial cold autoinflammatory syndrome, *MWS* Muckle-Wells syndrome, *CINCA* chronic infantile neurologic cutaneous and articular syndrome, *NOMID* neonatal-onset multisystem inflammatory disease downregulation of genes involved in epigenetic regulation, encompassing genes encoding histone proteins, histone-modifying genes, and genes associated with DNA methylation. Furthermore, transcriptome analysis of skin lesions revealed a downregulated expression of miR-29c and miR-103-2, which have been implicated in tissue fibrosis formation and the modulation of mRNA stabilization, respectively [30, 31]. Additionally, there was an elevated expression of other skin-specific miRNAs, including miR-9-1, miR-199a-2, miR-203, and miR-320a, which contribute to the skin's inflammatory response and the pathogenesis of psoriatic disease [32].

10.4 Tumor Necrosis Factor Receptor-Associated Periodic Syndrome (TRAPS)

Tumor necrosis factor (TNF) receptor-associated periodic syndrome (TRAPS) is an autosomal dominant hereditary disease caused by mutations in the TNFRSF1A gene, which encodes TNF-receptor 1 (TNFR1) [26]. The disease is marked by recurrent episodes of long-lasting fever associated with various systemic manifestations, including serositis, migratory rash, arthritis, and periorbital oedema [26] (Table 10.3). The pathogenesis of the condition is intricate and not entirely elucidated, incorporating diverse mechanisms, among which epigenetic dysregulation has been implicated [33]. One of the investigated mechanisms suggests that mutations in the TNFRSF1A gene cause the formation of a misfolded protein that is retained within the endoplasmic reticulum (ER) [33, 34]. This event leads to the activation of inositol-requiring enzyme 1 (IRE-1), an ER-associated endonuclease which acts as an ER sensor stressor. IRE-1 targets miRNAs, limiting protein production [33–36]. Specifically, IRE1 cleaves miR-155 and miR-146, which act as regulators of TLR4-mediated cellular response to LPS, suppressing the expression of NF-kB's target genes such as interleukins IL-6, IL-8, IL-1, and tumor necrosis factor alpha (TNF-α) [37, 38]. Harrison et al. [36] demonstrated decreased levels of miR-155 and miR-146 in dermal fibroblasts from patients with TRAPS, both before and after stimulation with LPS. MiR-155 and miR-146 are not the only miRNAs found to be dysregulated in TRAPS. In a study that compared 15 patients with TRAPS and 8 controls, the levels of six miRNAs (miR-134, miR-17-5p, miR-498, miR-451a, miR-572, miR-92a-3p) were significantly lower in TRAPS patients [39]. Moreover, 4 miRNAs (miR-150-3p, miR-92a-3p, miR-22-3p, miR-30d-5p) exhibited differential expression between patients treated with anakinra and untreated patients. Notably, the levels of miR-92a-3p and miR-150-3p increased to levels comparable with those observed in healthy subjects after anakinra treatment [39]. Additionally, miRNA levels were found to correlate with clinical and laboratory findings: reduced levels of miR-92b were associated with an increased number of fever attacks per year, whereas elevated levels of miR-377-5p correlated with higher concentrations of serum amyloid A [39]. Within this group of miRNAs, miR-150, miR-92, and miR-17 have been shown to play pivotal roles as regulators determining specific lineage decisions in the adaptive immune system [39, 40].

Table 10.3 Clinical features reported in TRAPS

General and organ-specific manifestations	Clinical features
Family history	Sporadic or autosomal dominant
Duration of fever/episode	1–3 weeks, continuous
Cutaneous	Centrifugal migratory erythema, edematous plaques, annular serpiginous patches, urticaria-like lesions
Musculoskeletal	Muscle cramps, migratory myalgia, fasciitis, arthralgia, oligo or monoarthritis, sacroileitis
Gastrointestinal	Abdominal pain, peritonitis-like tenderness, vomiting
Ocular	Periorbital edema, conjunctivitis, ocular pain, uveitis
Respiratory	Chest pain, pleurisy
Central nervous	Headache, aseptic meningitis, optic neuritis, behavioral abnormalities
Urogenital	Urethral strictures, scrotal pain
Cardiovascular	Pericarditis, myocarditis, ventricular tachycardia, restrictive cardiomyopathy
Lymphatic	Swollen and painful lymph nodes
Renal	Amyloidosis-related nephrotic syndrome
Autonomic nervous system	Orthostatic hypotension, bowel disturbance

10.5 Systemic Juvenile Idiopathic Arthritis (sJIA)/Still's disease

Juvenile idiopathic arthritis (JIA) is a heterogeneous group of chronic inflammatory diseases with onset before the age of 16. Among its subtypes, systemic juvenile idiopathic Arthritis (sJIA)/Still's disease stands out as a distinct entity due to its unique clinical and pathogenetic features. Unlike other JIA subtypes, sJIA is characterized by systemic inflammation, sharing clinical and immunological features with autoinflammatory syndromes.

The pathogenesis of sJIA remains not completely understood, but available evidence suggests immune dysregulation with a prominent role in the innate response. This involves neutrophils, monocytes, macrophages, and natural killer cells, along with the production of elevated levels of proinflammatory cytokines, particularly interleukin (IL)-1, IL-6, and IL-18 [41, 42]. While genetics has been implicated in sJIA susceptibility, recent research has increasingly focused on the role of epigenetics including microRNA (miRNA), as interesting modifiers of gene expression programs.

In the study by Sun et al. [43], the authors identified specific miRNA for sJIA and assessed their involvement in cytokines regulation. Two miRNAs, miR-26a and miR-145, were identified as significantly upregulated in sJIA. Particularly, plasma miR-26a levels discriminated sJIA patients from healthy controls and other rheumatic diseases. Positive correlations between miR-26a levels and IL-6 concentrations were also observed. Previous studies suggested that miR-26a may play an anti-inflammatory role [44, 45], implying the increased levels of miR-26a could be induced by an excessive IL-6 production in sJIA patients. The author therefore suggested that miR-26a might participate in a signaling pathway associated with proinflammatory cytokine expression.

Other studies have explored the role of epigenetics in the JAK/STAT signaling pathway in sJIA [46]. Li et al. demonstrated significantly lower expression levels of miR-19a and miR-21 in sJIA patients compared with the control group. Existing literature indicates that miR-21 regulates the interaction of Th1 and Th2 cells [47], while miR-19a serves a negative regulatory role in inflammation [48]. In this study, the authors highlighted the role of miR-19a and miR-21 in the activation of the JAK/STAT signaling pathway through the expression of IL-6, TNF-α, and SOCS3.

Jiang et al. [49] recently identified an additional mechanism interacting with the JAK/STAT pathway in sJIA by exploring the role of the long noncoding RNAs (lncRNAs), specifically metastasis-associated lung adenocarcinoma transcript 1 (MALAT1). The authors demonstrated that MALAT1 promotes the production of pro-inflammatory cytokine and matrix-metalloproteinases (MMPs) through JAK/STAT signaling pathway in sJIA.

Kamiya et al. [50] identified miRNAs as potential disease markers in JIA. Particularly, miR-223 serum levels were significantly higher in clinically active sJIA than in controls. Additionally, in systemic-onset JIA and polyarthritis patients, miR-223 and miR-16 levels correlated with erythrocyte sedimentation rate and matrix metalloproteinase-3. MiR-223 is expressed in myeloid cells, including neutrophils and macrophages, and was found to be an important modulator of myeloid differentiation [51]. It is known that children with sJIA exhibit increased levels of circulating monocytes with upregulated expression of monocyte differentiation genes, along with monocyte-derived proinflammatory cytokines, contributing to the pathophysiology of sJIA [52]. Monocytes in active sJIA display a distinct functional phenotype, showing features of both classical "M1" proinflammatory macrophages and alternatively activated macrophages "M2". An increasing number of studies have indicated molecular mechanisms involved in the regulation of macrophage polarization, and among these, miRNAs appear as critical modulators of monocyte polarization. Schulert et al. [53] studied monocyte miRNA expression profiling in patients with sJIA, identifying 110 miRNAs, including miR-155, miR-146a, miR-23a, and miR-27a, more expressed in monocytes from clinically active compared to inactive sJIA patients. MiRNA-125a-5p, known for driving M2-like macrophage polarization [54], was highly upregulated in children with active sJIA, correlating with serum ferritin levels, platelet and peripheral blood white cell counts, and systemic disease features.

In 2018, a study by Do et al. expanded the knowledge on the role of epigenetics in regulating and polarizing monocyte activity [55]. Two miRNAs, miR-125a-5p and miR-181c, previously found to be elevated in active sJIA, significantly reduced macrophage CD163 expression. CD163 directly mediates some anti-inflammatory functions of monocytes and macrophages by sequestrating toxic pro-inflammatory complexes.

Li et al. [56] focused on miR-146a expression in sJIA investigating its function in monocyte polarization. In their study, miR-146a expression was highly upregulated in monocytes from patients with active disease. MiR-146a levels were higher in monocytes polarized with M2 conditions, correlating with systemic features and inflammatory markers.

10.6 Chronic Non-bacterial Osteomyelitis (CNO)

CNO represents an autoinflammatory disease that primarily affects bones [57, 58]. The condition predominantly occurs in children and adolescents, presenting with a diverse clinical spectrum [57]. Some patients present with monofocal and self-limiting lesions. On the other hand, a distinct subgroup develops persistently active and recurrent bone inflammation affecting multiple sites, leading to the designation of chronic recurrent multifocal osteomyelitis (CRMO) [57, 58]. The bone involvement manifests with symptoms like pain, tenderness, and swelling over the affected areas. In certain cases, a subset of patients experiences additional inflammatory manifestations, including, but not limited, to palmoplantar pustulosis, severe acne, inflammatory bowel diseases, and peripheral or axial arthritis. The precise pathogenetic mechanisms of CNO remain unknown [57, 58]. However, it is recognized that its pathogenesis involves an imbalance of pro- and anti-inflammatory cytokines and regulatory signals [5, 58]. Specifically, diminished expression of immunoregulatory cytokines such as IL-10 and IL-19, alongside a dysregulated activation of the inflammasome complex proteins, emerges as a central aspect of the pathomechanism and these events have been directly attributed to epigenetic dysregulation [5, 58, 59].

IL-10 and IL-19 are members of the IL-10 cytokine family, contributing to the regulation of both innate and adaptive immune responses [60]. These two cytokines act by negatively affecting the expression of the pro-inflammatory cytokine IL-10 family, such as IL-20, and limiting pro-inflammatory reactions to prevent tissue damage [60]. The dysregulation of IL-10 and IL-19 cytokines contributes to the development of autoinflammatory and autoimmune diseases [10]. The expression of IL-10 and IL-19 is induced by the toll-like receptor (TLR)-4 activation, which in turn activates mitogen-activated protein kinases, extracellular signal reactive kinases (ERK) [5, 61, 62]. ERK, in turn, contributes to histone phosphorylation at histone 3 serine residue 10 (H3S10), resulting in an open chromatin structure at promoter regions of IL-10 and IL-19, thereby favoring their expression [5, 61, 62]. Monocytes from patients with CNO demonstrated diminished expression of IL-10 subsequent to TLR-4 activation induced by lipopolysaccharide (LPS) [59]. This observed phenomenon has been associated with attenuated ERK activation, resulting in compromised phosphorylation of histone H3S10 and subsequently impaired chromatin structure at the promoter regions of IL-10 and IL-19 [59]. On the contrary, monocytes isolated from patients with CNO exhibited an upregulation of pro-inflammatory IL-20 following stimulation with LPS. This dysregulation of IL-20 has been linked to a decrease in DNA methylation at the IL-20 gene promoter [63]. Furthermore, differential DNA methylation has been observed in both the IL-10 intronic enhancer element (I-SRE) and the IL-19 gene [6, 63].

Another dysregulated pathway in patients with CNO is represented by the hyperactivation of proteins forming the inflammasome complex. Elevated levels of inflammatory cytokines have been correlated with the unregulated activation of the NLRP3 inflammasome [64]. Monocytes isolated from CNO patients exhibit elevated mRNA levels of caspase-1, ASC and IL-1β, and this alteration has been associated with site-specific DNA hypomethylation in the vicinity of the corresponding genes, namely NLRP3 and PYCARD [63–65].

10.7 Behçet's Syndrome (BS)

Behçet's syndrome (BS) is a complex autoinflammatory disorder characterized by recurrent oral and genital ulcers, ocular inflammation, and skin lesions (Table 10.4). The distribution of BS is not uniform across the globe, and certain populations, particularly those in the Mediterranean region, Middle East, and East Asia, have a higher prevalence of the disease. This geographic distribution suggests a genetic component to the syndrome [66]. Genetic factors play a significant role in the development of BS. The human leukocyte antigen (HLA) system, located on chromosome 6, is involved in regulating the immune system and is strongly associated with BS susceptibility. The HLA-B51 allele, located within the major histocompatibility complex (MHC) class I region, is the most strongly associated genetic risk factor for BS. Carriers of the HLA-B51 allele have a risk of BS more than five times higher than non-carriers [67]. However, HLA-B51 accounts for less than 20% of the genetic risk, indicating that other genetic factors contribute to the development of the syndrome [68].

Table 10.4 Overall frequencies of clinical manifestations of Behçet syndrome

Manifestations	Frequency (%)
Common/overlapping	
– Oral aphthous ulcers	97–100
– Papular-pustular lesions	70–90
– Erythema nodosum-like lesions	40–60
– Arthritis	40–50
Less common/overlapping	
– Skin pathergy reaction	30–80
– Genital ulcers	0–90
– Superficial thrombophlebitis	15–30
– Gastrointestinal involvement	2–30
Distinctive features	
– Uveitis	30–50
– Deep vein thrombosis	10–15
– Arterial aneurysms	5
– Neurologic involvement	5–10

Other susceptibility loci within the HLA region, such as MICA, HLA-C, HLA-B27 (though with a weak correlation), and HLA-A*26, have also been identified. Furthermore, non-HLA genes have been implicated in BS susceptibility, including genes involved in proinflammatory and regulatory responses (e.g., STAT4, IL23R–IL12RB2 intergenic region, IL10, IL1A, IL1B, IRF-8, TNF promoter region), antigen presentation (e.g., ERAP1), and chemotaxis (e.g., CCR1 and CCR3) [66]. Variants of MEFV and mutations in toll-like receptor (TLR) genes have also been found in some patients with BS [69]. Additionally, epigenetic factors have been implicated in the pathogenesis of BS [70–72].

A study conducted by Hughes et al. delved into the investigation of DNA methylation abnormalities in BS, aiming to elucidate their potential contribution to its pathogenesis. Sixteen untreated male BS patients were compared with matched healthy controls, with additional samples collected post-treatment and remission from 12 of the BS patients [70]. Genome-wide DNA methylation patterns were analyzed in monocytes and CD4+ T cells using the Infinium HumanMethylation450 BeadChip array, revealing differential methylation at 383 CpG sites in monocytes and 125 CpG sites in CD4+ T cells between BS patients and controls. Bioinformatic analysis highlighted aberrant DNA methylation patterns among genes regulating cytoskeletal dynamics, suggesting their involvement in BD pathogenesis. Remarkably, treatment of BS led to the reversal of aberrant DNA methylation observed between patients and controls, indicating the responsiveness of these epigenetic changes to treatment [70].

In addition, activation histone marks have been overrepresented in natural killer cells, monocytes, lymphocytes, and neutrophils from patients with BS [71].

Emmi et al. investigated circulating miRNA (ci-miRNA) profiles in BS patients compared to healthy controls (HCs), systemic lupus erythematosus (SLE), and giant cell arteritis (GCA) patients [72]. Microarray screening of 16 BS patients and 18 HCs identified 29 deregulated ci-miRNAs, which were validated using poly(T) adaptor PCR (PTA-PCR) in a cohort of 30 BS patients and 30 HCs, alongside SLE and GCA patient groups. The identified ci-miRNAs, including miR-224-5p, miR-206, and miR-653-5p, discriminated BS from HCs, SLE, and GCA. Functional annotation analyses revealed pathways related to thrombo-inflammatory mechanisms, and the expression levels of ci-miRNAs correlated with leucocyte reactive oxygen species production and plasma lipid peroxidation [72]. Overall, the etiology of BS involves a complex interplay of genetic predisposition, immune dysregulation, and epigenetic modifications. Further research is needed to fully understand the mechanisms underlying this multifactorial disease.

References

1. Ben-Chetrit E, Gattorno M, Gul A, Kastner DL, Lachmann HJ, Touitou I, Ruperto N, Paediatric Rheumatology International Trials Organisation (PRINTO) and the AIDs Delphi study participants (2018) Consensus proposal for taxonomy and definition of the autoinflammatory diseases (AIDs): a Delphi study. Ann Rheum Dis 77:1558–1565
2. Nigrovic PA, Lee PY, Hoffman HM (2020) Monogenic autoinflammatory disorders: conceptual overview, phenotype, and clinical approach. J Allergy Clin Immunol 146:925–937

3. Martinez-Quiles N, Goldbach-Mansky R (2018) Updates on autoinflammatory diseases. Curr Opin Immunol 55:97–105
4. Akbaba TH, Sag E, Balci-Peynircioglu B, Ozen S (2020) Epigenetics for clinicians from the perspective of pediatric rheumatic diseases. Curr Rheumatol Rep 22:46
5. Surace AEA, Hedrich CM (2019) The role of epigenetics in autoimmune/inflammatory disease. Front Immunol 10:1525
6. Álvarez-Errico D, Vento-Tormo R, Ballestar E (2017) Genetic and epigenetic determinants in autoinflammatory diseases. Front Immunol 8:318
7. Schnappauf O, Chae JJ, Kastner DL, Aksentijevich I (2019) The pyrin inflammasome in health and disease. Front Immunol 7(10):1745
8. Chae JJ, Wood G, Masters SL, Richard K, Park G, Smith BJ, Kastner DL (2006) The B302 domain of pyrin, the familial Mediterranean fever protein, interacts directly with caspase-1 to modulate IL-1β production. Proc Natl Acad Sci USA 103(26):9982–9987
9. Ibrahim JN, Jéru I, Lecron JC, Medlej-Hashim M (2017) Cytokine signatures in hereditary fever syndromes (HFS). Cytokine Growth Factor Rev 33:19–34
10. Zekry ME, Sallam AAM, AbdelHamid SG, Zarouk WA, El-Bassyouni HT, El-Mesallamy HO (2023) Genetic and epigenetic regulation of MEFV gene and their impact on clinical outcome in auto-inflammatory familial Mediterranean fever patients. Curr Issues Mol Biol 45(1):721–737
11. Ulum YA, Peynircioglu BB, Batu E et al (2015) MEFV gene methylation pattern analysis in familial Mediterranean fever patients with altered expression levels. Pediatr Rheumatol 13(1):P113
12. Kirectepe AK, Kasapcopur O, Arisoy N, Celikyapi Erdem G, Hatemi G, Ozdogan H, Tahir Turanli E (2011) Analysis of MEFV exon methylation and expression patterns in familial Mediterranean fever. BMC Med Genet 7(12):105
13. Caldiran FY, Deveci K, Cacan E (2023) NLRP13 inflammasome complex is hypermethylated in familial Mediterranean fever and global methylation correlates with the disease severity. Ann Hum Genet 87(3):115–124
14. Vento-Tormo R, lvarez-Errico D, Garcia-Gomez A et al (2017) DNA demethylation of inflammasome- associated genes is enhanced in patients with cryopyrin-associated periodic syndromes. J Allergy Clin Immunol 139(1):202–211. e6
15. Akkaya-Ulum YZ, Balci-Peynircioglu B, Karadag O et al (2017) Alteration of the microRNA expression profile in familial mediterranean fever patients. Clin Exp Rheumatol 35 Suppl 108(6):90–94
16. Balci-Peynircioglu B, Akkaya-Ulum YZ, Akbaba TH, Tavukcuoglu Z (2019) Potential of miRNAs to predict and treat inflammation from the perspective of familial Mediterranean fever. Inflamm Res 68(11):905–913
17. Amarilyo G, Pillar N, Ben-Zvi I, Weissglas-Volkov D, Zalcman J, Harel L, Livneh A, Shomron N (2018) Analysis of microRNAs in familial Mediterranean fever. PLoS One 13(5):e0197829
18. Wada T, Toma T, Matsuda Y, Yachie A, Itami S, Taguchi YH, Murakami Y (2017) Microarray analysis of circulating microRNAs in familial Mediterranean fever. Mod Rheumatol 27(6):1040–1046
19. Kahraman CY, Egin ME, Tatar A, Turkez H, Mardinoglu A (2021) The assessment of selected miRNA profile in familial mediterranean fever. Biomed Res Int 13:6495700
20. Koga T, Migita K, Sato T et al (2018) MicroRNA- 204-3p inhibits lipopolysaccharide-induced cytokines in familial Mediterranean fever via the phosphoinositide 3-kinase γ pathway. Rheumatology (Oxford) 57(4):718–726
21. Demir F, Çebİ AH, Kalyoncu M (2019) Assessment of circulating microribonucleic acids in patients with familial mediterranean fever. Arch Rheumatol 35(1):52–59
22. Latsoudis H, Mashreghi MF, Grün JR et al (2017) Differential expression of miR-4520a associated with pyrin mutations in familial mediterranean fever (FMF). J Cell Physiol 232(6):1326–1336

23. Akkaya-Ulum YZ, Akbaba TH, Tavukcuoglu Z, Chae JJ, Yilmaz E, Ozen S, Balci-Peynircioglu B (2021) Familial Mediterranean fever-related miR-197-3p targets IL1R1 gene and modulates inflammation in monocytes and synovial fibroblasts. Sci Rep 11(1):685
24. Hortu HO, Karaca E, Sozeri B et al (2019) Evaluation of the effects of miRNAs in familial Mediterranean fever. Clin Rheumatol 38(3):635–643
25. Karpuzoglu EM, Kisla Ekinci RM, Balci S, Bisgin A, Yilmaz M (2021) Altered expression of apoptosis related, circulating cell-free miRNAs in children with familial Mediterranean fever: a cross-sectional study. Rheumatol Int 41(1):103–111
26. Lachmann HJ (2017) Periodic fever syndromes. Best Pract Res Clin Rheumatol 31:596–609
27. Booshehri LM, Hoffman HM (2019) CAPS and NLRP3. J Clin Immunol 39:277–286
28. Vento-Tormo R, Álvarez-Errico D, Garcia-Gomez A et al (2017) DNA demethylation of inflammasome-associated genes is enhanced in patients with cryopyrin-associated periodic syndromes. J Allergy Clin Immunol 139:202–211.e6
29. Aubert P, Suárez-Fariñas M, Mitsui H et al (2012) Homeostatic tissue responses in skin biopsies from NOMID patients with constitutive overproduction of IL-1β. PLoS One 7:e49408
30. van Rooij E, Sutherland LB, Thatcher JE, DiMaio JM, Naseem RH, Marshall WS, Hill JA, Olson EN (2008) Dysregulation of microRNAs after myocardial infarction reveals a role of miR-29 in cardiac fibrosis. Proc Natl Acad Sci USA 105:13027–13032
31. Liao Y, Lönnerdal B (2010) Global microRNA characterization reveals that miR-103 is involved in IGF-1 stimulated mouse intestinal cell proliferation. PLoS One 5:e12976
32. Sonkoly E, Wei T, Janson PCJ et al (2007) MicroRNAs: novel regulators involved in the pathogenesis of psoriasis? PLoS One 2:e610
33. Cudrici C, Deuitch N, Aksentijevich I (2020) Revisiting TNF receptor-associated periodic syndrome (TRAPS): current perspectives. Int J Mol Sci 21:3263
34. Dickie LJ, Aziz AM, Savic S et al (2012) Involvement of X-box binding protein 1 and reactive oxygen species pathways in the pathogenesis of tumour necrosis factor receptor-associated periodic syndrome. Ann Rheum Dis 71:2035–2043
35. Tirasophon W, Lee K, Callaghan B, Welihinda A, Kaufman RJ (2000) The endoribonuclease activity of mammalian IRE1 autoregulates its mRNA and is required for the unfolded protein response. Genes Dev 14:2725–2736
36. Harrison SR, Scambler T, Oubussad L, Wong C, Wittmann M, McDermott MF, Savic S (2018) Inositol-requiring enzyme 1-mediated downregulation of MicroRNA (miR)-146a and miR-155 in primary dermal fibroblasts across three TNFRSF1A mutations results in Hyperresponsiveness to lipopolysaccharide. Front Immunol 9:173
37. Schulte LN, Westermann AJ, Vogel J (2013) Differential activation and functional specialization of miR-146 and miR-155 in innate immune sensing. Nucleic Acids Res 41:542–553
38. Saba R, Sorensen DL, Booth SA (2014) MicroRNA-146a: a dominant, negative regulator of the innate immune response. Front Immunol 5:578
39. Lucherini OM, Obici L, Ferracin M et al (2013) First report of circulating microRNAs in tumour necrosis factor receptor-associated periodic syndrome (TRAPS). PLoS One 8:e73443
40. Ghafouri-Fard S, Niazi V, Taheri M (2021) Role of miRNAs and lncRNAs in hematopoietic stem cell differentiation. Noncoding RNA Res 6:8–14
41. de Jager W, Hoppenreijs EP, Wulffraat NM, Wedderburn LR, Kuis W, Prakken BJ (2007) Blood and synovial fluid cytokine signatures in patients with juvenile idiopathic arthritis: a cross-sectional study. Ann Rheum Dis 66:589–598
42. Lotito AP, Campa A, Silva CA, Kiss MH, Mello SB (2007) Interleukin 18 as a marker of disease activity and severity in patients with juvenile idiopathic arthritis. J Rheumatol 34:823–830
43. Sun J, Feng M, Wu F, Ma X, Lu J, Kang M, Liu Z (2016) Plasma miR-26a as a diagnostic biomarker regulates cytokine expression in systemic juvenile idiopathic arthritis. J Rheumatol 43:1607–1614
44. Jiang C, Zhu W, Xu J et al (2014) MicroRNA-26a negatively regulates toll-like receptor 3 expression of rat macrophages and ameliorates pristane induced arthritis in rats. Arthritis Res Ther 16:R9

45. Wei C, Kim I-K, Kumar S et al (2013) NF-κB mediated miR-26a regulation in cardiac fibrosis. J Cell Physiol 228:1433–1442
46. Li HW, Xie Y, Li F, Sun GC, Chen Z, Zeng HS (2016) Effect of miR-19a and miR-21 on the JAK/STAT signaling pathway in the peripheral blood mononuclear cells of patients with systemic juvenile idiopathic arthritis. Exp Ther Med 11:2531–2536
47. Lu TX, Hartner J, Lim E-J et al (2011) MicroRNA-21 limits in vivo immune response-mediated activation of the IL-12/IFN-γ pathway, Th1 polarization, and the severity of delayed-type hypersensitivity. J Immunol 187:3362–3373
48. Gantier MP, Stunden HJ, McCoy CE et al (2012) A miR-19 regulon that controls NF-κB signaling. Nucleic Acids Res 40:8048–8058
49. Jiang H, Zhu M, Wang H, Liu H (2021) Suppression of lncRNA MALAT1 reduces pro-inflammatory cytokines production by regulating miR-150-5p/ZBTB4 axis through JAK/STAT signal pathway in systemic juvenile idiopathic arthritis. Cytokine 138:155397
50. Kamiya Y, Kawada J, Kawano Y, Torii Y, Kawabe S, Iwata N, Ito Y (2015) Serum microRNAs as potential biomarkers of juvenile idiopathic arthritis. Clin Rheumatol 34:1705–1712
51. Chen CZ, Li L, Lodish HF, Bartel DP (2004) MicroRNAs modulate hematopoietic lineage differentiation. Science 303:83–86
52. Macaubas C, Nguyen K, Deshpande C et al (2010) Distribution of circulating cells in systemic juvenile idiopathic arthritis across disease activity states. Clin Immunol 134:206–216
53. Schulert GS, Fall N, Harley JB, Shen N, Lovell DJ, Thornton S, Grom AA (2016) Monocyte microRNA expression in active systemic juvenile idiopathic arthritis implicates miR-125a-5p in polarized monocyte phenotypes. Arthritis Rheumatol Hoboken NJ 68:2300–2313
54. Banerjee S, Xie N, Cui H, Tan Z, Yang S, Icyuz M, Abraham E, Liu G (2013) microRNA let-7c regulates macrophage polarization. J Immunol Baltim Md 1950 190:6542–6549
55. Do T, Tan R, Bennett M, Medvedovic M, Grom AA, Shen N, Thornton S, Schulert GS (2018) MicroRNA networks associated with active systemic juvenile idiopathic arthritis regulate CD163 expression and anti-inflammatory functions in macrophages through two distinct mechanisms. J Leukoc Biol 103:71–85
56. Li D, Duan M, Feng Y, Geng L, Li X, Zhang W (2016) MiR-146a modulates macrophage polarization in systemic juvenile idiopathic arthritis by targeting INHBA. Mol Immunol 77:205–212
57. Zhao Y, Ferguson PJ (2018) Chronic nonbacterial osteomyelitis and chronic recurrent multifocal osteomyelitis in children. Pediatr Clin N Am 65:783–800
58. Hofmann SR, Kapplusch F, Girschick HJ, Morbach H, Pablik J, Ferguson PJ, Hedrich CM (2017) Chronic recurrent multifocal osteomyelitis (CRMO): presentation, pathogenesis, and treatment. Curr Osteoporos Rep 15:542–554
59. Hofmann SR, Morbach H, Schwarz T, Rösen-Wolff A, Girschick HJ, Hedrich CM (2012) Attenuated TLR4/MAPK signaling in monocytes from patients with CRMO results in impaired IL-10 expression. Clin Immunol 145:69–76
60. Hofmann SR, Rösen-Wolff A, Tsokos GC, Hedrich CM (2012) Biological properties and regulation of IL-10 related cytokines and their contribution to autoimmune disease and tissue injury. Clin Immunol 143:116–127
61. Lucas M, Zhang X, Prasanna V, Mosser DM (2005) ERK activation following macrophage FcgammaR ligation leads to chromatin modifications at the IL-10 locus. J Immunol 175:469–477
62. Zhang X, Edwards JP, Mosser DM (2006) Dynamic and transient remodeling of the macrophage IL-10 promoter during transcription. J Immunol 177:1282–1288
63. Hofmann SR, Kubasch AS, Ioannidis C, Rösen-Wolff A, Girschick HJ, Morbach H, Hedrich CM (2015) Altered expression of IL-10 family cytokines in monocytes from CRMO patients result in enhanced IL-1β expression and release. Clin Immunol 161:300–307
64. Scianaro R, Insalaco A, Bracci Laudiero L, De Vito R, Pezzullo M, Teti A, De Benedetti F, Prencipe G (2014) Deregulation of the IL-1β axis in chronic recurrent multifocal osteomyelitis. Pediatr Rheumatol Online J 12:30

65. Brandt D, Sohr E, Pablik J et al (2018) CD14+ monocytes contribute to inflammation in chronic nonbacterial osteomyelitis (CNO) through increased NLRP3 inflammasome expression. Clin Immunol 196:77–84
66. Emmi G, Bettiol A, Hatemi G, Prisco D (2024) Behçet's syndrome. Lancet S0140–6736(23):02629-6
67. Burillo-Sanz S, Montes-Cano M-A, García-Lozano J-R et al (2019) Behçet's disease and genetic interactions between HLA-B*51 and variants in genes of autoinflammatory syndromes. Sci Rep 9(2777):22
68. Remmers EF, Cosan F, Kirino Y et al (2010) Genome-wide association study identifies variants in the MHC class I, IL10, and IL23RIL12RB2 regions associated with Behçet's disease. Nat Genet 42:698–702
69. Seyahi E, Ugurlu S, Amikishiyev S, Gul A (2023) Behçet disease, familial Mediterranean fever and MEFV variations: more than just an association. Clin Immunol 251:109630
70. Hughes T, Ture-Ozdemir F, Alibaz-Oner F, Coit P, Direskeneli H, Sawalha AH (2014) Epigenome-wide scan identifies a treatment-responsive pattern of altered DNA methylation among cytoskeletal remodeling genes in monocytes and CD4+ T cells from patients with Beh. et's disease. Arthritis Rheumatol 66:1648–1658
71. Ortiz-Fernandez L, Sawalha AH (2021) Genetics of Behcet's disease: functional genetic analysis and estimating disease heritability. Front Med (Lausanne) 8:625710
72. Emmi G, Bagni G, Lastraioli E et al (2022) A unique circulating miRNA profile highlights thrombo-inflammation in Behçet's syndrome. Ann Rheum Dis 81(3):386–397

Epigenetics and Immune-Mediated Diseases

11

A. A. Senatore, P. Currao, E. Arnaudo, R. Castagnoli, and S. Savasta

Contents

11.1 Introduction 179
11.2 Immuno-Mediated Diseases 181
11.3 Systemic Lupus Erythematosus 181
11.4 Rheumatoid Arthritis 183
11.5 Type 1 Diabetes 184
11.6 Neuro-Immuno-Mediated Diseases 185
References 188

11.1 Introduction

Autoimmune diseases are complex pathologies characterized by a dysfunction of the immune system, which mistakenly attacks "self" antigens, resulting in cellular and tissue damage. Although these conditions are mainly caused by a genetic predisposition, other factors, such as the environment, hormones, and infections, are also believed to be decisive. Epigenetics is the study of the molecular modifications dictated by these latter factors, which regulate gene expression without altering the DNA sequence [1].

A. A. Senatore · E. Arnaudo · R. Castagnoli
Pediatric Clinic, IRCCS Policlinico San Matteo, Department of Clinical-Surgical, Diagnostic and Pediatric Sciences, University of Pavia, Pavia, Italy
e-mail: antonioandrea.senatore01@universitadipavia.it; elena.arnaudo01@universitadipavia.it; riccardo.castagnoli@unipv.it

P. Currao · S. Savasta (✉)
Pediatric Clinic and Rare Diseases, "A. Cao" Hospital for Microcytic Anemia, Department of Medical Sciences and Public Health, University of Cagliari, Cagliari, Italy
e-mail: salvatore.savasta@unica.it

L. Emmi, V. Fanos (eds.), *Epigenetics and Rare Diseases*, Rare Diseases of the Immune System, https://doi.org/10.1007/978-3-032-02902-7_11

Known epigenetic mechanisms include DNA methylation, histone modification, and the action of non-coding RNAs (ncRNAs).

DNA methylation involves the addition of a methyl group at cytosine C5 in the context of CpG dinucleotides [2]. There are four enzymes responsible for methylation and they are called methyltransferases (in mammals DNMT3A, DNMT3B, DNMT3L, DNMT1), and their activity can depend on certain toxins, dietary components such as folate and choline, and, finally, environmental pollutants that produce reactive oxygen species [1]. The source of methyl used by methyltransferases is reactive methyl bound to the sulfur of S-adenosyl methionine (SAM) [3]. The methylation process reduces access to DNA for transcription factors and RNA polymerases, thus repressing transcription. However, this is a reversible process through the action of the TET family of dioxygenases (ten-eleven translocation): these are enzymes that convert 5-methylcytosine (5mC) into 5-hydroxymethylcytosine (5hmC) [4]. Recent studies have shown that DNA methylation causes alterations in the gene expression of some cells, particularly T cells, and therefore consequently determines the onset of various autoimmune diseases [3]. DNA methylation plays a role in both the formation and activation of T cells. Methylation at the Cd4 locus contributes primarily to the formation of CD4+ cells [5] and in the thymus the interaction between the methyltransferase DNMT1 and the transcription factor FOXP3 (Forkhead Box P3) determines the formation of T reg cells, a heterogeneous population of CD4 positive T cells characterized by a high expression of CD25 and CD127 [6]. Finally, methylation is essential for the production of interleukin 2 (IL2) and for the development of the different subpopulations of "helper" T lymphocytes (Th1, Th2, Th17) [7].

Histone modification can consist of methylation, citrullination, or acetylation. The latter is favored by histone acetyltransferase (HAT) and deacetylase (HDACi) enzymes, which respectively add and remove acetyl groups at the histone level based on environmental stimuli such as hormones or inflammatory cytokines, thus altering chromatin structure and gene expression [1]. For example, the acetylation of lysine 27 of histone 3 (H3K27ac) correlates with increased transcriptional activity [8]. Many studies have shown that some enzymes involved in deacetylation are involved in the transcription of genes linked to autoimmunity and in the regulation of T-cell immune and inflammatory processes, such as the V(D)J recombination process [1, 9].

Finally, non-coding RNAs include two types of RNA, microRNAs (miRs) and long non-coding RNAs (lncRNAs), which derive from both introns and exons and are distinguished by their length; microRNAs are made up of 20–24 nucleotides, while long non-coding RNAs are made up of about 300 [1]. Various external factors can influence the transcription of non-coding RNA, which play different roles by binding to particular areas of DNA (miRs bind to 3' UTR regions) or to proteins (long non-coding RNA bind to histone-modifying proteins or to transcription factors) [10]; they regulate the formation of heterochromatin, histone methylation and DNA methylation, determine the silencing of some genes and also activate complexes that regulate the development and differentiation of various immune cells [3]. In particular, long non-coding RNAs stimulate the proliferation and differentiation of erythroid progenitors by acting on factors such as GATA1, TAL1, and KLF1, and promote granulocyte differentiation through the action of HOTAIRM1, which acts as a cell cycle regulator [11–13].

Below, we will describe the potential role of epigenetic alterations in the pathogenesis of some pathological entities, classifying them according to the area of the immune system involved, therefore mainly in disorders concerning humoral or cellular immunity.

11.2 Immuno-Mediated Diseases

Immune-mediated diseases are a group of disorders that result from a malfunction of the immune system, which loses the ability to distinguish the body's own structures, triggering abnormal humoral and cell-mediated responses against its own tissues. These conditions are characterized by an alteration of the immune system that increases the risk of complications and, in many cases, reduces life expectancy.

A distinctive element of these pathologies is the production of autoantibodies, including anti-nucleus antibodies (ANA), considered fundamental serological markers.

Recent studies suggest that there are biological pathways shared between these conditions, which could pave the way for new insights into common pathogenetic mechanisms. These present similarities in several aspects, including: a higher incidence in women, common symptoms (arthritis, pulmonary and vascular diseases), serological characteristics (ANA and anti-Ro52/TRIM21 antibodies), immunological alterations (excessive activation of type I interferon and dysfunctions in CD4+ T lymphocytes, in particular in Th17 and Treg populations) and shared genetic traits.

Numerous studies have highlighted the key role of epigenetic alterations in the immune mechanisms involved in this group of pathologies, in particular at the level of CD4+ T cells and co-stimulatory proteins that compromise the regulation and functioning of the immune system, contributing to the progression of the disease [3].

11.3 Systemic Lupus Erythematosus

Systemic lupus erythematosus (SLE) is the autoimmune disease most studied in relation to epigenetic modifications. It mainly affects women and is primarily attributed to a dysregulation of T lymphocytes, which makes this pathology extremely complex to manage. SLE is characterized by the production of autoantibodies directed against nuclear antigens, causing damage to various organs and tissues, mainly the kidneys and the vascular system.

In recent years, research has focused on lncRNAs and circular RNAs (circRNAs) as new biomarkers for the diagnosis and prognosis of SLE, in particular for monitoring tissue damage. Some studies have shown that the expression of some lncRNAs is significantly increased in monocyte-derived dendritic cells (moDCs) and is closely correlated with a worse disease prognosis.

CircRNAs mainly act as competitive endogenous RNAs (ceRNAs) for miRNAs, influencing various biological processes [14]. For example, in patients with SLE, a reduction in the levels of some circRNAs has been observed that are thought to inhibit apoptosis and reduce the production of IL-2 [15]. CircIBTK, on the other hand, seems to act as a sponge for miR-29b, regulating DNA demethylation and the AKT signaling pathway [16].

A recent study has revealed a new role for circRNAs in the pathogenesis of SLE. PKR kinase is a double-stranded RNA receptor involved in the antiviral response that appears to be regulated by circRNAs. It has been observed that circRNAs form double-helix structures, preferentially binding to PKR and inhibiting its activation. In patients with SLE, the overall reduction of circRNAs has been associated with an increase in PKR phosphorylation, suggesting a possible connection with the activation of the type I interferon pathway [17].

RNA modifications also play a crucial role in regulating gene expression. Among these, N6-methyladenosine (m6A) is the most common, but others such as N1-methyladenosine (m1A), N5-methylcytosine (m5C), and pseudouridine (Ψ) can also have a significant impact [18]. For example, it has been observed that ALKBH5, an enzyme involved in the removal of m6A methylation, is associated with the presence of anti-dsDNA and anti-nucleosome autoantibodies in patients with SLE, suggesting that RNA modifications may influence the progression of the disease [19].

Cellular apoptosis also appears to be a process involved. A genetic variant located in the enhancer element of RNA associated with SLE (SLEAR) reduces the expression of this RNA in the peripheral blood monocytes of patients, contributing to the apoptosis of Jurkat cells in vitro. This happens because SLEAR activates genes with anti-apoptotic function, such as BCL2 and XIAP [20]. Similarly, increased expression of lincRNA-p21 has been associated with upregulation of the pro-apoptotic genes PUMA and BAX in patients with LN and in mouse models of the disease.

A study conducted on monozygotic and dizygotic twins discordant for SLE revealed that more marked epigenetic modifications occur in B cells compared to other cell types, with widespread hypermethylation in gene promoters [21]. A subgroup of B cells (CD27–CD11c + T-BET+CXCR5–) is particularly expanded in patients with SLE and its activation is driven by Toll-like receptor 7 (TLR7) stimulation and IL-21. Even T cells are subject to epigenetic modifications in SLE, with alterations in DNA methylation and histone modifications in genes involved in the disease, including CD11a, CD40L, CD70, and perforin [22].

The dysregulation of the type I interferon pathway is closely linked to the pathogenesis of SLE and increases susceptibility to viral infections, tumors, and autoimmune disorders. IFN-α and IFN-β are produced mainly by innate immune cells in response to the recognition of pathogenic stimuli by pattern recognition receptors (PRRs). Increased IFN pathway activity in patients with SLE has been correlated with disease severity, with a higher prevalence in Asian and Hispanic populations [23].

In SLE, immune complexes, endogenous nucleic acids, and excess apoptotic cells are recognized by receptors such as TLR7, TLR9, RIG-I, and cGAS, contributing to the abnormal activation of the immune response. Recent studies have shown that some microRNAs, such as miR574, let-7b and miR21, contained in the plasma exosomes of patients with SLE, can act as new ligands for TLR7, stimulating plasmacytoid dendritic cells (pDC) and amplifying the IFN response [24].

Even some lncRNAs can directly modulate the IFN pathway. For example, MALAT1, which is reduced in the PBMCs of patients with SLE, seems to inhibit IRF3, a key factor in the response to IFN [25].

11.4 Rheumatoid Arthritis

Rheumatoid arthritis is a chronic inflammatory disease that causes progressively destructive arthritis, with extensive damage to the joints. Among the epigenetic mechanisms involved in the disease are modifications in the methylation status of both fibroblasts and T and B immune cells [26]. Some studies have shown that T cells from patients with rheumatoid arthritis have generalized hypomethylation, a characteristic similar to that observed in systemic lupus erythematosus [27].

One of the most commonly used drugs in the treatment of RA, methotrexate (MTX), is known to induce demethylation of the FOXP3 gene promoter, thus promoting the accumulation of regulatory T cells (Treg). However, given its action as a folic acid antagonist, MTX could influence the course of the disease through other mechanisms that are not yet fully understood. Some studies have also shown that treatment with MTX can restore normal methylation status in peripheral blood mononuclear cells [28].

The first large-scale epigenomic studies date back to 2001, when Nakano et al. compared the synovial fibroblasts of patients with RA and osteoarthritis and found hypermethylation in about 60% of the 1859 genomic regions with different methylation status [29]. Subsequently, a study by de La Rica et al. identified new genes with altered methylation patterns, including IL6R, CAPN8, and DPP4. By analyzing the correlation between DNA methylation in RA synovial fibroblasts, miRNA expression, and transcriptomic data, it was observed that numerous identified genes showed inverse regulation and that several CpG sites were hypermethylated and associated with reduced expression of the corresponding miRNAs [30].

Although histone modifications have been studied to a lesser extent than DNA methylation, their involvement in the pathogenesis of rheumatoid arthritis has nevertheless been documented. The altered activity of histone deacetylases (HDAC) in peripheral blood polymorphonuclear cells of RA patients does not appear to be influenced by conventional anti-TNF drug treatments. However, the use of the selective HDAC2/3 inhibitor (MI192) has been shown to reduce IL-6 production in a dose-dependent manner, suggesting a possible innovative therapeutic approach for the disease [31].

A study conducted by Wendling et al. analyzed the link between Sirt1 activity/expression and IL-3 levels in peripheral polymorphonuclear cells, showing how an increase in IL-3 serum concentration was associated with a reduction in Sirt1 expression and an increase in the rate of apoptosis in RA patients [32]. Other studies have shown that the administration of TNF-α in the synovial fibroblasts of RA patients induces an increase in HDAC1 expression [33–35].

In synovial tissues affected by RA, the balance between histone acetyltransferase (HAT) and HDAC activity is severely altered. Increased HAT activity leads to excessive acetylation of histone proteins, contributing to the progression of the disease [36]. This phenomenon has been correlated with increased IL-6 expression in synovial fibroblasts, due to hyperacetylation of histone H3 [37].

In recent years, interest in the role of microRNAs in the pathogenesis of RA has also grown considerably. In particular, upregulation of miR-146a in response to TNF-α has been observed in affected patients, while miR-363 and miR-498 were found to be downregulated in CD4+ cells [38]. However, subsequent studies have reported a reduction of miR-146a and miR-155 in Treg cells after T cell stimulation in RA patients [39].

Another important aspect concerns the expression of miR-126a, which was found to be increased, determining hypomethylation of the CD11a and CD70 promoters and, consequently, excessive expression of these genes [40]. Finally, it has been shown that overexpression of miR-21 favors the accumulation of Treg cells in the synovial fibroblasts of patients affected by the disease [41].

11.5 Type 1 Diabetes

In type 1 diabetes (T1D), hyperglycemia is mainly due to the destruction of the β cells of the pancreatic islets and the action of T cells. Autoantibodies, including those against insulin (IAA), glutamic acid decarboxylase (GAD), and islet antigen 2A (IA-2A), can appear several years before the clinical diagnosis of T1D. Individuals with first-degree relatives with T1D and who present multiple islet autoantibodies are considered to be at high risk of developing the disease.

As with other autoimmune diseases, DM1 is also characterized by abnormalities in T-cell function. Numerous studies have shown that alterations in microRNAs can compromise the function of regulatory T cells (Tregs), affecting their ability to maintain immune tolerance. In subjects at high risk of T1D, compared to healthy subjects, upregulation of miR-26a is observed in activated Tregs, with a reduction in EZH2 expression [42]. In addition, miRNA181a in CD4+ T cells modulates the PTEN/PI3K/AKT pathway, resulting in a reduction in Foxp3 expression [43]. This same pathway is involved in the increase of precursor Tfh cells, regulated by miRNA92a. Inhibition of miRNA181a and miRNA92a has been shown to be effective in reducing islet autoimmunity in murine and humanized models through the use of antagomirs (anti-miR, antisense oligonucleotides complementary to their specific target miRNA) [44].

Several studies have shown that β-cell-specific CD8+ T cells have a long lifespan. Combined analysis of whole genome sequencing has shown that the DNA methylation profile of β-cell-specific CD8+ T cells is similar to the phenotype of memory T stem cells with multi- and totipotent capacity [45].

During the course of the disease, CD8+ T cells recognize the C-peptide of preproinsulin and destroy pancreatic β cells. It has been shown that endoplasmic reticulum (ER) stress of β cells, induced by IFN-γ and IL-1β, can promote the production of autoantigenic peptides by reducing the expression of miR-17–5p [46].

Epigenetic alterations precede the alteration of the glycemic balance and are associated with the production of autoantibodies and the degeneration of β cells, suggesting their possible use as predictive indicators of the onset of DM1. One of the most analyzed genes from an epigenetic point of view is INS, responsible for the production of preproinsulin [47]. It has been observed that β cells have an increase in unmethylated CpG sites in INS, and their detection in the bloodstream could represent a sign of β cell destruction.

In addition to INS, the glucokinase gene (GCK) and the chromatin target gene PRMT1 (CHTOP) have also been proposed as epigenetic biomarkers specific to pancreatic islets. Several miRNAs, including miR-204 and miR-101–3p, have been identified in the serum of subjects at high risk of developing DM1 [48–50]. With the onset of clinical symptoms, some miRNAs circulating in the plasma have been correlated with a reduction in C-peptide levels [51], while others have been indicated as predictors of an early graft-versus-host-disease (GVHD) or a favorable prognosis. Others, however, have been associated with an accelerated progression towards terminal renal failure [52, 53].

From the point of view of pathogenetic mechanisms, in DM1 miR-150 and miR-21, activated by the transcription factor NF-κB, have been shown to inhibit inflammation and apoptosis of β cells, acting on PUMA and PDCD4 respectively. On the contrary, an increase in some lncRNAs in patients with DM1 can promote β-cell death and Treg dysfunction.

On the biological therapy front, the use of nanodrugs containing miRNAs has shown promising results, stimulating β-cell proliferation by reducing PTEN expression [54–56].

11.6 Neuro-Immuno-Mediated Diseases

The term “neuro-immuno-mediated diseases” refers to a group of diseases that affect the central nervous system [CNS] and are characterized by a specific immuno-inflammatory activation against the latter.

The etiology and pathogenesis underlying nanodrugs in these clinical conditions are extremely complex and complex. Often, the anamnestic history of nanodrugs reveals an external trigger (e.g., an infection) that determines the activation of the immune system, which in turn affects the CNS [56]. For these pathologies, it is becoming increasingly important to realize that not only genetics play a fundamental role, but also epigenetics can greatly influence phenotypic manifestations.

The term “epigenetics” refers to those factors and mechanisms capable of modulating gene transcription based on external factors and without altering the composition of DNA [3]. Furthermore, in these pathologies, there is often the presence of several altered genetic elements, and these can be altered to a different extent even within the same pathology. This, together with the probable co-presence of associated environmental factors, which influence both the genetics and the epigenetics of the patient in various ways, contributes to the considerable variety of symptoms of these pathologies. Furthermore, the association of different genetic mutations and different environmental factors that can intervene in one patient rather than another explains why the same pathology (such as, for example, multiple sclerosis) can present itself in a different way from one individual to another [56, 57].

The epigenetic mechanisms that can influence gene transcription are different (DNA methylation, chromatic remodeling, histone alterations), and indirectly influence the regulation of the immune mechanisms that underlie the neurological pathologies of our interest [3]. For this reason, projects have been created to precisely identify the various gene loci linked to different autoimmune pathologies, also considering non-coding loci that deal with epigenetic mechanisms, which can influence or even be the cause of the aforementioned autoimmune pathologies [57].

Various studies have also focused on the onset and effect that excessive alcohol consumption during adolescence can have on neurological diseases, including neuro-immunological ones. In fact, it has been noted that alcohol can cause alterations at both the genetic and epigenetic level, with lasting effects even in the long term and which can lead to severe conditions even in adulthood [58].

Most studies in this field have focused on multiple sclerosis (MS), which serves as a model to explain some of the mechanisms involved in neuro-immunological diseases and the factors that can influence them. Numerous studies have focused on the search for specific loci and gene variants that may be linked to an increased likelihood of developing MS. This research has revealed both various genes directly linked to the development of the disease (associated with alterations of the immune system), and genes indirectly involved, that have epigenetic regulatory effects, such as certain genes involved in the regulation of DNA methylation [59].

From the various studies carried out, it has emerged that the main exogenous risk factors that are statistically significantly associated with the development of MS are EBV infections and cigarette smoking, the precise genetic or epigenetic mechanisms of which are still poorly understood [60]. Another study analyzed vitamin D deficiency in children, and noted that insufficient levels of this vitamin correlated with a greater probability of developing MS. Factors that could be the cause of an increased risk of relapse have also been implicated (such as, for example, some specific gene variants, particular bacteria of the intestinal microbiota, daily salt intake), but studies are still needed to confirm these relationships [61].

Endogenous factors, such as adolescent obesity, have also been implicated as a possible triggering factor or co-factor, influencing specific HLA loci in subjects already genetically predisposed to developing MS [62].

All the previously mentioned exogenous factors would have an effect on the increase or decrease in DNA methylation and acetylation, regulating numerous genes involved in inflammatory and immune processes, as well as influencing axonal degeneration, apoptosis of neural cells, and intracellular dysfunction (especially mitochondrial), leading to the phenotypic expression of the disease. Some mRNAs have also been implicated in the pathogenetic onset of MS. Based on all these discoveries, some specific therapies have also been directed towards the epigenetics of this disease, such as histone deacetylase inhibitors, which are showing increasingly promising results in preclinical studies of various diseases, both immunological and neurological. In particular, these therapies are based on the assumption that epigenetic modifications are reversible, and therefore acting on them can restore their previous function, thus acting upstream on the genetic modifications that caused the pathological status [63, 64].

Various studies have been carried out on MS, also focusing on other pathologies. One example is neuromyelitis optica (NMO), whose clinical picture presents optic neuritis, alterations affecting the brain stem and transverse myelitis, and whose pathogenesis is determined by the presence of anti-aquaporin 4 antibodies. In this pathology it has been noted that breastfeeding and early exposure to peers, thus contracting the normal viral infections of developmental age, may represent protective factors against the development of this neuro-immunological pathology, influencing the genome in a still unclear way through likely epigenetic mechanisms [65].

The clinical manifestations of these pathologies are extremely varied and complex, ranging from forms that affect a single specific area of the CNS, to forms that affect multiple areas, with or without a consensual alteration of the state of consciousness [56]. Another possible classification is based on the difference between forms that affect the white matter (acquired demyelinating forms) and forms that affect the gray matter (whose pathogenesis is usually mediated by antibodies) [56].

Fundamental for the classification of these complex pathologies are radiological investigations aimed at producing neuro-imaging, which can provide important clues and point towards a specific pathology, and bio-humoral investigations, carried out both on blood, but even more so by taking cerebrospinal fluid (CSF), aimed at searching for auto-antibodies that can identify a specific condition affecting the CNS [56].

There are many therapeutic approaches. These include symptomatic treatments, such as the support that is sometimes necessary in a Pediatric Intensive Care Unit, or the management of the main complications that can occur in addition to the main condition (e.g., superinfections or venous thrombosis). Then there are specific therapies for the basic immune condition, such as the use of high doses of corticosteroids or the infusion of intravenous immunoglobulins. In the most severe cases, it is possible to reach the point of using truly immunosuppressive therapies [56].

We can conclude by stating that neuro-immune-mediated diseases represent a group of complex pathologies that are increasingly better known and studied. The task of the clinician is not only to learn to recognize them early in order to treat them better, but also to understand how to prevent them, by investigating the genetics of individual patients, studying what the environmental "triggers" might be that act on predisposed individuals and researching increasingly personalized therapies, based on the patient's specific mutations and on the epigenetics that characterizes them.

References

1. Danieli MG, Casciaro M, Paladini A, Bartolucci M, Sordoni M, Shoenfeld Y, Gangemi S (2024) Exposome: epigenetics and autoimmune diseases. Autoimmun Rev 23(6):103584. https://doi.org/10.1016/j.autrev.2024.103584. Epub 2024 Aug 2 PMID: 39097180
2. Romano R, Cillo F, Moracas C, Pignata L, Nannola C, Toriello E, De Rosa A, Cirillo E, Coppola E, Giardino G, Brunetti-Pierri N, Riccio A, Pignata C (2022) Epigenetic alterations in inborn errors of immunity. J Clin Med 11(5):1261. https://doi.org/10.3390/jcm11051261. PMID: 35268351; PMCID: PMC8910960
3. Mazzone R, Zwergel C, Artico M, Taurone S, Ralli M, Greco A, Mai A (2019) The emerging role of epigenetics in human autoimmune disorders. Clin Epigenetics 11(1):34. https://doi.org/10.1186/s13148-019-0632-2. PMID: 30808407; PMCID: PMC6390373
4. Surace AEA, Hedrich CM (2019) The role of epigenetics in autoimmune/inflammatory disease. Front Immunol 10:1525. https://doi.org/10.3389/fimmu.2019.01525. PMID: 31333659; PMCID: PMC6620790
5. Sellars M, Huh JR, Day K, Issuree PD, Galan C, Gobeil S, Absher D, Green MR, Littman DR (2015) Regulation of DNA methylation dictates Cd4 expression during the development of helper and cytotoxic T cell lineages. Nat Immunol 16(7):746–754. https://doi.org/10.1038/ni.3198. Epub 2015 Jun 1. PMID: 26030024; PMCID: PMC4474743
6. Piotrowska M, Gliwiński M, Trzonkowski P, Iwaszkiewicz-Grzes D (2021) Regulatory T cells-related genes are under DNA methylation influence. Int J Mol Sci 22(13):7144. https://doi.org/10.3390/ijms22137144. PMID: 34281195; PMCID: PMC8267835
7. Feng Y, Arvey A, Chinen T, van der Veeken J, Gasteiger G, Rudensky AY (2014) Control of the inheritance of regulatory T cell identity by a cis element in the Foxp3 locus. Cell 158(4):749–763. https://doi.org/10.1016/j.cell.2014.07.031. PMID: 25126783; PMCID: PMC4151558
8. Smith ZD, Meissner A (2013) DNA methylation: roles in mammalian development. Nat Rev Genet 14(3):204–220. https://doi.org/10.1038/nrg3354. Epub 2013 Feb 12 PMID: 23400093
9. Selimyan R, Gerstein RM, Ivanova I, Precht P, Subrahmanyam R, Perlot T, Alt FW, Sen R (2013) Localized DNA demethylation at recombination intermediates during immunoglobulin heavy chain gene assembly. PLoS Biol 11(1):e1001475. https://doi.org/10.1371/journal.pbio.1001475. Epub 2013 Jan 29. PMID: 23382652; PMCID: PMC3558432
10. Panni S, Lovering RC, Porras P, Orchard S (2020) Non-coding RNA regulatory networks. Biochim Biophys Acta Gene Regul Mech 1863(6):194417. https://doi.org/10.1016/j.bbagrm.2019.194417. Epub 2019 Sep 4 PMID: 31493559
11. Tian X, Tian J, Tang X, Ma J, Wang S (2016) Long non-coding RNAs in the regulation of myeloid cells. J Hematol Oncol 9(1):99. https://doi.org/10.1186/s13045-016-0333-7. PMID: 27680332; PMCID: PMC5041333
12. Zhang X, Weissman SM, Newburger PE (2014) Long intergenic non-coding RNA HOTAIRM1 regulates cell cycle progression during myeloid maturation in NB4 human promyelocytic leukemia cells. RNA Biol 11(6):777–787. https://doi.org/10.4161/rna.28828. Epub 2014 Apr 24. PMID: 24824789; PMCID: PMC4156508
13. Dahariya S, Paddibhatla I, Kumar S, Raghuwanshi S, Pallepati A, Gutti RK (2019) Long non-coding RNA: classification, biogenesis and functions in blood cells. Mol Immunol 112:82–92. https://doi.org/10.1016/j.molimm.2019.04.011. Epub 2019 May 9 PMID: 31079005
14. Tsai CY, Shen CY, Liu CW, Hsieh SC, Liao HT, Li KJ, Lu CS, Lee HT, Lin CS, Wu CH, Kuo YM, Yu CL (2020) Aberrant non-coding RNA expression in patients with systemic lupus erythematosus: consequences for immune dysfunctions and tissue damage. Biomolecules 10(12):1641. https://doi.org/10.3390/biom10121641. PMID: 33291347; PMCID: PMC7762297
15. Li LJ, Zhu ZW, Zhao W, Tao SS, Li BZ, Xu SZ, Wang JB, Zhang MY, Wu J, Leng RX, Fan YG, Pan HF, Ye DQ (2018) Circular RNA expression profile and potential function of hsa_circ_0045272 in systemic lupus erythematosus. Immunology 155(1):137–149. https://doi.org/10.1111/imm.12940. Epub 2018 May 23. PMID: 29700819; PMCID: PMC6099170

16. Wang X, Zhang C, Wu Z et al (2018) CircIBTK inhibits DNA demethylation and activation of AKT signaling pathway via miR-29b in peripheral blood mononuclear cells in systemic lupus erythematosus. Arthritis Res Ther 20:118. https://doi.org/10.1186/s13075-018-1618-8
17. Liu CX, Li X, Nan F, Jiang S, Gao X, Guo SK, Xue W, Cui Y, Dong K, Ding H, Qu B, Zhou Z, Shen N, Yang L, Chen LL (2019) Structure and degradation of circular RNAs regulate PKR activation in innate immunity. Cell 177(4):865–880.e21. https://doi.org/10.1016/j.cell.2019.03.046. Epub 2019 Apr 25 PMID: 31031002
18. Roundtree IA, Evans ME, Pan T, He C (2017) Dynamic RNA modifications in gene expression regulation. Cell 169(7):1187–1200. https://doi.org/10.1016/j.cell.2017.05.045. PMID: 28622506; PMCID: PMC5657247
19. Luo Q, Fu B, Zhang L, Guo Y, Huang Z, Li J (2020) Decreased peripheral blood ALKBH5 correlates with markers of autoimmune response in systemic lupus erythematosus. Dis Markers 2020:8193895. Published 2020 Jun 25. https://doi.org/10.1155/2020/8193895
20. Fan Z, Chen X, Liu L, Zhu C, Xu J, Yin X, Sheng Y, Zhu Z, Wen L, Zuo X, Zheng X, Zhang Y, Xu J, Huang H, Zhou F, Sun L, Luo J, Zhang D, Chen X, Cui Y, Hao Y, Cui Y, Zhang X, Chen R (2020) Association of the Polymorphism rs13259960 in SLEAR with predisposition to systemic lupus erythematosus. Arthritis Rheumatol 72(6):985–996. https://doi.org/10.1002/art.41200. Epub 2020 Apr 26 PMID: 31930717.
21. Ulff-Møller CJ, Asmar F, Liu Y, Svendsen AJ, Busato F, Grønbaek K, Tost J, Jacobsen S (2018) Twin DNA methylation profiling reveals flare-dependent interferon signature and B cell promoter Hypermethylation in systemic lupus erythematosus. Arthritis Rheumatol 70(6):878–890. https://doi.org/10.1002/art.40422. Epub 2018 May 9 PMID: 29361205
22. Lu Q (2013) The critical importance of epigenetics in autoimmunity. J Autoimmun 41:1–5. https://doi.org/10.1016/j.jaut.2013.01.010. Epub 2013 Feb 1 PMID: 23375849
23. Lanata CM, Paranjpe I, Nititham J, Taylor KE, Gianfrancesco M, Paranjpe M, Andrews S, Chung SA, Rhead B, Barcellos LF, Trupin L, Katz P, Dall'Era M, Yazdany J, Sirota M, Criswell LA (2019) A phenotypic and genomics approach in a multi-ethnic cohort to subtype systemic lupus erythematosus. Nat Commun 10(1):3902. https://doi.org/10.1038/s41467-019-11845-y. Erratum in: Nat Commun 2020 Feb 27;11(1):1164. https://doi.org/10.1038/s41467-020-15097-z. PMID: 31467281; PMCID: PMC6715644
24. Salvi V, Gianello V, Busatto S, Bergese P, Andreoli L, D'Oro U, Zingoni A, Tincani A, Sozzani S, Bosisio D (2018) Exosome-delivered microRNAs promote IFN-α secretion by human plasmacytoid DCs via TLR7. JCI Insight 3(10):e98204. https://doi.org/10.1172/jci.insight.98204. PMID: 29769437; PMCID: PMC6012509
25. Liu W, Wang Z, Liu L, Yang Z, Liu S, Ma Z, Liu Y, Ma Y, Zhang L, Zhang X, Jiang M, Cao X (2020) LncRNA *Malat1* inhibition of TDP43 cleavage suppresses IRF3-initiated antiviral innate immunity. Proc Natl Acad Sci U S A 117(38):23695–23706. https://doi.org/10.1073/pnas.2003932117. Epub 2020 Sep 9. PMID: 32907941; PMCID: PMC7519350
26. Glossop JR, Emes RD, Nixon NB, Packham JC, Fryer AA, Mattey DL, Farrell WE (2016) Genome-wide profiling in treatment-naive early rheumatoid arthritis reveals DNA methylome changes in T and B lymphocytes. Epigenomics 8(2):209–224. https://doi.org/10.2217/epi.15.103. Epub 2015 Nov 10 PMID: 26556652.
27. Richardson B, Scheinbart L, Strahler J, Gross L, Hanash S, Johnson M (1990) Evidence for impaired T cell DNA methylation in systemic lupus erythematosus and rheumatoid arthritis. Arthritis Rheum 33(11):1665–1673. https://doi.org/10.1002/art.1780331109. PMID: 2242063
28. de Andres MC, Perez-Pampin E, Calaza M et al (2015) Assessment of global DNA methylation in peripheral blood cell subpopulations of early rheumatoid arthritis before and after methotrexate. Arthritis Res Ther 17:233. https://doi.org/10.1186/s13075-015-0748-5
29. Nakano K, Whitaker JW, Boyle DL, Wang W, Firestein GS (2013) DNA methylome signature in rheumatoid arthritis. Ann Rheum Dis 72(1):110–117. https://doi.org/10.1136/annrheumdis-2012-201526. Epub 2012 Jun 26. PMID: 22736089; PMCID: PMC3549371
30. de la Rica L, Urquiza JM, Gómez-Cabrero D, Islam AB, López-Bigas N, Tegnér J, Toes RE, Ballestar E (2013) Identification of novel markers in rheumatoid arthritis through integrated

analysis of DNA methylation and microRNA expression. J Autoimmun 41:6–16. https://doi.org/10.1016/j.jaut.2012.12.005. Epub 2013 Jan 8. PMID: 23306098

31. Gillespie J, Savic S, Wong C, Hempshall A, Inman M, Emery P, Grigg R, McDermott MF (2012) Histone deacetylases are dysregulated in rheumatoid arthritis and a novel histone deacetylase 3-selective inhibitor reduces interleukin-6 production by peripheral blood mononuclear cells from rheumatoid arthritis patients. Arthritis Rheum 64(2):418–422. https://doi.org/10.1002/art.33382. PMID: 21952924.
32. Zhu X, Wu L, Mo X et al (2020) Identification of PBMC-expressed miRNAs for rheumatoid arthritis. Epigenetics 15(4):386–397. https://doi.org/10.1080/15592294.2019.1676613
33. Yang G, Wu D, Zeng G, Jiang O, Yuan P, Huang S, Zhu J, Tian J, Weng Y, Rao Z (2015) Correlation between miR-126 expression and DNA hypomethylation of CD4+ T cells in rheumatoid arthritis patients. Int. J Clin Exp Pathol 8(8):8929–8936. PMID: 26464634; PMCID: PMC4583866
34. Kmiołek T, Rzeszotarska E, Wajda A, Walczuk E, Kuca-Warnawin E, Romanowska-Próchnicka K, Stypinska B, Majewski D, Jagodzinski PP, Pawlik A, Paradowska-Gorycka A (2020) The interplay between transcriptional factors and MicroRNAs as an important factor for Th17/Treg balance in RA patients. Int J Mol Sci 21(19):7169. https://doi.org/10.3390/ijms21197169. PMID: 32998457; PMCID: PMC7583886
35. Kurowska-Stolarska M, Alivernini S, Melchor EG, Elmesmari A, Tolusso B, Tange C, Petricca L, Gilchrist DS, Di Sante G, Keijzer C, Stewart L, Di Mario C, Morrison V, Brewer JM, Porter D, Milling S, Baxter RD, McCarey D, Gremese E, Lemke G, Ferraccioli G, McSharry C, McInnes IB (2017) MicroRNA-34a dependent regulation of AXL controls the activation of dendritic cells in inflammatory arthritis. Nat Commun 8:15877. https://doi.org/10.1038/ncomms15877. PMID: 28639625; PMCID: PMC5489689
36. Yang P, Zhang M, Wang X, Xu AL, Shen M, Jiang B, Zhou X, Zhou L (2020) MicroRNA let-7g-5p alleviates murine collagen-induced arthritis by inhibiting Th17 cell differentiation. Biochem Pharmacol 174:113822. https://doi.org/10.1016/j.bcp.2020.113822. Epub 2020 Jan 25 PMID: 31987855.
37. Wu YH, Liu W, Xue B, Zhang L, Liu XY, Liu B, Wang Y, Cai Y, Duan R (2016) Upregulated expression of microRNA-16 correlates with Th17/Treg cell imbalance in patients with rheumatoid arthritis. DNA Cell Biol 35(12):853–860. https://doi.org/10.1089/dna.2016.3349. Epub 2016 Nov 22 PMID: 27875659.
38. Dong L, Wang X, Tan J, Li H, Qian W, Chen J, Chen Q, Wang J, Xu W, Tao C, Wang S (2014) Decreased expression of microRNA-21 correlates with the imbalance of Th17 and Treg cells in patients with rheumatoid arthritis. J Cell Mol Med 18(11):2213–2224. https://doi.org/10.1111/jcmm.12353. Epub 2014 Aug 28. PMID: 25164131; PMCID: PMC4224555
39. Xie M, Wang J, Gong W, Xu H, Pan X, Chen Y, Ru S, Wang H, Chen X, Zhao Y, Li J, Yin Q, Xia S, Zhou X, Liu X, Shao Q (2019) NF-κB-driven miR-34a impairs Treg/Th17 balance via targeting Foxp3. J Autoimmun 102:96–113. https://doi.org/10.1016/j.jaut.2019.04.018. Epub 2019 May 23 PMID: 31130368.
40. Xiang HY, Pan F, Yan JZ, Hong LQ, Zhang LH, Liu YH, Feng X, Cai CS (2018) Upregulation of miR-498 suppresses Th17 cell differentiation by targeting STAT3 in rheumatoid arthritis patients. Sheng Li Xue Bao 70(2):167–174. Chinese PMID: 29691581
41. Friday SC, Fox DA (2016) Phospholipase D enzymes facilitate IL-17- and TNFα-induced expression of proinflammatory genes in rheumatoid arthritis synovial fibroblasts (RASF). Immunol Lett 174:9–18. https://doi.org/10.1016/j.imlet.2016.04.001. Epub 2016 Apr 4 PMID: 27058440.
42. Zhang Y, Feng ZP, Naselli G, Bell F, Wettenhall J, Auyeung P, Ellis JA, Ponsonby AL, Speed TP, Chong MM, Harrison LC (2016) MicroRNAs in CD4(+) T cell subsets are markers of disease risk and T cell dysfunction in individuals at risk for type 1 diabetes. J Autoimmun 68:52–61. https://doi.org/10.1016/j.jaut.2015.12.006. Epub 2016 Jan 11 PMID: 26786119
43. Serr I, Scherm MG, Zahm AM, Schug J, Flynn VK, Hippich M, Kälin S, Becker M, Achenbach P, Nikolaev A, Gerlach K, Liebsch N, Loretz B, Lehr CM, Kirchner B, Spornraft M, Haase B, Segars J, Küper C, Palmisano R, Waisman A, Willis RA, Kim WU, Weigmann

B, Kaestner KH, Ziegler AG, Daniel C (2018) A miRNA181a/NFAT5 axis links impaired T cell tolerance induction with autoimmune type 1 diabetes. Sci Transl Med 10(422):eaag1782. https://doi.org/10.1126/scitranslmed.aag1782. Erratum in: Sci Transl Med. 2019 Feb 20;11(480):eaaw9304. https://doi.org/10.1126/scitranslmed.aaw9304. PMID: 29298866; PMCID: PMC5828501

44. Serr I, Fürst RW, Ott VB et al (2016) miRNA92a targets KLF2 and the phosphatase PTEN signaling to promote human T follicular helper precursors in T1D islet autoimmunity. Proc Natl Acad Sci USA 113(43):E6659–E6668. https://doi.org/10.1073/pnas.1606646113. PMID: 27791035; PMCID: PMC5087025
45. Abdelsamed HA, Zebley CC, Nguyen H, Rutishauser RL, Fan Y, Ghoneim HE, Crawford JC, Alfei F, Alli S, Ribeiro SP, Castellaw AH, McGargill MA, Jin H, Boi SK, Speake C, Serti E, Turka LA, Busch ME, Stone M, Deeks SG, Sekaly RP, Zehn D, James EA, Nepom GT, Youngblood B (2020) Beta cell-specific $CD8^+$ T cells maintain stem cell memory-associated epigenetic programs during type 1 diabetes. Nat Immunol 21(5):578–587. https://doi.org/10.1038/s41590-020-0633-5. Epub 2020 Mar 30. PMID: 32231298; PMCID: PMC7183435
46. Thomaidou S, Kracht MJL, van der Slik A, Laban S, de Koning EJ, Carlotti F, Hoeben RC, Roep BO, Zaldumbide A (2020) β-Cell stress shapes CTL immune recognition of Preproinsulin signal peptide by posttranscriptional regulation of endoplasmic reticulum aminopeptidase 1. Diabetes 69(4):670–680. https://doi.org/10.2337/db19-0984. Epub 2020 Jan 2 PMID: 31896552
47. Wang Z, Xie Z, Lu Q, Chang C, Zhou Z (2017) Beyond genetics: what causes type 1 diabetes. Clin Rev Allergy Immunol 52(2):273–286. https://doi.org/10.1007/s12016-016-8592-1. PMID: 27878451
48. Syed F, Tersey SA, Turatsinze JV, Felton JL, Kang NJ, Nelson JB, Sims EK, Defrance M, Bizet M, Fuks F, Cnop M, Bugliani M, Marchetti P, Ziegler AG, Bonifacio E, Webb-Robertson BJ, Balamurugan AN, Evans-Molina C, Eizirik DL, Mather KJ, Arslanian S, Mirmira RG (2020) Circulating unmethylated CHTOP and INS DNA fragments provide evidence of possible islet cell death in youth with obesity and diabetes. Clin Epigenetics 12(1):116. https://doi.org/10.1186/s13148-020-00906-5. PMID: 32736653; PMCID: PMC7393900
49. Xu G, Thielen LA, Chen J, Grayson TB, Grimes T, Bridges SL Jr, Tse HM, Smith B, Patel R, Li P, Evans-Molina C, Ovalle F, Shalev A (2019) Serum miR-204 is an early biomarker of type 1 diabetes-associated pancreatic beta-cell loss. Am J Physiol Endocrinol Metab 317(4):E723–E730. https://doi.org/10.1152/ajpendo.00122.2019. Epub 2019 Aug 13. PMID: 31408375; PMCID: PMC6842918
50. Santos AS, Cunha Neto E, Fukui RT, Ferreira LRP, Silva MER (2019) Increased expression of circulating microRNA 101-3p in type 1 diabetes patients: new insights into miRNA-regulated pathophysiological pathways for type 1 diabetes. Front Immunol 10:1637. https://doi.org/10.3389/fimmu.2019.01637. PMID: 31396209; PMCID: PMC6665278
51. Garavelli S, Bruzzaniti S, Tagliabue E et al (2020) Plasma circulating miR-23~27~24 clusters correlate with the immunometabolic derangement and predict C-peptide loss in children with type 1 diabetes. Diabetologia 63:2699–2712. https://doi.org/10.1007/s00125-020-05237-x
52. Roels S, Costa OR, Tersey SA, Stangé G, De Smet D, Balti EV, Gillard P, Keymeulen B, Ling Z, Pipeleers DG, Gorus FK, Mirmira RG, Martens GA (2019) Combined analysis of GAD65, miR-375, and Unmethylated insulin DNA following islet transplantation in patients with T1D. J Clin Endocrinol Metab 104(2):451–460. https://doi.org/10.1210/jc.2017-02520. PMID: 30203041; PMCID: PMC6310912
53. Pezzolesi MG, Satake E, McDonnell KP, Major M, Smiles AM, Krolewski AS (2015) Circulating TGF-β1-regulated miRNAs and the risk of rapid progression to ESRD in type 1 diabetes. Diabetes 64(9):3285–3293. https://doi.org/10.2337/db15-0116. Epub 2015 Apr 30. PMID: 25931475; PMCID: PMC4542435
54. Huang Y-N, Chiang S-L, Lin Y-J, Liu S-C, Li Y-H, Liao Y-C, Lee M-R, Su P-H, Tsai F-J, Hung H-C et al (2021) Long, noncoding RNA SRA induces apoptosis of ?-cells by promoting the IRAK1/LDHA/lactate pathway. Int J Mol Sci 22:1720. https://doi.org/10.3390/ijms22041720

55. Wang P, Liu Q, Zhao H, Bishop JO, Zhou G, Olson LK, Moore A (2020) miR-216a-targeting theranostic nanoparticles promote proliferation of insulin-secreting cells in type 1 diabetes animal model. Sci Rep 10(1):5302. https://doi.org/10.1038/s41598-020-62269-4. PMID: 32210316; PMCID: PMC7093482
56. Wells E, Hacohen Y, Waldman A, Tillema JM, Soldatos A, Ances B, Benseler S, Bielekova B, Dale RC, Dalmau J, Gaillard W, Gorman M, Greenberg B, Hyslop A, Pardo CA, Tasker RC, Yeh EA, Bar-Or A, Pittock S, Vanderver A, Banwell B; Attendees of the International Neuroimmune Meeting. Neuroimmune disorders of the central nervous system in children in the molecular era. Nat Rev Neurol. 2018;14(7):433-445. https://doi.org/10.1038/s41582-018-0024-9. Erratum in: Nat Rev Neurol. 2018 Dec;14(12):749. https://doi.org/10.1038/s41582-018-0077-9. PMID: 29925924
57. Farh KK, Marson A, Zhu J, Kleinewietfeld M, Housley WJ, Beik S, Shoresh N, Whitton H, Ryan RJ, Shishkin AA, Hatan M, Carrasco-Alfonso MJ, Mayer D, Luckey CJ, Patsopoulos NA, De Jager PL, Kuchroo VK, Epstein CB, Daly MJ, Hafler DA, Bernstein BE (2015) Genetic and epigenetic fine mapping of causal autoimmune disease variants. Nature 518(7539):337–343. https://doi.org/10.1038/nature13835. Epub 2014 Oct 29. PMID: 25363779; PMCID: PMC4336207
58. Crews FT, Coleman LG Jr, Macht VA, Vetreno RP (2023) Targeting persistent changes in Neuroimmune and epigenetic signaling in adolescent drinking to treat alcohol use disorder in adulthood. Pharmacol Rev 75(2):380–396. https://doi.org/10.1124/pharmrev.122.000710
59. Andlauer TF, Buck D, Antony G, Bayas A, Bechmann L, Berthele A, Chan A, Gasperi C, Gold R, Graetz C, Haas J, Hecker M, Infante-Duarte C, Knop M, Kümpfel T, Limmroth V, Linker RA, Loleit V, Luessi F, Meuth SG, Mühlau M, Nischwitz S, Paul F, Pütz M, Ruck T, Salmen A, Stangel M, Stellmann JP, Stürner KH, Tackenberg B, Then Bergh F, Tumani H, Warnke C, Weber F, Wiendl H, Wildemann B, Zettl UK, Ziemann U, Zipp F, Arloth J, Weber P, Radivojkov-Blagojevic M, Scheinhardt MO, Dankowski T, Bettecken T, Lichtner P, Czamara D, Carrillo-Roa T, Binder EB, Berger K, Bertram L, Franke A, Gieger C, Herms S, Homuth G, Ising M, Jöckel KH, Kacprowski T, Kloiber S, Laudes M, Lieb W, Lill CM, Lucae S, Meitinger T, Moebus S, Müller-Nurasyid M, Nöthen MM, Petersmann A, Rawal R, Schminke U, Strauch K, Völzke H, Waldenberger M, Wellmann J, Porcu E, Mulas A, Pitzalis M, Sidore C, Zara I, Cucca F, Zoledziewska M, Ziegler A, Hemmer B, Müller-Myhsok B (2016) Novel multiple sclerosis susceptibility loci implicated in epigenetic regulation. Sci Adv 2(6):e1501678. https://doi.org/10.1126/sciadv.1501678. PMID: 27386562; PMCID: PMC4928990
60. Belbasis L, Bellou V, Evangelou E, Ioannidis JP, Tzoulaki I (2015) Environmental risk factors and multiple sclerosis: an umbrella review of systematic reviews and meta-analyses. Lancet Neurol 14(3):263–273. https://doi.org/10.1016/S1474-4422(14)70267-4. Epub 2015 Feb 4 PMID: 25662901
61. Waubant E, Ponsonby AL, Pugliatti M, Hanwell H, Mowry EM, Hintzen RQ (2016) Environmental and genetic factors in pediatric inflammatory demyelinating diseases. Neurology 87(9 Suppl 2):S20–S27. https://doi.org/10.1212/WNL.0000000000003029. PMID: 27572857
62. Alfredsson L, Olsson T (2019) Lifestyle and environmental factors in multiple sclerosis. Cold Spring Harb Perspect Med 9(4):a028944. https://doi.org/10.1101/cshperspect.a028944. PMID: 29735578; PMCID: PMC6444694
63. Küçükali Cİ, Kürtüncü M, Çoban A, Çebi M, Tüzün E (2015) Epigenetics of multiple sclerosis: an updated review. NeuroMolecular Med 17(2):83–96. https://doi.org/10.1007/s12017-014-8298-6. Epub 2014 Mar 21 PMID: 24652042
64. Falkenberg KJ, Johnstone RW (2014) Histone deacetylases and their inhibitors in cancer, neurological diseases and immune disorders. Nat Rev Drug Discov 13(9):673–691. https://doi.org/10.1038/nrd4360. Epub 2014 Aug 18 PMID: 25131830
65. Graves J, Grandhe S, Weinfurtner K et al (2014) Protective environmental factors for neuromyelitis optica. Neurology 83(21):1923–1929. https://doi.org/10.1212/WNL.0000000000001001

Epigenetics and Inherited Metabolic Disorders

12

Catia Cavicchi, Jean-Louis Guéant, and Amelia Morrone

Contents

12.1 Introduction to Inherited Metabolic Disorders 194
12.2 General Overview on the Role of Epigenetics in IEMs 194
12.3 Episignatures May Contribute to the Clinical Heterogeneity, Pathophysiology, Prognosis, and Treatment Monitoring of IEMs 200
12.3.1 Lysosomal Storage Disorders 200
12.3.2 Wilson Disease 205
12.3.3 Phenylketonuria 206
12.3.4 Homocystinuria-Megaloblastic Anemia cblG Type 207
12.4 Epimutation Causative of IEM 208
12.5 Epigenetics for the Treatment of IEMs: Future Perspectives 212
References 214

C. Cavicchi
Laboratory of Molecular Genetics of Neurometabolic Diseases, Department of Neurosciences and Medical Genetics, Meyer Children's Hospital IRCCS, Florence, Italy
e-mail: catia.cavicchi@meyer.it

J.-L. Guéant
Inserm UMRS 1256 NGERE – Nutrition, Genetics, and Environmental Risk Exposure, University of Lorraine, Nancy, France

National Center of Inborn Errors of Metabolism, University Regional Hospital Center of Nancy, Nancy, France
e-mail: jean-louis.gueant@univ-lorraine.fr

A. Morrone (✉)
Laboratory of Molecular Genetics of Neurometabolic Diseases, Department of Neurosciences and Medical Genetics, Meyer Children's Hospital IRCCS, Florence, Italy

Department of NEUROFARBA, University of Florence, Florence, Italy
e-mail: amelia.morrone@meyer.it; amelia.morrone@unifi.it

L. Emmi, V. Fanos (eds.), *Epigenetics and Rare Diseases*, Rare Diseases of the Immune System, https://doi.org/10.1007/978-3-032-02902-7_12

12.1 Introduction to Inherited Metabolic Disorders

Inborn errors of metabolism (IEMs) are a large and heterogeneous group of genetic disorders caused by single-gene defects that impair the functioning of a metabolic pathway. This impairment often results in the accumulation of toxic substrates or a deficiency of essential metabolites and clinical symptoms. Although IEMs are considered rare, they occur in 1 in every 2500 births. They are typically transmitted as autosomal recessive traits, but some have X-linked, dominant or mitochondrial (matrilineal) inheritance. A few cases occur sporadically.

Recently, the "International Classification of Inherited Metabolic Disorders" (ICIMD) was published, which groups IEMs according to protein defect and biochemical characteristics (www.icimd.org) [1]. There are 24 categories with 124 groups, for a total of 1450 disorders. The ICIMD database reflects the complexity of these genetic disorders concerning their underlying pathophysiology as well as their biochemical workup and molecular analysis. Based on the time of onset and clinical course, IEMs are often grouped into three forms: acute neonatal onset forms, acute late onset forms, and forms with late onset and a chronic and progressive course. IEMs can present at any age, but most diagnoses occur in the pediatric population. Newborn screening (NBS) programs are now active for more than 40 IEMs, which has led to earlier diagnoses, prompt treatment, and significant reductions in mortality and disease burden.

12.2 General Overview on the Role of Epigenetics in IEMs

Like other Mendelian disorders, in a number of IEMs, correlations between genotypes and phenotypes are not clear-cut, and patients with the same pathogenic variants can present with different disease severities and symptoms. Even monozygotic twins affected by the same IEM can present different phenotypes. These findings have long suggested that factors that do not depend on genetics affect phenotypes. Indeed, it is likely that the cells and organs of genetically affected individuals adapt to different and specific environmental scenarios, for instance, in response to altered intracellular/extracellular metabolite levels or diet. Given the phenotypical variability of IEMs, there are likely to be multiple and complementary factors that can influence their clinical presentation and course. As yet these factors have not been extensively explored and thus epigenetic research into IEMs is a promising and welcome departure.

The main epigenetic mechanisms recognized so far as being involved in IEMs are DNA methylation, post-translational modifications of histones, mainly histone acetylation or methylation, and noncoding RNAs, mainly micro-RNAs (miRNAs). All of these influence gene expression regulation. DNA methylation of a gene promoter generally down-regulates gene expression, histone acetylation facilitates transcriptional activation by leading to a more relaxed chromatin structure and histone methylation may have both a repressing or activating role. MiRNAs act as post-transcriptional gene regulators.

Metabolic processes are directly linked to covalent DNA and histone modifications, as many metabolite products act as substrates for DNA or histone modifications or modulate the activity of DNA-modifying and histone-modifying enzymes [2]. The main examples that support this link are: a) S-adenosylmethionine (SAM) and S-adenosylhomocysteine (SAH) regulate DNA and histone methylation; b) acetyl-CoA is a key node in metabolism that also serves in histone acetylation; c) nicotinamide adenine dinucleotide (NAD^+) is required as a co-substrate by sirtuin-dependent histone deacetylases; d) mitochondrial oxidative metabolism is closely associated with histone demethylation. Besides methylation or acetylation, histones undergo multiple other covalent modifications. Among them, the more relevant for IEMs could perhaps be citrullination, succinylation, malonylation, O-glycosylation, and ADP-ribosylation.

Although next-generation sequencing (NGS) facilitates genetic diagnoses in IEMs by analyzing several genes at once and allowing for the detection of both DNA single-nucleotide variants and copy number variants (CNVs), it can be inconclusive in some patients who nevertheless have a clear-cut clinical phenotype. This suggests that the causal role of epigenetics is underestimated in IEMs. A recent large-scale study on the prevalence and distribution of rare epigenetic changes in the human genome supports this hypothesis. The study described the methylation profiles of 23,116 individuals and identified more than 4,400 unique autosomal epivariations [3]. 384 of these were associated with the outlier expression of genes linked to human diseases, suggesting that they may contribute to the mutational spectra underlying many Mendelian disorders, including IEMs [4]. In the future, epigenome profiling could complement sequence-based approaches in the diagnosis of IEMs.

Comprehension of the epigenetic mechanisms associated with IEMs can also facilitate treatment monitoring and provide new therapeutic opportunities. Epigenetic changes are dynamic and, unlike genetic variants, can be reversed, which may represent a significant advantage in therapeutic settings.

The main IEMs with clear-cut epigenetic components are summarized in Table 12.1.

Table 12.1 Main IEMs with clear-cut epigenetic components

IEM (OMIM entry, gene, inheritance)	Category, Group (according to the ICIMD)	Epigenetic modification	Main epigenetic findings	Significance of the epigenetic changes	Reference
Fabry disease (#301500, *GLA*, XL)	Disorders of complex molecule degradation, Sphingolipid degradation	DNA methylation	Conflicting findings on correlation between X-inactivation pattern obtained by the *HUMARA* assay and clinical phenotype; proven correlation between methylation of the *GLA* non-mutated allele and its mRNA expression with disease severity	Contribution to phenotypic heterogeneity in heterozygous females	[5]
			Altered DNA methylation in the promoter of the *HUMARA* gene in cell lines of patients; differential methylated CpG sites in the genome associated with abnormal expression of at least 21 genes potentially involved in relevant pathways for the disease (including the hypomethylated *COL4A1* and *COL4A2*)	Biomarker for pathophysiology; contribution to phenotypic heterogeneity	[6]
		MicroRNA	10 plasma miRNAs specific for Fabry patients; miR-199a-5p and miR-126-3p involved in endothelial dysfunction; miR-423-5p and miR-451a likely correlated with cardiac involvement; levels of miR-199a-5p and miR-126-3p and miR-423-5p were ERT responsive	Epimarker for diagnosis, prognosis, and ERT efficacy	[7]
			miR-1307-5p, miR-21-5p, miR-152-5p and miR-26a-5p were down-regulated in the serum of affected patients after ERT; male patients treated with ERT showed down-regulation of miR-19a-3p and miR-486-5p	Epimarker for ERT efficacy; contribution to phenotypic heterogeneity	[8]
		Mitochondrial-related microRNAs	Dysregulation of circulating Mito-miRNAs involved in metabolism, biogenesis, oxidative damage, and apoptosis of mitochondria; overexpression of miR-181c and miR-338 induced the down-regulation of mitochondrial complex I and IV, respectively	Biomarker for mitochondrial involvement	[9]
Gaucher disease (#230800, *GBA*, AR)	Disorders of complex molecule degradation, Sphingolipid degradation	MicroRNA	In fibroblasts of affected patients, miR-127-5p down-regulated *SCARB-2* gene, causing improper trafficking of the GBA enzyme; miR-16-5p and miR-195-5p were able to up-regulate GBA activity	Biomarker for pathophysiology; potential therapy targets	[10]
			Differential expression of miRNAs and mRNAs in brain of a neuropathic Gaucher mouse model	Biomarker for neuropathic form	[11]

Homocystinuria-megaloblastic anemia, cblG type (#250940, *MTR*, AR)	Disorders of amino acid metabolism, Sulfur-containing amino acids	DNA methylation	Differentially methylated CpGs of several genes involved in eye development, cone differentiation, and retinoid metabolism in retina tissues from the *Mtr*-cKO mouse model: *Rarα* and *Epha2* (hypermethylated); *Vegfa* and *Nr2e3* (hypomethylated); significant decreased expression of *Lrat*, *Rpe65*, *Rdh5*, and *Pax6* genes	Biomarker for retinopathy	[30]
Krabbe disease (#245200, *GALC*, AR)	Disorders of complex molecule degradation, Sphingolipid degradation	MicroRNA	Down-regulation of miR-219 in oligodendrocytes of the twitcher mouse model associated with cell-autonomous developmental defects, apoptotic death, and psychosine accumulation; restoring of these pathogenic findings by the exogenous supplementation of miR-219	Biomarker for pathophysiology	[13]
Methylmalonic aciduria and homocystinuria, cblC type (#277400, *MMACHC*, AR)	Disorders of vitamin and cofactor metabolism, Cobalamin	Histone methylation and DNA methylation	Hypermethylation of the CpG islands belonging to the *MMACHC-CCDC163P* and the *TESK2* promoters triggered by a splicing variant in the nearby *PRDX1* gene which caused aberrant antisense transcription through the active promoters, deposition of H3K36me3 mark, and finally silencing of both *MMACHC* and *TESK2* genes	Causative of epi-cblC disease	[4, 31, 35, 37]
Mucopolysaccharidosis I (#607014, #607015 and #607016, *IDUA*, AR)	Disorders of complex molecule degradation, Glycosaminoglycan degradation	MicroRNA	Differential expression of miR-20b and miR-106b in the cerebellum of MPSI mouse model correlated with reduction of *Neu1* expression; overexpression of miR-20b-5p; underexpression of miR-20b-3p and miR-106b-5p; alteration of miR-20b-3p/ miR-20b-5p ratio	Biomarker for neuropathology	[14]
Mucopolysaccharidoses IIIB and IVA (#252920, *NAGLU*, AR and #253000, *GALNS*, AR)	Disorders of complex molecule degradation, Glycosaminoglycan degradation	DNA methylation and histone modifications	Global DNA hypomethylation and increased histone acetylation (H3K14ac) in cell lines for both diseases; decreased histone methylation (H3K9me3) in MPS IIIB cells	Biomarker for pathogenesis; contribution to phenotypic heterogeneity	[15]
Multiple sulfatase deficiency (#272200, *SUMF1*, AR)	Disorders of complex molecule degradation, Sphingolipid degradation	MicroRNA	Up-regulation of miR-95 in human cell lines responsible for reduction of SUMF1 protein levels and sulfatase activity resulting in a severe block in autophagy-mediated degradation	Biomarker for disease pathophysiology	[17]
Niemann-Pick C (#257220, #607625, *NPC1* and *NPC2*, AR)	Disorders of complex molecule degradation, Other	DNA methylation	Decreased expression of DNMT3A and methyl-CpG domain binding proteins, reduced 5-methylcytosine immunoreactivity in the molecular and Purkinje cell layers, demethylation of genome-wide repetitive LINE-1 elements, and hypermethylation in specific promoter regions of single-copy genes in NPC1-deficient cerebellum of mouse model at early stages of the disease	Biomarker for neuropathophysiology	[18]

(continued)

Table 12.1 (continued)

IEM (OMIM entry, gene, inheritance)	Category, Group (according to the ICIMD)	Epigenetic modification	Main epigenetic findings	Significance of the epigenetic changes	Reference
Phenylketonuria (PKU); Hyperphenylalaninemia, non-PKU mild (#261600, *PAH, AR*)	Disorders of amino acid metabolism, Phe and Tyr	DNA methylation	In brain samples of PKU patients: enrichment of differentially methylated loci (predominantly hypermethylated); many loci encoded miRNAs. nverse correlation between the expression levels of miRNAs and target genes; promoter hypermethylation of genes encoding structural proteins and transporters resulting in gene silencing. In leukocytes: gene body hypermethylation was more marked in therapy noncompliant patients	Biomarker for neuropathophysiology and relative Phe exposure	[24]
			Extensive pattern of aberrant DNA methylation in brain tissues of hyperphenylalaninemic animals (PAH^{enu2} mice): hyper- and hypomethylated gene coding regions, gene body methylation of miRNAs, and promoter hypermethylation of genes with synaptic involvement resulting in down-regulation of their expression		[25]
		Long noncoding RNA	Depletion of the human lncRNA *HULC* and the mouse *Pair* correlated with the clinical and biochemical features of PKU; *HULC* modulated PAH enzyme activity (as a "therapeutic chaperone") by facilitating PAH-substrate and PAH-cofactor interactions; GalNAc-*HULC* mimics reduced excessive Phe and improved Phe tolerance in $Pair^{-/-}$ and $Pair^{R408W/R408W}$ mice	Biomarker for liver damage; possible therapeutic strategy	[26]
Pompe disease (#232300, *GAA*, AR)	Disorders of complex molecule degradation, Other	MicroRNA	Differential expression of several miRNAs in mouse model tissues and plasma of affected patients, potentially affecting the expression of genes involved in autophagy, muscle regeneration, and muscle atrophy; infantile patients showed higher levels of miR-133a which decreased during ERT	Biomarker for phenotype severity and ERT monitoring	[16]

Wilson disease (WD) (#277900, *ATP7B*, AR)	Disorders of trace elements and metals, Copper	DNA methylation	Global DNA hypomethylation, due to copper overload and copper-mediated inflammation, in tx-j mouse livers at early stages of disease which is associated with down-regulation of the hepatic Sahh and Dnmt3b enzymes; reduced transcript levels of selected genes involved in endoplasmic reticulum stress, lipid synthesis, and fatty acid oxidation	Biomarker for early liver damage	[19]
			DNA methylation patterns and transcriptional changes were influenced by maternal provision of methyl-group donor choline and copper chelator penicillamine in tx-j mice during pregnancy: choline prevented aberrant fetal hepatic gene expression in offspring; sexually dimorphic responses to treatments	Biomarker for treatment monitoring	[20]
			DNA methylation changes affecting oxidative stress pathways (*Txn1* and *Prdx1* genes) were corrected by postnatal choline supplementation in tx-j mice; TXN system-related genes were differentially methylated also in liver samples of affected patients	Biomarker for disease pathophysiology and treatment monitoring	[21]
			In liver samples: 969 hypermethylated and 871 hypomethylated DMRs which were exclusive for WD patients and could differentiate early from severe fibrosis; hypermethylation of the *HDAC5*. In blood: DMRs which could differentiate WD patients, healthy subjects, subjects with other liver disorders, early from advanced stages of disease, and organ phenotypes; hypermethylation of the *HDAC5*	Biomarker for diagnosis, progression, and organ-specific pathogenesis	[22]
		Histone acetylation and methylation	Increased histone acetylation (H3K9ac and H3K27ac marks) and decreased histone methylation (H3K9me3 and H3K27me3 marks) in tx-j mice; down-regulation of HDAC4 and HDAC5 enzymes and up-regulation of HAT1; aberrant expression of numerous genes involved in energy metabolism and cellular stress/development	Biomarker for disease progression	[23]

Abbreviations: *AR* autosomal recessive, *CCDC163P* coiled-coil domain containing 163, pseudogene, *COL4A1 and COL4A2* collagen type IV alpha-1 and alpha-2, *DMRs* differentially methylated regions, *DNMT3A and Dnmt3b* DNA methyltransferase 3 alpha and beta, *Epha2*, ephrin receptor A2, *ERT* enzyme replacement therapy, *GalNAc-HULC* N-acetylgalactosamine-*HULC*, *HAT1* histone acetyltransferase 1, *HDAC4* histone deacetylase 4, *HDAC5* histone deacetylase 5, *HULC* hepatocellular carcinoma up-regulated long noncoding RNA, *HUMARA* human androgen receptor, *ICIMD* international classification of inborn metabolic disorders, *LINE-1* long interspersed nuclear element-1, lncRNA, long noncoding RNA, *Lrat* lecithin retinol acyltransferase, *Nr2e3* nuclear receptor subfamily 2 group E member 3, *Pair* PAH-activating lincRNA, *Pax6* paired box gene 6, *Phe* phenylalanine, *PRDX1* peroxiredoxin 1, *Rara* retinoic acid receptor alpha, *Rpe65* retinoid isomerohydrolase, *Rdh5* retinol dehydrogenase 5, *Sahh* S-adenosylhomocysteine hydrolase, *SCARB-2* scavenger receptor class B member 2, *TESK2* testis-associated actin-modeling kinase 2, *Txn1* thioredoxin 1, *Tyr* tyrosine, *Vegfa* vascular endothelial growth factor A, *XL* X-linked

12.3 Episignatures May Contribute to the Clinical Heterogeneity, Pathophysiology, Prognosis, and Treatment Monitoring of IEMs

In this review, we used the term "Episignatures" for epigenomic alterations associated with a disease and "epimarkers" for episignatures which meet all the criteria of a biomarker.

Research on epigenomic alterations of IEMs in humans and in animals indicates that episignatures could help to shed light on the pathophysiology, differences in clinical phenotype and prognosis of affected patients, as well as be useful for the monitoring of treatment efficacy. Herein, we report the IEMs for which putative epivariations have been identified.

12.3.1 Lysosomal Storage Disorders

Lysosomal storage disorders (LSDs) are rare IEMs which cause lysosomal protein alteration due to monogenic defects. They are characterized by a wide spectrum of phenotypes, with onset in utero, in neonates, or later; sometimes onset is in adulthood. Despite the monogenetic nature of LSDs, clinical manifestations are heterogeneous and clear-cut genotype/phenotype correlations do not appear to exist. There are often discordant phenotypes among patients sharing the same genotype, even within the same family. Studies on the epigenetic mechanisms involved in LSDs, particularly miRNAs and DNA methylation, suggest that epigenetic factors may significantly modulate pathways governing lysosomal functions.

12.3.1.1 Fabry Disease

GLA pathogenic gene variants cause Fabry disease (FD), an X-linked sphingolipidosis characterized by total or partial deficiency of the lysosomal enzyme alpha-galactosidase A (GLA). GLA deficiency leads to the progressive accumulation of globotriaosylceramide (Gb3) and its metabolites in different organs and tissues, resulting in multisystemic clinical manifestations, including hypertrophic cardiomyopathy and renal insufficiency. Genetic heterogeneity, intrafamilial heterogeneity, and conflicting data on several variants associated with in vitro *GLA* enzyme deficiency or unknown significance (VUS) have been reported. It is often difficult to make phenotypic correlations with *GLA* variants, even in male patients within the same family. Male monozygotic twins expressing different progressions of FD have also been described. Phenotype expression in heterozygous FD females has been hypothesized to depend on X-chromosome inactivation (XCI) patterns. XCI is accompanied by changes in DNA methylation at CpG islands with transcriptional inactivation of several genes on one of the two X-chromosomes. However, there is conflicting data regarding the correlation between XCI and clinical phenotype in FD heterozygous females [5].

Using Fabry patient cellular models, Shen et al. [6] hypothesized that dysregulated DNA methylation plays a role in the development of FD. They found that changes in GLA activity and Gb3 accumulation led to altered DNA methylation in the promoter of the androgen receptor gene (*AR*; *HUMARA* according to the old nomenclature), which was associated with up-regulation of AR transcription in FD human cells. As AR encodes a ligand-dependent transcription factor that modulates the expression of androgen-regulated genes, an up-regulation of AR significantly impacts the expression of several genes. Genome-wide methylation analysis showed differentially methylated CpG site regions in the genome, including 15 relevant signaling pathways for FD. Integrative analysis of DNA methylation and RNA-seq led to the identification of 21 genes with both differential mRNA expression and methylation. Among them, collagen type IV alpha 1 and alpha 2 genes (*COL4A1* and *COL4A2*) exhibited an increased expression of transcripts, which correlated with their decreased methylation status. The authors hypothesized that *COL4A1* and *COL4A2* contribute to vasculopathy in FD. The high methionine levels detected in FD cells suggested that abnormal methionine metabolism affects DNA methylation homeostasis and may play a role in the pathogenic mechanisms determining the phenotypic heterogeneity of FD. However, there wasn't a direct correlation between the levels of methionine and the SAM/SAH ratio in FD cells [6].

MiRNAs could also play a role in the diagnosis, prognosis, and treatment of FD. Cammarata et al. [7] found that FD patients share a common miRNA plasma profile, including both up- and down-regulated miRNAs. Ten miRNAs discriminated between FD patients and controls. Specifically, the up-regulated miR199a-5p and miR-126-3p, involved in endothelial dysfunction, differentiated FD patients from subjects with left ventricular hypertrophy related to other disorders. The miR-423-5p (down-regulated) and the miR-451a (up-regulated) are likely to be associated with cardiac involvement. Patients who underwent enzyme replacement therapy (ERT) showed variations in plasma levels of miR199a-5p, miR-126-3p, and miR-423-5p, a finding that could be used for ERT monitoring [7]. In 2019, *Xiao* et al. [8] confirmed that circulating miRNAs in FD patients could be epimarkers for estimating ERT efficacy.

Gb3 and its derivate lysoGb3 are considered to be key markers in FD pathophysiology and are the targets of FD treatments. Current treatments cannot fully reverse the clinical manifestations of FD and other approaches should be investigated. For example, *Gambardella* et al. [9] investigated mitochondrial-related microRNAs (mito-miRNAs) as potential biomarkers of mitochondrial involvement in FD. Their data showed a dysregulation of mito-miRNAs in FD patients, which were involved in metabolism, biogenesis, oxidative damage, and apoptosis of mitochondria. Up-regulation of mito-miRNAs affecting the electron transport chain inhibited its expression and/or assembly. For instance, the up-regulated miR-181c and miR-338 induced down-regulation of the mitochondrial transport chain complexes I and IV, respectively [9].

12.3.1.2 Gaucher Disease

Gaucher disease (GD), a rare autosomal recessive LSD, is one of the most common sphingolipidoses caused by variants in the *GBA* gene, resulting in a deficiency of beta-glucocerebrosidase (GBA). One important modifier of GBA is the lysosomal integral membrane protein type 2 (LIMP-2) that is encoded by the *SCARB-2* gene. LIMP-2 is the specific receptor for GBA trafficking from the endoplasmic reticulum to the lysosome.

Siebert et al. [10] demonstrated that miRNA (miR-127-5p) down-regulates both the transcript and protein levels of the *SCARB-2* gene through interactions with its 3'UTR with consequent down-regulation of GBA activity correlated with improper GBA trafficking. Two additional miRNAs (miR-16-5p and miR-195-5p) up-regulate GBA activity by enhancing both expression and protein levels of the enzyme [10]. A neuropathic GD mouse model has revealed differential expression of miRNAs in the brain, primarily in the mesencephalic region [11]. In summary, miRNAs can modify GBA activity in GD patient cells and may also function as modifiers in GD with an impact on the biological mechanisms that lead to the neuropathic form.

GBA gene variants in GD patients or heterozygous carriers are the most common genetic risk factor for developing Parkinson disease. Clinical evaluation of sibling pairs discordant for Parkinsonism and with GD suggests that there may be other genetic, epigenetic, and/or environmentalfactors involved [12].

12.3.1.3 Krabbe Disease

Krabbe disease (KD) is an autosomal recessive LSD caused by a deficiency of the galactosylceramidase (GALC) enzyme, in which miRNAs may play a role in the brain pathophysiology. GALC catalyzes the breakdown of galactosylceramide, a major myelin lipid in developing oligodendrocytes (OLs). GALC deficiency leads to the accumulation of the cytotoxic psychosine (galactosylsphingosine), resulting in nervous system degeneration and inflammation. Most patients show early-onset demyelination with the death of apoptotic OL and exitus within 2 years of age. However, pathophysiological mechanisms in KD are still poorly understood.

The microRNA miR-219 is a regulator of OL development, myelination, and remyelination. By using twitcher mice, an authentic mouse model of KD, Inamura et al. [13] demonstrated that the expression and functional activity of miR-219 were reduced in the mouse OLs. They also showed that defects of cell-autonomous development, apoptotic death, and psychosine accumulation were restored by the exogenous supplementation of miR-219 [13]. We still do not fully understand the processes regulating miR-219 expression and which target genes are involved in KD pathophysiology.

12.3.1.4 Mucopolysaccharidosis I

Pathogenic variants in the *IDUA* gene are responsible for α-iduronidase (IDUA) enzyme deficiency, resulting in Mucopolysaccharidosis type I (MPS I), an autosomal recessive LSD. IDUA deficiency leads to ubiquitous accumulation of glycosaminoglycans (GAGs) and to multisystemic progressive disease. Clinical manifestations are widely heterogeneous and severely affected patients present developmental delay, cognitive impairment, and other systemic alterations. Secondary accumulation and abnormal distribution of gangliosides, mainly GM2 and GM3, have been described in the neuronal cells of MPS I patients and animal models. A reduced expression of neuraminidase (Neu1) was observed in the cerebellum of MPS I mice. Neu1 enzyme acts in a lysosomal multienzyme complex (LMC) with the β-galactosidase enzyme and the protective protein/cathepsin A and is involved in stepwise degradation of gangliosides.

Pereira et al. [14] analyzed LMC gene expression and the miRNA profiles of the miR-17 family, predicted to bind at least two LMC components, in the cerebellum of MPS I. They confirmed the significant underexpression of Neu1, whereas a normal expression of the other two LMC genes was present. The down-expression of Neu1 was linked to miR-20b and miR-106b which were differentially expressed in MPS I mice. By comparing the expression of both 3p and 5p strands for each miRNA in MPS I and control mice, the authors found that in MPS I mice, miR-20b-5p was overexpressed, while miR-20b-3p and miR-106b-5p were underexpressed. In addition, in the cerebellum of MPS I mice, the ratio between miR-20b-3p and miR-20b-5p was altered. The results of this work showed that miR-20b and miR-106-b directly contribute to the down-regulation of Neu1 expression and could function as modulators in the pathogenic cascades that underlie MPS I neuropathology [14].

12.3.1.5 Mucopolysaccharidoses IIIB and IVA

Mucopolysaccharidosis type IIIB (MPS IIIB or Sanfilippo B) and mucopolysaccharidosis IVA (MPS IVA or Morquio A) are autosomal recessive LSDs caused by pathogenetic variants in the *NAGLU* and *GALNS* genes, respectively. MPS IIIB results in a deficiency of the N-alpha-acetylglucosaminidase (NAGLU) that leads to the accumulation of the glycosaminoglycan heparan sulfate. The disease is characterized by progressive degeneration of the central nervous system and severe intellectual disability. MPS IVA is due to a deficiency of the N-acetylgalactosamine-6-sulfatase that causes the accumulation of keratan sulfate and chondroitin 6-sulfate. MPS IVA is characterized by skeletal dysplasia and patients exhibit mild-to-severe phenotypes.

Evidence for epigenetic factors has recently been reported in MPS IIIB and IVA by Vargas-Lopez et al. [15]. The authors describe the potential impact of DNA methylation and histone modifications on the phenotypic heterogeneity and pathogenesis observed in MPS IIIB and IVA. DNA methylation analysis showed global DNA hypomethylation in cell lines for both diseases, which may indicate a potential activation of normally silenced genes. DNA hypomethylation was linked to oxidative stress through the one-carbon cycle (i.e., decreased availability of methyl groups). Analysis of histone modifications showed an increased histone acetylation (H3K14ac, transcriptionally active chromatin marker) in both MPS IIIB and IVA cells and a reduced histone methylation (H3K9me3, transcriptionally silent heterochromatin marker) in MPS IIIB cells [15].

12.3.1.6 Pompe Disease

Glycogenosis type 2 or Pompe disease is caused by pathogenic variants in the *GAA* gene leading to deficiency of the lysosomal alpha-glucosidase (GAA) enzyme and the accumulation of glycogen mainly in the heart and skeletal muscles. Clinical heterogeneity is present with a phenotypic continuum starting from infantile onset. Identification of biomarkers is useful for diagnosis, prognosis, and treatment.

Tarallo et al. [16] analyzed miRNA expression by small RNA-seq in mouse model tissues (muscle and heart) and in the plasma of Pompe patients (GAA patients). They found differential expression of several miRNAs both in mouse models and in GAA patients. Sixteen of these dysregulated miRNAs were common between GAA patients and the mouse model. These miRNAs were predicted to regulate the expression of genes involved in pathways for Pompe disease pathophysiology, such as autophagy, muscle regeneration, and muscle atrophy. miR-133a, previously described as a biomarker for Duchenne/Becker muscular dystrophy, was significantly up-regulated in GAA patients, with higher levels in infants compared with late-onset patients. These findings suggested a correlation between miR-133a levels and phenotype severity. Importantly, in three infantile patients, the levels of miR-133a decreased after they started ERT, suggesting that miR-133a could represent a marker for ERT monitoring [16].

12.3.1.7 Multiple Sulfatase Deficiency

Sulfatases are enzymes that catalyze the hydrolysis of sulfate esters, such as GAGs and other macromolecules in lysosomes. Their catalytic residue, C-alpha-formylglycine (FGly), is post-translationally generated from a cysteine by sulfatase-modifying factor 1 (SUMF1). Genetic lesions in the *SUMF1* gene lead to a multiple sulfatase deficiency (MSD), a severe LSD.

Frankel et al. [17] identified the miRNA miR-95 in human cell lines that reduced SUMF1 protein levels and sulfatase activity, suppressing the proteoglycan catabolism and lysosomal function. Of note, the overexpression of miR-95 inhibited autophagy-mediated degradation with accumulation of autophagosomes and autophagic substrates in the cytosol. The authors demonstrated that *SUMF1* is a direct endogenous target of miR-95 as it interacts with its 3'UTR region. They also hypothesized that, by targeting miR-95 in cells from MSD patients, residual SUMF1 expression can be increased, allowing a reactivation of sulfatase activity and increased clearance of sulfated GAGs [17].

12.3.1.8 Niemann-Pick Type C

Clinical heterogeneity has also been reported in Niemann-Pick disease type C (NPC), a severe heterogeneous neurovisceral storage disease characterized by endolysosomal storage of unesterified cholesterol in the brain, liver, spleen, and other tissues. Both the *NPC1* and *NPC2* genes, responsible for NPC, encode proteins involved in the transport of cholesterol and lipids in cells. However, in most cases, NPC is caused by mutations in *NPC1*. The pathogenic mechanisms leading to neuronal dysfunction caused by an NPC protein defect are not well understood.

Kennedy et al. [18] reported imbalances in amino acid metabolism in the cerebellum of pre-symptomatic and early symptomatic NPC-1 mouse models, mainly associated with an increase in glutamine oxidation and alterations in one-carbon metabolism. The alterations in one-carbon metabolism impacted both folate and methylation pathways, which provide one carbon units for DNA methylation and influenced the epigenetic regulation of gene expression. Decreased expression of DNA methyltransferase 3a (DNMT3A) and methyl-CpG-domain binding proteins, reduced 5-methylcytosine immunoreactivity, demethylation of genome-wide repetitive LINE-1 elements, and hypermethylation in specific promoter regions of single-copy genes were observed in the early stages of the disease [18]. Overall, these changes may have functional consequences for neurotransmitter and nitrogen homeostasis and in the epigenetic regulation of gene expression, contributing to neurological decline and neurodegeneration in NPC disease.

12.3.2 Wilson Disease

Wilson disease (WD) is an autosomal recessive disorder caused by pathogenic variants in the *ATP7B* copper transporter gene. Consequent copper overload results in a variable and unpredictable clinical phenotype involving hepatic, neurologic, and psychiatric symptoms, without a firm genotype-phenotype correlation. One factor that may contribute to the pathogenesis of WD and explain the variability of affected patients is epigenetic regulation of gene expression.

Epigenetic changes in WD are triggered by copper overload, responses to disease progression, and environmental factors such as diet. It is well known that copper accumulation alters methionine metabolism by inhibiting the SAH hydrolase which leads to an accumulation of SAH that inhibits DNA methyltransferases. Global alterations in DNA methylation and expression of genes involved in liver injury in the tx-j mouse model of WD have been reported by Medici et al. [19, 20] who observed a similar pattern of differentially methylated genes in the liver of human WD patients, especially in genes of the thioredoxin (TXN) system, which is an important mediator of antioxidative responses [21]. In a subsequent study, the researchers better defined the epigenomic signature of WD patients by aiming at relevant disease mechanisms [22]. In liver specimens, they found 969 hypermethylated and 871 hypomethylated DMRs, which were exclusive to WD and could differentiate early from severe fibrosis in WD patients. WD-hypermethylated liver DMRs were enriched in liver-specific enhancers and transcription factor binding sites, suggesting that methylation changes have functional relevance by affecting gene expression regulation. DMRs detected in the blood reflected a subset of WD-specific liver DMRs and discriminated between WD patients, healthy subjects, and subjects with other liver diseases. Blood DMRs were also able to differentiate between stages of WD and between hepatic and neurologic phenotypes. These findings demonstrate that DNA methylation contributes to the progression and pathogenesis of WD and could be clinically useful for its diagnostic and prognostic potential.

Post-translational modifications of histones may also have a role in the pathogenesis of WD. The histone deacetylase 5 (HDAC5) gene has been significantly hypermethylated in WD liver and blood samples [22]. Sarode et al. [23] examined the potential link between copper overload, histone acetylation, and gene expression in the tx-j mouse model. They found reduced levels of histone deacetylase 4 and 5 (HDAC4/5) in 12- and 24-week tx-j mice compared with control animals. This reduction worsened with copper overload and was due to HDAC phosphorylation by the AMP-activated protein kinase that targeted HDAC4/5 proteins for cytosolic degradation. HDAC4 and HDAC5 down-regulation correlated with increased histone acetylation (H3K9ac and H3K27ac marks) and impaired methylation (H3K9me3 and H3K27me3). These findings suggest that a reduction in HDAC4/5 impacts the acetylation/methylation balance and leads to transcriptional reprogramming with consequent aberrant expression of numerous genes during WD progression. Chromatin immunoprecipitation sequencing (ChIP-Seq) integrated with RNA-seq showed altered epigenetic regulation and transcript levels of genes involved in energy metabolism and liver regeneration. The most relevant targets regulated by histone acetylation in tx-j mice were Pparα and Pparγ, two isoforms of peroxisome proliferator-activated receptors (PPARs), which are key regulators of lipid, glucose, and mitochondrial energy metabolisms.

12.3.3 Phenylketonuria

Phenylalanine hydroxylase (PAH)-deficient phenylketonuria (PKU) is one of the most well-known and frequent IEM (global prevalence 1:23,930 live births). It is known that overloaded phenylalanine (Phe) causes brain damage; however, the mechanisms underlying Phe toxicity in the brain are poorly defined.

An initial methylome analysis performed on offspring born of PHA^{enu2} mouse dams, used as a model of maternal PKU (MPKU), found that in utero Phe toxicity altered DNA methylation in the brain and had downstream impacts upon gene expression.

Dobrowolski et al. [24] investigated whether DNA methylation could also occur in PKU patients as a response to Phe overexposure. They performed methylated DNA immunoprecipitation sequencing (MeDIP-Seq) and expression studies on brain tissues from classic PKU patients and on blood from well- and poorly controlled PKU patients. A comparison of findings obtained from PKU and control brain tissues showed an enrichment of differentially methylated loci in the PKU group with a prevalence of hypermethylated loci (63%). Approximately 35% of these loci were noncoding RNA genes, mainly miRNAs. Both down- and up-regulated miRNAs were identified and there was an inverse relationship between miRNA expression and target gene expression. Differential promoter methylation in PKU brain tissues involved hypermethylation, which was correlated with the down-regulation of several genes encoding structural proteins and transporters. These results highlight that both overexpression of inhibitory miRNAs and promoter hypermethylation were the epigenetic mechanisms underlying the down-regulation of genes, which are potentially involved in the pathophysiology of PKU. Significant differences in methylation were also observed in the leukocytes of well- and poorly controlled patients, with gene body hypermethylation more skewed toward therapy noncompliant patients. DNA methylation patterns in leukocytes appear to be influenced by Phe exposure and have the potential to serve as a biomarker for relative Phe exposure [24]. Altered DNA methylation patterns associated with Phe exposure in humans resembled those of the PHA^{enu2} mouse model [25].

The noncoding RNA transcript from *HULC* (hepatocellular carcinoma up-regulated long noncoding RNA) associates with the human PAH enzyme and modulates its activity [26]. *HULC* depletion leads to reduced PAH enzymatic activities in human induced pluripotent stem cell-differentiated hepatocytes. Mechanistically, HULC facilitates PAH-substrate and PAH-cofactor interactions as a therapeutic chaperone.

12.3.4 Homocystinuria-Megaloblastic Anemia cblG Type

Homocystinuria-megaloblastic anemia cblG type is a rare cobalamin-related remethylation defect caused by biallelic pathogenic variants in the *MTR* gene, encoding the methionine synthase (MS). Affected patients show elevated plasma homocysteine (Hcy) and a wide spectrum of symptoms, including neurological manifestations, severe megaloblastic anemia, and ophthalmological abnormalities, which share similarities with those from methylmalonic aciduria and homocystinuria cblC type (cblC disease) due to *MMACHC* mutations and impaired intracellular processing of cobalamins (vitamin B12 isoforms) [27–29]. The current standard treatment is hydroxocobalamin injections, which improve metabolic and hematological parameters but have little to no effect on neurological and ocular symptoms. MS catalyzes the methyl-cobalamin dependent remethylation of homocysteine to methionine and plays a pivotal role in the methionine cycle of one-carbon metabolism, which is involved in generating the universal methyl donor and end-product (SAM and SAH) of epigenetic transmethylation reactions.

Hcy and its metabolites are known to impair epigenetic control of gene expression mediated by DNA and histone methylation. A recent epigenomic-wide study exploring alterations of DNA methylation in an original murine model with conditional Mtr deletion in retinal tissue (Mtr-cKO mouse model) demonstrated that cblG ocular disease is associated with a dramatic reduction of the SAM/SAH ratio and impaired DNA methylation at specific CpGs, predominantly clustered within CpG islands of promotors [30]. DNA methylation changes affected genes involved in eye development, cone differentiation, and retinoid metabolism. The authors reported promoter hypermethylation of the retinoic acid receptor alpha gene (*Rarα*), which is a known regulator of expression of the lecithin retinol acyltransferase gene (*Lrat*), and a reduced expression of *Lrat* and two other pivotal genes for eye retinoid metabolism, the retinoid isomerohydrolase (*Rpe65*) and the retinol dehydrogenase 5 (*Rdh5*). Down-regulation of the paired box gene 6 (*Pax6*), considered the master regulator of the eye, correlated with a severe delay in the opening of both eyes on postnatal day 14 observed in Mtr-cKO mice. These findings showed that altered

DNA methylation contributes to ocular pathophysiology in cblG. Ocular symptoms are also typical of other IEMs affecting the cobalamin metabolism pathway, such as cblC disease, which is the most common among these disorders. It is likely that cblG and cblC diseases share similar epigenomic changes for eye pathophysiology.

12.4 Epimutation Causative of IEM

An epimutation is a stable alteration of DNA methylation (and/or post-translational histone modification) transmitted from parents to offspring and directly involved in the underlying molecular mechanisms of the disease. "Constitutional" epimutations are defined as being derived from the germline and should be present in all tissues of an individual. In contrast, "somatic" epimutations arise in cells from a specific somatic tissue. A primary epimutation results from a direct disruption of epigenetic programming (e.g., DNA methylation or histone modification patterns) that is subsequently propagated as such by epigenetic inheritance (demonstrated in animals but not in humans). A secondary epimutation is initiated by a genetic defect that then leads to disruption of normal epigenetic programming that can be propagated by either genetic or epigenetic inheritance, or both [4]. To our knowledge, epi-cblC is the unique IEM caused by a clear-cut epimutation which produces the silencing of the *MMACHC* gene, which is mutated in cblC disease. Like cblG, cblC presents with both severe neurological and systemic metabolic abnormalities. Biallelic pathogenic variants in the *MMACHC* gene cause hyperhomocysteinemia, high levels of methylmalonic acid in plasma and urine, and decreased or normal production of methionine. Since this metabolic pattern is common to other inherited cobalamin disorders, multigene panel testing of at least 12 genes is needed to make differential diagnoses. Nevertheless, some affected cases remain without a conclusive molecular diagnosis, which has led researchers to look at other potential determining factors.

In 2018, *Gueant* et al. discovered an *MMACHC* epimutation in 7 cblC patients harboring only one heterozygous genetic variant in the *MMACHC* gene and named this disease "epi-cblC" [31]. They considered it a new disease in regard to the genetic/epigenetic presentation and clinical manifestations, with digenic epimutations in the *MMACHC* and *TESK2* genes secondary to a mutation in the *PRDX1* flanking gene [4]. The *MMACHC* epimutation consisted of 32 hypermethylated CpGs inside a CpG island encompassing the promoter and first exon of *MMACHC* and caused the silencing of the *MMACHC* gene by a mechanism of transcriptional interference (i.e., *in cis* suppression of one transcriptional process by another). Transcriptional interference in epi-cblC arises from an aberrant antisense transcription encompassing the *CCDC163P*/*MMACHC* promoter, which is triggered by a sequence variation in *PRDX1*, an antisense gene located at the 3′ end of *MMACHC*. Specifically, the *PRDX1* c.515-1G > T and c.515-2A > T variants affect the canonical splice acceptor site of intron 5 and cause the skipping of the last exon and the polyA transcription termination signal of *PRDX1*. Consequently, an aberrant extension of *PRDX1* antisense transcription occurs through the *MMACHC*,

CCDC163P, and *TESK2* genes (Fig. 12.1a). Antisense transcription may regulate gene expression after transcription initiation by direct collision of RNA polymerases in the bidirectional promoter, which matches *cis*-hypermethylation of the CpG islands in the *MMACHC* promoter with subsequent *MMACHC* repression (Fig. 12.1b). The epimutation is located in a trio of genes with the specific configuration *R1-F2-R3* (reverse *CCDC163P*-forward *MMACHC*-reverse *PRDX1*), which produces the silencing of *MMACHC* mediated by the *PRDX1* aberrant anti-sense transcription through the bidirectional promoter common to *MMACHC* and *CCDC163P* (Fig. 12.1b). Under this configuration, the writing of the epimutation is strongly influenced by the ratio of the transcription level of *R3* vs *F2* [4]. The *PRDX1*/*MMACHC* (*R3*/*F2*) ratio is very high in many tissues, including germ cells [4], making the *MMACHC* epimutation a constitutional epimutation. This explains that the epimutation is present in sperm and maintained through generations. The epivariation in the *CCDC163P*/*MMACHC* promoter, with a population frequency of ~1:950, is one of the most frequent among the 384 promoter hypermethylation events that have been identified in disease-causing OMIM genes in the epigenome-wide dataset of 23,116 individuals [3].

Epigenetic silencing caused by aberrant transcription has been previously described for Lynch syndrome (*MSH2* gene) and α-thalassemia (*HBA2* gene), in

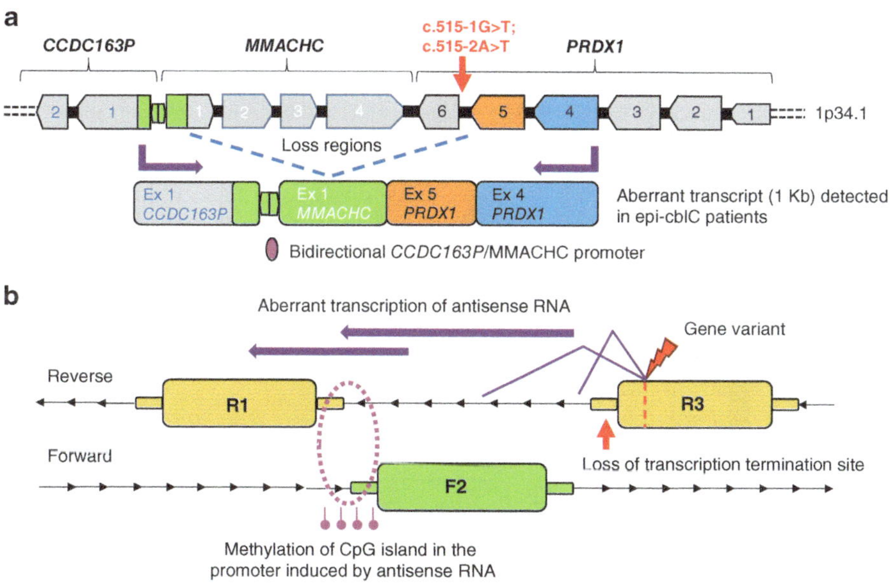

Fig. 12.1 (**a**) The *PRDX1* c.515-1G > T variant causes aberrant antisense transcription through the *MMACHC* and *CCDC163P* genes by skipping the last exon and the polyA transcription termination signal of *PRDX1*. The aberrant 1 Kb transcript detected in epi-*cblC* patients encompasses the *PRDX1* exons 4 and 5, part of exon 1 and promoter of *MMACHC*, and part of exon 1 of *CCDC163P*. (**b**) Schematic representation of the trio of genes composed by *CCDC163P-MMACHC-PRDX1* with the specific configuration *R1-F2-R3* and consequences of aberrant transcriptional elongation on F2 allele expression triggered by a splicing variant in R3

which the cause of transcriptional elongation is a genomic deletion (in epi-cblC it is a *PRDX1* splice site variant). These epimutations were somatic and erased in sperm [4, 32, 33].

ChIP-seq studies showed a H3K36me3 histone mark covering the *MMACHC* promoter in fibroblasts from epi-cblC patients compared to control cells. SETD2 is the single histone-lysine N-methyltransferase known to write the H3K36me3 mark, also in germ cells. As this histone mark may recruit DNA methyltransferases, it is likely that H3K36me3 subsequently triggers the de novo methylation of the *MMACHC* promoter (Fig. 12.2) [31].

So far, 21 epi-cblC cases have been described in the literature [31, 34–36]. Among them, eleven epi-cblC probands were reported in Italy by revisiting inconclusive cases with combined increased methylmalonic acid and homocysteine [35]. Ten cases had a monoallelic *MMACHC* epimutation due to the heterozygous *PRDX1* c.515-1G > T in association with a monoallelic *MMACHC* genetic variant. One patient identified at NBS, in which *MMACHC* sequencing did not find a pathogenic variant, had a biallelic *MMACHC* epimutation, due to the homozygous *PRDX1* c.515-1G > T transmitted by both parents (Fig. 12.3a, b). The biallelic epimutation correlated with a complete silencing of *MMACHC* in the patient's fibroblasts (Fig. 12.3c), a DNA full-methylated profile in the patient and a hemimethylated profile in his parents (Fig. 12.3d). We identified two other cases with homozygous *MMACHC* epimutation caused by the c.515-1G > T variant, including one in France and one in Italy (data unpublished). Retrospective analyses of the Italian cohort indicated that epi-cblC disease was more frequent than thought, accounting for about 13% of cblC cases diagnosed by NBS in the Tuscany and Umbria regions and

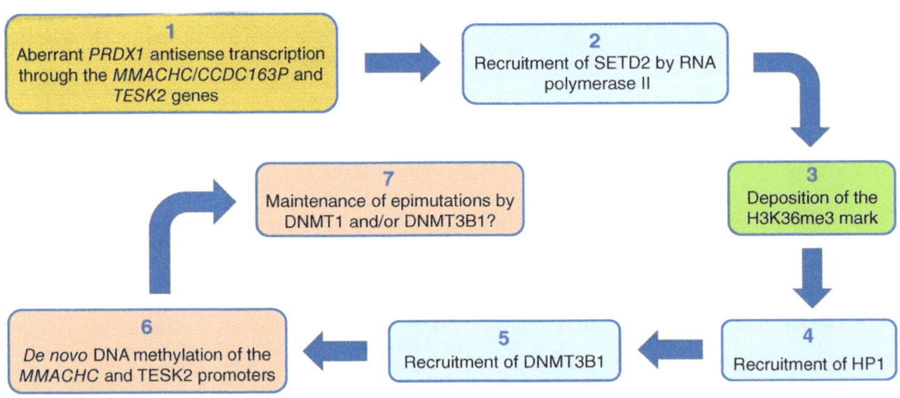

Fig. 12.2 Proposed scheme for the pathomechanism that produces the writing of the epimutation in epi-*cblC*. High levels of antisense transcript caused by the *PRDX1* variant produce the H3K36me3 histone mark. This mark is specifically synthesized by the SET domain-containing 2 (SETD2) enzyme, a histone lysine methyltransferase which catalyzes the trimethylation of lysine 36 in histone H3 by a direct interaction with RNA polymerase II during elongation. In turn, the H3K36me3 mark binds the heterochromatin protein 1 (HP1), which recruits the DNA Methyltransferase 3 beta, isoform 1 (DNMT3B1). Finally, DNMT3B1 induces de novo methylation of the CpG island of the *MMACHC* promoter. Whether DNA Methyltransferase 1 (DNMT1) and DNMT3B independently maintain the promoter methylation or collaborate themselves is yet unclear

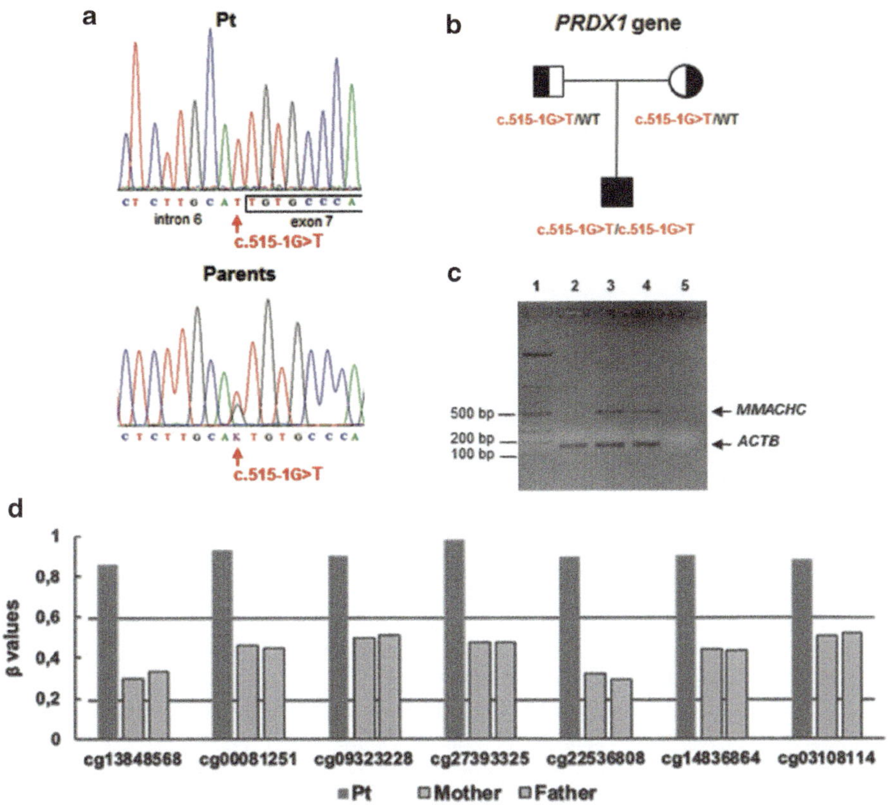

Fig. 12.3 Genetic and epigenetic results of the homozygous *PRDX1* patient (Pt). (**a**) *PRDX1* Sanger sequencing showing the c.515-1G > T pathogenic variant in DNA samples of Pt (at a homozygous level) and his parents (at a heterozygous level). (**b**) Pedigree of the family. (**c**) Multiplex RT-PCR showing amplification of *MMACHC* and *ACTB* cDNAs (fragments: 512 bp and 174 bp, respectively). *ACTB* was used as a housekeeping gene. Lane 1: molecular marker; lane 2: Pt; lane 3 and 4: normal controls; lane 5: no cDNA template. (**d**) Methylation levels of the *CCDC163P/MMACHC* bidirectional promoter in Pt and his parents. All the reported CpG probes exhibit a full-methylated profile in Pt (β values above 0.6) and a hemimethylated profile in the parents (β values between 0.2 and 0.6)

for 11% of the 104 genetically characterized Italian cblC probands [35]. These findings suggest that *MMACHC* epimutation/*PRDX1* mutation analyses should be part of routine genetic testing for all patients presenting with a metabolic phenotype that combines methylmalonic aciduria and homocystinuria.

In 2022, *Oussalah* et al. showed that the aberrant transcription triggered by *PRDX1* splicing variants also produces an epimutation in the promoter of *TESK2*, an antisense gene neighboring *CCDC163P* [37]. In epi-cblC cases, the authors observed a dramatic increase in *PRDX1* aberrant transcripts, together with a reduction of both *MMACHC* and *TESK2* mRNAs and an accumulation of H3K36me3 mark. They showed that the readthrough transcription of *PRDX1* is also prolonged through the *TESK2* promoter, with the consequent silencing of *TESK2* by the H3K36me3 deposition (Fig. 12.2).

Current knowledge on epi-cblC disease suggests that the clinical manifestations of epi-cblC patients are similar to those of canonical cblC patients [35]. However, the two epi-cblC homozygous patients appear to have a more severe clinical phenotype than *MMACHC*:c.271dup homozygotes, especially in regard to a more severe hypomethioninemia, which is known to correlate with poorer neurodevelopmental outcomes if concomitant with a higher peak of Hcy in the neonatal period.

In epi-cblC, the main phenotype of the disease is due to an epimutation in the *CCDC163P/MMACHC* promoter, but the silencing of *TESK2* and functional consequences of the *PRDX1* splicing variant on the *PRDX1* protein may significantly influence the disease course. *PRDX1* encodes peroxiredoxin 1, an antioxidant enzyme that reduces hydrogen peroxide into water through a 2-cysteine catalytic mechanism involving the Cys52 residue and the Cys173. As Cys173 is located in the last exon of *PRDX1*, which is skipped due to *PRDX1* splicing variants, a PRDX1 enzymatic defect is expected in homozygous patients. As PRDX1 protein has been shown to inhibit the c-Abl and c-Myc oncoproteins and to enhance natural killer cell cytotoxicity, its functional impairment could potentially lead to an increased risk of cancer. *TESK2* encodes a protein kinase called testis-associated actin-modeling kinase 2 (TESK2), mainly expressed in the testis and prostate, which plays an important role in spermatogenesis. A low expression of TESK2 protein has been associated with poor survival in patients with lung adenocarcinoma and premetastatic stage in human lung-to-brain metastasis. The potential clinical consequences of *TESK2* silencing on spermatogenesis and an increased cancer risk related to the functional defect of both TESK2 and PRDX1 proteins suggest that epi-cblC should be considered a clinical entity that differs from cblC.

12.5 Epigenetics for the Treatment of IEMs: Future Perspectives

Available treatments for IEMs cannot completely resolve clinical symptoms; thus, developing new therapeutic approaches is essential to improving and personalizing patient care. Increasing evidence links IEMs to epigenetic changes that are primarily reversible and could be targeted to counteract aberrant gene expression.

Drugs able to correct epigenomic alterations ("epi-drugs"), administered alone or in combination with existing treatments, could significantly improve survival rates and quality of life for patients. To our knowledge, no preclinical or clinical trials are currently in progress for epigenetic therapy in IEMs. To date, only a few studies on epigenetic therapeutic approaches have been reported in vitro, using animal models and a very limited number of patients. Herein, we report some of these studies and discuss potential future directions for research into epigenetic therapies for IEMs.

In Fabry disease, ERT is still the standard of care for the classic form, but its beneficial effects are reduced in cases with advanced tissue injury, in the presence of oxidative stress and persistent abnormal immune responses linked to progressive Gb3 storage. Apabetalone is a clinical-stage oral inhibitor of bromodomain and extra-terminal domain (BET) proteins, which has been shown to improve the inflammation pathway related to cardiovascular and kidney disease. BET proteins are epigenetic "readers" acting as regulators in the transcription of inflammatory genes by binding to acetylated histones and other nuclear proteins needed for the regulation of gene expression. In stimulated peripheral blood mononuclear cells, apabetalone dose-dependently decreased the production of chemokine monocyte chemoattractant protein-1 (MCP-1) and interleukin-12 (IL-12), countering the inflammatory state and oxidative stress in Fabry patients receiving ERT [38]. Apabetalone could potentially be used in combination with ERT in Fabry patients.

Cerebral X-linked adrenoleukodystrophy (CALD), caused by *ABCD1* gene variants, is a neurodegenerative disease characterized by the accumulation of very long chain fatty acids (VLCFAs) in peroxisomes, leading to devastating cerebral inflammatory demyelination. Neuroinflammation can be halted by hematopoietic stem cell transplantation (HSCT) or gene therapy (HSCGT), but these therapies have limited effects in patients with more advanced disease. *ABCD2*, the closest homolog of *ABCD1,* is epigenetically repressed by histone modifications (H3K4me3, H3K27ac, and H3K36me3 marks) in monocytes/macrophages, which are the most severely affected HSC-derived immune cells. It has been reported that Vorinostat, a histone deacetylase inhibitor (HDACi) used as an anti-cancer agent, compensates for the metabolic defect in X-ALD monocytes/macrophages by enhancing *ABCD2* expression, peroxisomal ß-oxidation and, in one CALD patient, ameliorating VLCFA accumulation [39].

Additional HDACis could potentially be used as epigenetic treatments for IEMs. For instance, Trichostatin (TSA) has been shown to rescue histone acetylation changes in cells from patients affected by MPS IVA and Niemann-Pick type C [40]. Panobinostat is also able to normalize cholesterol homeostasis in fibroblasts from Niemann-Pick type C patients [40].

As miRNAs are deregulated in PKU and several LSDs, their modulation could potentially be used in novel therapeutic strategies [40].

In Wilson disease, some studies suggest that supplementation with the methyl-donor choline may recover impaired DNA methylation and histone acetylation/methylation imbalance, improving gene expression [22, 23]. The consequent transcriptional reprogramming counteracts the hepatic presentation and progression of the disease. This makes choline, either alone or combined with the copper chelator D-penicillamine, a potential adjunctive treatment for Wilson disease.

A new emerging field for the treatment of IEMs is CRISPR-based epigenome editing, which combines epigenetic effectors with a nuclease-null deactivated (or "dead") Cas9 protein to directly modify the epigenomic signature in a site-specific manner, leaving the primary DNA sequence intact. Depending on their functions, epigenetic effectors act as "writers" (enzymes that deposit modifications: DNA methyltransferases, histone and lysine/methyltransferases) or "erasers" (enzymes that delete modifications: histone deacetylases/demethylases and enzymes involved in removing DNA methylation) [40, 41]. Epigenomic editing has several advantages, such as the enhancement of residual activity, functional compensation, precise and temporal control of endogenous expression, targeting several genes simultaneously and, importantly, greater safety than gene editing, due to the maintenance of intact DNA sequences and reversible effects. However, despite the promising clinical potential of epigenomic editing, there remain obstacles to its introduction into clinical practice, such as inefficient delivery, immune response, and off-target effects.

The in vivo development of CRISPR-based epigenome editing is an exciting development in the potential treatment of IEMs with an X-linked inheritance, such as X-linked adrenoleukodystrophy, ornithine transcarbamylase deficiency, Danon disease, Mucopolysaccharidosis type II, Fabry disease, etc. In affected females, these diseases arise from the expression of a faulty allele on the active X chromosome, while the inactive X chromosome contains a functional copy of the gene. Epigenetic editing could reactivate the silent functional allele from the inactive X chromosome by complementing the mutant allele. Recent evidence obtained from studies on human embryonic stem cells and derived neurons from patients with Rett syndrome provides proof-of-concept for the use of DNA methylation editing to treat X-linked diseases.

Acknowledgments We gratefully thank the AMMeC (Associazione Malattie Metaboliche Congenite, Italia) for their continuing support.

References

1. Ferreira CR, Rahman S, Keller M, Zschocke J, ICIMD Advisory Group (2021) An international classification of inherited metabolic disorders (ICIMD). J Inherit Metab Dis [Internet]. 44:164–177. Available from: http://www.ncbi.nlm.nih.gov/pubmed/33340416
2. Li X, Egervari G, Wang Y, Berger SL, Lu Z (2018) Regulation of chromatin and gene expression by metabolic enzymes and metabolites. Nat Rev Mol Cell Biol [Internet] 19:563–578. Available from: https://www.nature.com/articles/s41580-018-0029-7
3. Garg P, Jadhav B, Rodriguez OL, Patel N, Martin-Trujillo A, Jain M et al (2020) A survey of rare epigenetic variation in 23,116 human genomes identifies disease-relevant Epivariations and CGG expansions. Am J Hum Genet [Internet] 107:654–669. Available from: https://linkinghub.elsevier.com/retrieve/pii/S0002929720302883
4. Guéant J-L, Siblini Y, Chéry C, Schmitt G, Guéant-Rodriguez R-M, Coelho D et al (2022) Epimutation in inherited metabolic disorders: the influence of aberrant transcription in adjacent genes. Hum Genet [Internet]. 141:1309–1325. Available from: https://link.springer.com/10.1007/s00439-021-02414-9
5. Di Risi T, Vinciguerra R, Cuomo M, Della Monica R, Riccio E, Cocozza S et al (2021) DNA methylation impact on Fabry disease. Clin Epigenetics [Internet]. 13:24. Available from: http://www.ncbi.nlm.nih.gov/pubmed/33531072
6. Shen J-S, Balaji U, Shigeyasu K, Okugawa Y, Jabbarzadeh-Tabrizi S, Day TS et al (2022) Dysregulated DNA methylation in the pathogenesis of Fabry disease. Mol Genet Metab reports [Internet] 33:100919. Available from: http://www.ncbi.nlm.nih.gov/pubmed/36186841
7. Cammarata G, Scalia S, Colomba P, Zizzo C, Pisani A, Riccio E et al (2018) A pilot study of circulating microRNAs as potential biomarkers of Fabry disease. Oncotarget [Internet] 9:27333–27345. Available from: http://www.ncbi.nlm.nih.gov/pubmed/29937989

8. Xiao K, Lu D, Hoepfner J, Santer L, Gupta S, Pfanne A et al (2019) Circulating microRNAs in Fabry disease. Sci Rep [Internet] 9:15277. Available from: http://www.ncbi.nlm.nih.gov/pubmed/31649303
9. Gambardella J, Fiordelisi A, Sorriento D, Cerasuolo F, Buonaiuto A, Avvisato R et al (2023) Mitochondrial microRNAs are dysregulated in patients with Fabry disease. J Pharmacol Exp Ther [Internet] 384:72–78. Available from: http://www.ncbi.nlm.nih.gov/pubmed/35764328
10. Siebert M, Westbroek W, Chen Y-C, Moaven N, Li Y, Velayati A et al (2014) Identification of miRNAs that modulate glucocerebrosidase activity in Gaucher disease cells. RNA Biol [Internet] 11:1291–1300. Available from: http://www.ncbi.nlm.nih.gov/pubmed/25584808
11. Dasgupta N, Xu Y-H, Li R, Peng Y, Pandey MK, Tinch SL et al (2015) Neuronopathic Gaucher disease: dysregulated mRNAs and miRNAs in brain pathogenesis and effects of pharmacologic chaperone treatment in a mouse model. Hum Mol Genet [Internet] 24:7031–7048. Available from: http://www.ncbi.nlm.nih.gov/pubmed/26420838
12. Lopez G, Steward A, Ryan E, Groden C, Wiggs E, Segalà L et al (2020) Clinical evaluation of sibling pairs with gaucher disease discordant for parkinsonism. Mov Disord [Internet] 35:359–365. Available from: http://www.ncbi.nlm.nih.gov/pubmed/31785030
13. Inamura N, Go S, Watanabe T, Takase H, Takakura N, Nakayama A et al (2021) Reduction in miR-219 expression underlies cellular pathogenesis of oligodendrocytes in a mouse model of Krabbe disease. Brain Pathol [Internet] 31:e12951. Available from: http://www.ncbi.nlm.nih.gov/pubmed/33822434
14. Pereira VG, Queiroz MT, D'Almeida V (2016) Differential expression of microRNAs from miR-17 family in the cerebellum of mucopolysaccharidosis type I mice. Gene [Internet] 595:207–211. Available from: http://www.ncbi.nlm.nih.gov/pubmed/27720939
15. Vargas-López V, Prada LF, Alméciga-Díaz CJ (2024) Evidence of epigenetic landscape shifts in mucopolysaccharidosis IIIB and IVA. Sci Rep [Internet]. 14:3961. Available from: http://www.ncbi.nlm.nih.gov/pubmed/38368436
16. Tarallo A, Carissimo A, Gatto F, Nusco E, Toscano A, Musumeci O et al (2019) microRNAs as biomarkers in Pompe disease. Genet Med [Internet] 21:591–600. Available from: http://www.ncbi.nlm.nih.gov/pubmed/29997386
17. Frankel LB, Di Malta C, Wen J, Eskelinen E-L, Ballabio A, Lund AH (2014) A non-conserved miRNA regulates lysosomal function and impacts on a human lysosomal storage disorder. Nat Commun [Internet]. 5:5840. Available from: http://www.ncbi.nlm.nih.gov/pubmed/25524633
18. Kennedy BE, Hundert AS, Goguen D, Weaver ICG, Karten B (2016) Presymptomatic alterations in amino acid metabolism and DNA methylation in the cerebellum of a murine model of Niemann-pick type C disease. Am J Pathol [Internet] 186:1582–1597. Available from: http://www.ncbi.nlm.nih.gov/pubmed/27083515
19. Medici V, Shibata NM, Kharbanda KK, LaSalle JM, Woods R, Liu S et al (2013) Wilson's disease: changes in methionine metabolism and inflammation affect global DNA methylation in early liver disease. Hepatology [Internet] 57:555–565. Available from: https://onlinelibrary.wiley.com/doi/10.1002/hep.26047
20. Medici V, Kieffer DA, Shibata NM, Chima H, Kim K, Canovas A et al (2016) Wilson disease: epigenetic effects of choline supplementation on phenotype and clinical course in a mouse model. Epigenetics [Internet] 11:804–818. Available from: http://www.ncbi.nlm.nih.gov/pubmed/27611852
21. Mordaunt CE, Shibata NM, Kieffer DA, Czlonkowska A, Litwin T, Weiss KH et al (2018) Epigenetic changes of the thioredoxin system in the tx-j mouse model and in patients with Wilson disease. Hum Mol Genet [Internet]. 27:3854–3869. Available from: http://www.ncbi.nlm.nih.gov/pubmed/30010856
22. Mordaunt CE, Kieffer DA, Shibata NM, Członkowska A, Litwin T, Weiss K-H et al (2019) Epigenomic signatures in liver and blood of Wilson disease patients include hypermethylation of liver-specific enhancers. Epigenetics Chromatin [Internet] 12:10. Available from: http://www.ncbi.nlm.nih.gov/pubmed/30709419

23. Sarode GV, Neier K, Shibata NM, Shen Y, Goncharov DA, Goncharova EA et al (2021) Wilson disease: intersecting DNA methylation and histone acetylation regulation of gene expression in a mouse model of hepatic copper accumulation. Cell Mol Gastroenterol Hepatol [Internet] 12:1457–1477. Available from: https://linkinghub.elsevier.com/retrieve/pii/S2352345X21001144
24. Dobrowolski SF, Lyons-Weiler J, Spridik K, Biery A, Breck J, Vockley J et al (2015) Altered DNA methylation in PAH deficient phenylketonuria. Mol Genet Metab [Internet]. 115:72–77. Available from: https://linkinghub.elsevier.com/retrieve/pii/S109671921500092X
25. Dobrowolski SF, Lyons-Weiler J, Spridik K, Vockley J, Skvorak K, Biery A (2016) DNA methylation in the pathophysiology of hyperphenylalaninemia in the PAH(enu2) mouse model of phenylketonuria. Mol Genet Metab [Internet]. 119:1–7. Available from: http://www.ncbi.nlm.nih.gov/pubmed/26822703
26. Li Y, Tan Z, Zhang Y, Zhang Z, Hu Q, Liang K et al (2021) A noncoding RNA modulator potentiates phenylalanine metabolism in mice. Science (80-). Interne 373:662–673. Available from: https://www.science.org/doi/10.1126/science.aba4991
27. Huemer M, Diodato D, Schwahn B, Schiff M, Bandeira A, Benoist J et al (2017) Guidelines for diagnosis and management of the cobalamin-related remethylation disorders cblC, cblD, cblE, cblF, cblG, cblJ and MTHFR deficiency. J Inherit Metab Dis [Internet] 40:21–48. Available from: https://onlinelibrary.wiley.com/doi/10.1007/s10545-016-9991-4
28. Wiedemann A, Oussalah A, Lamireau N, Théron M, Julien M, Mergnac J-P et al (2022) Clinical, phenotypic and genetic landscape of case reports with genetically proven inherited disorders of vitamin B12 metabolism: a meta-analysis. Cell Reports Med [Internet] 3:100670. Available from: https://linkinghub.elsevier.com/retrieve/pii/S2666379122002026
29. Matmat K, Guéant-Rodriguez R-M, Oussalah A, Wiedemann-Fodé A, Dionisi-Vici C, Coelho D et al (2022) Ocular manifestations in patients with inborn errors of intracellular cobalamin metabolism: a systematic review. Hum Genet [Internet] 141:1239–1251. Available from: https://link.springer.com/10.1007/s00439-021-02350-8
30. Matmat K, Conart J-B, Graindorge P-H, El Kouche S, Hassan Z, Siblini Y et al (2023) A transgenic mice model of retinopathy of cblG-type inherited disorder of one-carbon metabolism highlights epigenome-wide alterations related to cone photoreceptor cells development and retinal metabolism. Clin Epigenetics [Internet]. 15:158. Available from: https://clinicalepigeneticsjournal.biomedcentral.com/articles/10.1186/s13148-023-01567-w
31. Guéant J-L, Chéry C, Oussalah A, Nadaf J, Coelho D, Josse T et al (2018) A PRDX1 mutant allele causes a MMACHC secondary epimutation in cblC patients. Nat Commun [Internet] 9:67. Available from: https://www.nature.com/articles/s41467-017-02306-5
32. Ligtenberg MJL, Kuiper RP, Geurts van Kessel A, Hoogerbrugge N (2013) EPCAM deletion carriers constitute a unique subgroup of Lynch syndrome patients. Fam Cancer [Internet] 12:169–174. Available from: http://link.springer.com/10.1007/s10689-012-9591-x
33. Tufarelli C, Stanley JAS, Garrick D, Sharpe JA, Ayyub H, Wood WG et al (2003) Transcription of antisense RNA leading to gene silencing and methylation as a novel cause of human genetic disease. Nat Genet [Internet] 34:157–165. Available from: https://www.nature.com/articles/ng1157
34. Pollini L, Tolve M, Nardecchia F, Galosi S, Carducci C, di Carlo E et al (2020) Multiple sclerosis and intracellular cobalamin defect (MMACHC/PRDX1) comorbidity in a young male. Mol Genet Metab Reports [Internet] 22:100560. Available from: https://linkinghub.elsevier.com/retrieve/pii/S2214426919302095
35. Cavicchi C, Oussalah A, Falliano S, Ferri L, Gozzini A, Gasperini S et al (2021) PRDX1 gene-related epi-cblC disease is a common type of inborn error of cobalamin metabolism with mono- or bi-allelic MMACHC epimutations. Clin Epigenetics [Internet]. 13:137. Available from: https://clinicalepigeneticsjournal.biomedcentral.com/articles/10.1186/s13148-021-01117-2
36. Zhang X, Chen Q, Song Y, Guo P, Wang Y, Luo S et al (2021) Epimutation of MMACHC compound to a genetic mutation in cblC cases. Mol Genet genomic Med [Internet] 9:e1625. Available from: http://www.ncbi.nlm.nih.gov/pubmed/33982424

37. Oussalah A, Siblini Y, Hergalant S, Chéry C, Rouyer P, Cavicchi C et al (2022) Epimutations in both the TESK2 and MMACHC promoters in the epi-cblC inherited disorder of intracellular metabolism of vitamin B12. Clin Epigenetics [Internet] 14:52. Available from: https://clinicalepigeneticsjournal.biomedcentral.com/articles/10.1186/s13148-022-01271-1
38. Fu L, Wasiak S, Tsujikawa LM, Rakai BD, Stotz SC, Wong NCW et al (2022) Inhibition of epigenetic reader proteins by apabetalone counters inflammation in activated innate immune cells from Fabry disease patients receiving enzyme replacement therapy. Pharmacol Res Perspect [Internet] 10:e00949. Available from: http://www.ncbi.nlm.nih.gov/pubmed/35417091
39. Zierfuss B, Weinhofer I, Kühl J-S, Köhler W, Bley A, Zauner K et al (2020) Vorinostat in the acute neuroinflammatory form of X-linked adrenoleukodystrophy. Ann Clin Transl Neurol [Internet] 7:639–652. Available from: http://www.ncbi.nlm.nih.gov/pubmed/32359032
40. Leal AF, Fnu N, Benincore-Flórez E, Herreño-Pachón AM, Echeverri-Peña OY, Alméciga-Díaz CJ et al (2023) The landscape of CRISPR/Cas9 for inborn errors of metabolism. Mol Genet Metab [Internet] 138:106968. Available from: http://www.ncbi.nlm.nih.gov/pubmed/36525790
41. Rutten MGS, Rots MG, Oosterveer MH (2020) Exploiting epigenetics for the treatment of inborn errors of metabolism. J Inherit Metab Dis [Internet]. 43:63–70. Available from: https://onlinelibrary.wiley.com/doi/10.1002/jimd.12093

Epigenetics and Psychiatric Diseases

13

M. Manchia, P. Paribello, and B. Chaumette

Contents

13.1 Introduction 220
13.1.1 DNA Methylation 222
13.1.2 Histone Acetylation and Methylation 224
13.1.3 Non-coding RNA 225
13.2 Epigenetic in Bipolar Disorder 226
13.2.1 Methylation Profile in BD 226
13.2.2 Histone Modifications in BD 226
13.2.3 ncRNA in BD 226
13.3 Epigenetic in Major Depressive Disorder 226
13.3.1 Methylation Profile in MDD 227
13.3.2 ncRNA in MDD 227
13.4 Epigenetics in Schizophrenia 228
13.4.1 Methylation Profile in SCZ 228
13.4.2 Histone Modifications in SCZ 229
13.4.3 ncRNA in SCZ 229
13.5 Conclusions 230
References 230

M. Manchia (✉) · P. Paribello
Section of Psychiatry, Department of Medical Sciences and Public Health, University of Cagliari, Cagliari, Italy

Unit of Clinical Psychiatry, University Hospital Agency of Cagliari, Cagliari, Italy
e-mail: mirkomanchia@unica.it

B. Chaumette
Université de Paris, Institute of Psychiatry and Neuroscience of Paris (IPNP), Paris, France

GHU Paris Psychiatrie et Neurosciences, Hôpital Sainte Anne, Paris, France

Department of Psychiatry, McGill University, Montreal, Canada

L. Emmi, V. Fanos (eds.), *Epigenetics and Rare Diseases*, Rare Diseases of the Immune System, https://doi.org/10.1007/978-3-032-02902-7_13

13.1 Introduction

Psychiatric disorders represent complex, vaguely defined groups of common conditions causing a particularly relevant disability burden in the general population [1, 2]. The etiology of psychiatric disorders remains to this day obscure. According to our current understanding of the pathophysiology of mental diseases, these are best conceptualized as a possible result of the interaction between biological and environmental factors [3]. Since the late eighteenth century in Europe and in the United States, the inheritance of mental disorders has been recognized as being particularly prominent. Indeed, the study of the inheritance of mental disorders has been argued to be the real origin of genetics in the medical field [4]. Prominent psychiatrists or alienists such as Esquirol, and the physician Thurnam started recording data regarding family trees for mental illness. Initially these efforts derived presumably from a genuine interest in describing the origin of mental illness and only at a later time turned to a more sinister direction, leading the way to eugenics with the tragic consequences this had during the following centuries [4]. **Inheritance** for severe mental illness such as schizophrenia has been estimated to be around 80%, with inheritance being defined as the cause of variation for the diagnosis in a given population [5]. The frequency of concordance of diagnosis between proband and relatives tends to increase in a non-linear fashion considering progressively closer relatives, a pattern typical of conditions with a multifactorial etiology [6]. This pattern suggests the presence of multiple, unknown elements interacting with inheritable factors in influencing the overall risk in a given individual (Fig. 13.1).

Common number variants account for a sizeable portion of the said inheritance for schizophrenia (24.4%) [9], with similar figures being also reported for bipolar disorder (with an estimated 70% heritability, which is explained by 25% by common genetic variants, respectively [10, 11].

According to our current nosological systems for psychiatric disorders, mental health conditions are defined based on three main elements: (1) signs, (2) symptoms, and (3) illness trajectories. Importantly, the current nosological systems are agnostic relative to the possible etiologies for the described mental health conditions. The underlying assumption would be that a specific set of clinical features so defined would be sufficient to distinguish between different mental health conditions. With a certain degree of variations in the criteria applied, this framework provided the basis for all major diagnostic classification systems available to date for psychiatric nosology [12]. Since the introduction of the Diagnostic and Statistical Manual of Mental Disorder—III Edition, this framework has offered a diagnostic construct with acceptable reliability, allowing for the advancement of the field in epidemiological studies and psychiatric research for the etiology of mental illness and for testing the efficacy of potential interventions [13]. Despite the usefulness of this framework for clinicians and researchers alike, it is increasingly evident how applying a categorical approach to natural phenomena may represent quite a significant constriction for understanding natural phenomena. First, specific signs and symptoms may not necessarily correlate with a specific diagnosis. In the twentieth century, much interest was devoted to classifying different types of fever with the

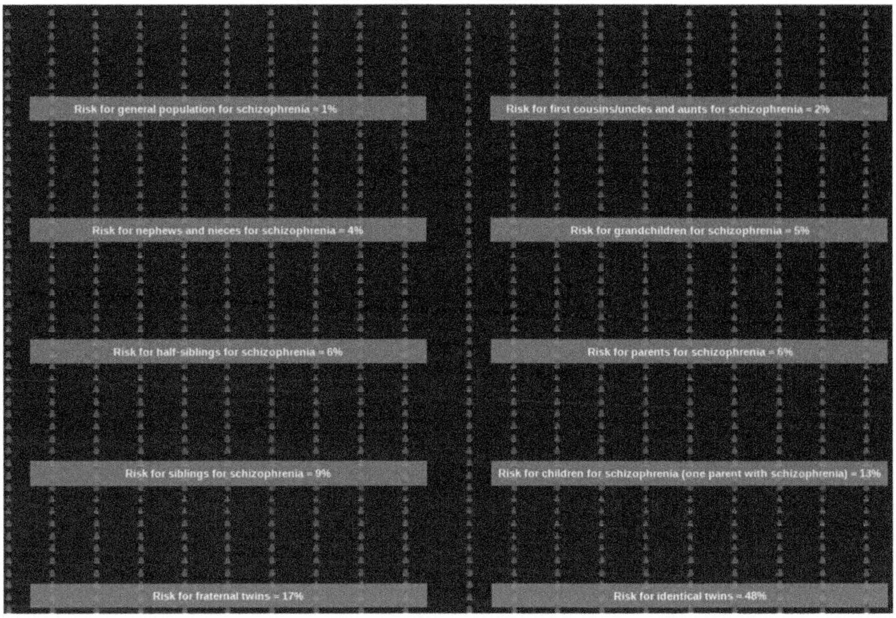

Fig. 13.1 Risk of receiving a diagnosis of schizophrenia for family members of probands with a formal schizophrenia diagnosis. Adapted from Gottesman 1991 fig. 10 - Schizophrenia Genesis: The Origins of Madness [7]. Waffle plot inspired by the code shared by Murray et al. 2021 [8]

hypothetical purpose of defining distinct pathological conditions. However, only in later times did it become apparent how this approach would not have resulted in a clear advancement in understanding underlying causes nor in focusing treatment therapeutic efforts as fever merely represents a non-specific reaction common to various conditions [12]. The specificity of the provided criteria is also lacking, with many symptoms and signs shared among different conditions. Not surprisingly, a corollary is represented by the frequent level of co-morbidity between different psychiatric conditions, which has been increasing steadily during the last decades [14]. The relative lack of reliability [15], stability in time [16], and the inherent heterogeneity for each psychiatry diagnosis [3, 17] represent additional elements that may represent additional hurdles in defining a clear and distinct phenotype for research and clinical purposes. Notwithstanding these limitations, several lines of evidence suggest a relevant level of inheritance for numerous psychiatric conditions, ranging from 80% for schizophrenia to 30% for major depressive disorder [18]. With this premise, considering the lack of specificity of diagnoses, high comorbidity, and shared symptomatology dimensions, it should be intuitive that the evidence for a relatively strong inheritability for numerous mental health disorders is also shared across different conditions [19]. Indeed, several lines of evidence suggest that common genetic architecture may underlie the risk of receiving a diagnosis of different conditions. However, genetic elements may only explain a relatively limited part of the observed overlap [19]. In this context, epigenetic elements have been increasingly analyzed to elucidate unknown elements that may underlie the observed inheritance and possibly represent a possible common denominator linking environmental exposure and biological elements. With epigenetics, we typically refer to a wide range of biological mechanisms regulating the gene expression level: transcriptional, post-transcriptional, translational, and post-translational (Fig. 13.2).

The three primary epigenetic processes are [20]:

- DNA methylation at CpG sites: This involves adding chemical groups to the DNA, particularly at cytosine-phosphate-guanine (CpG) sites, which can influence gene activity without altering the DNA sequence.
- Histone protein modifications: Chemical changes to histone proteins, which DNA wraps around, can affect how tightly or loosely DNA is packaged, thereby regulating gene expression.
- Non-coding RNA regulation: Non-coding RNAs modulate gene expression by interacting with DNA, RNA, or proteins to influence various cellular processes.

Epigenetic changes do not change the DNA sequence of the involved gene. However, by influencing gene expression, epigenetic change may dramatically impact health trajectories not only for the considered individual but potentially for their offspring too.

13.1.1 DNA Methylation

While a significant portion of the methylome remains relatively consistent within a given tissue type and among different individuals, approximately 30% of CpG sites exhibit substantial inter-individual variability. This variation is thought to reflect mainly genetic factors or gene-environment interactions and appears particularly relevant to study. For instance, epigenetic variation in DNA methylation (DNAm) at variable methylated regions (VMRs) can be influenced by several prenatal environmental factors (E), nearby genetic variants (cis-acting genotypes, G), or a combination of their additive (G + E) or interactive effects (GxE). The majority of VMRs are best explained by the combined influence of genetic and environmental factors. Additionally, CpG sites influenced by either genetic effects (G), their combination with environmental effects (G + E), or their interaction (GxE) are functionally distinct. Notably, genetic variants identified through GxE models are enriched in genome-wide association studies (GWAS) for complex disorders, highlighting their significance in the risk of receiving a specific diagnosis or in impacting on its clinical trajectory [21]. This approach may hardly address problems inherent to psychiatric diagnosis as we define it according to current nosological paradigms. Still, it may arguably widen our understanding of some of the inextricable interactions between the environment and the underlying biology in influencing mental health and the transgenerational transmission of mental illness. For example, past reports suggest that maternal methylome may indeed be associated with prenatal maternal depressive symptoms and that maternal methylome may impact both the placental and fetal methylome but only in

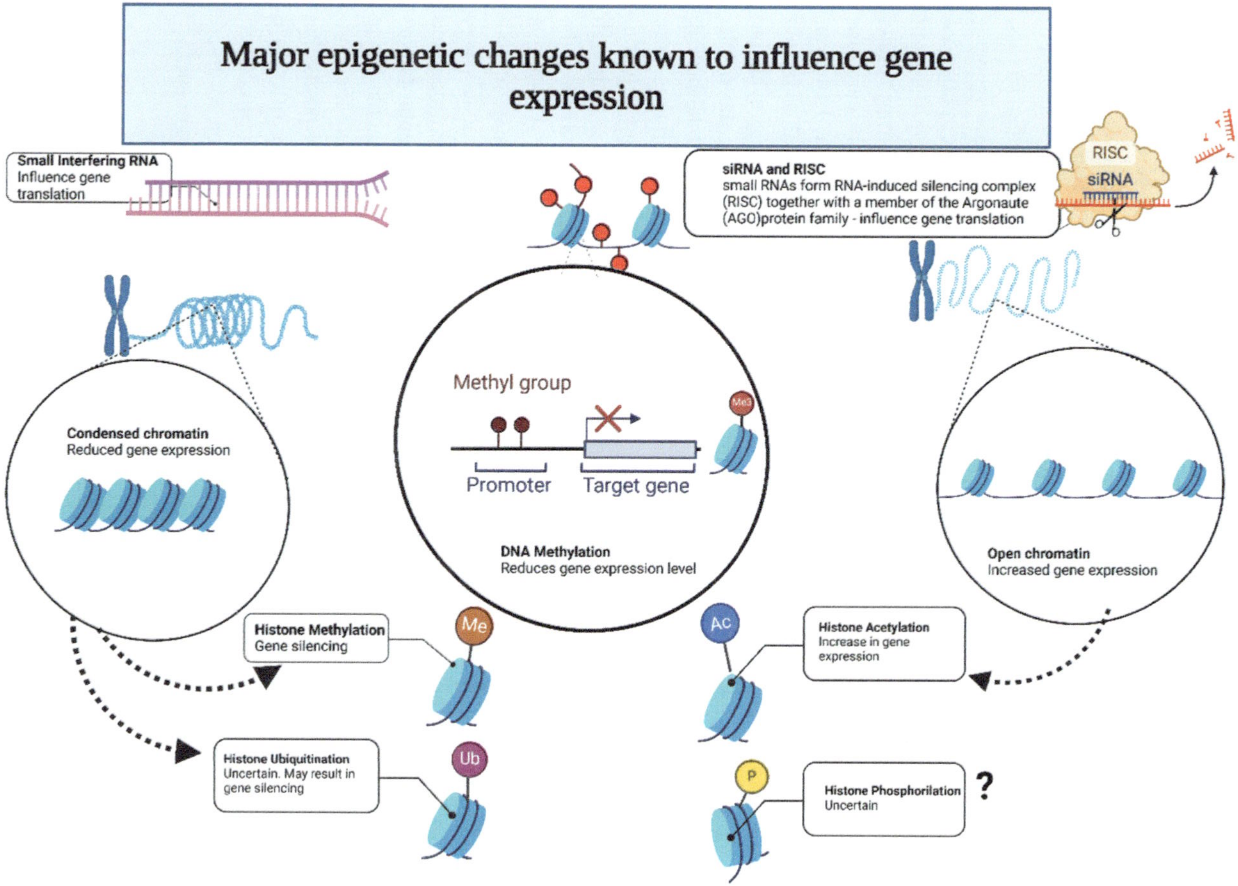

Fig. 13.2 Summary for some of the most frequently cited epigenetic changes. Image produced with BioRender

female fetuses. Importantly, maternal depressive symptoms (as defined according to the Edinburgh Postnatal Depression Scale – EPDS) scores did not vary by fetal sex, indicating a direct effect of fetal sex on the observed methylation patterns. This finding has been replicated in two independent cohorts, both in the Singaporean longitudinal birth cohort Growing Up in Singapore Towards healthy Outcomes (GUSTO) and in the UK-based birth cohort study Avon Longitudinal Study of Parents and Children (ALSPAC) [22]. Previous studies show sex-specific responses to maternal conditions, often with more pronounced impacts on male fetuses in terms of neurodevelopmental outcomes like ADHD, autism, and schizophrenia [22]. Overall, the available evidence suggests that there may be a relevant role played by the environment in influencing the epigenetic pattern in a given individual, and indeed, many factors, such as sex, may impact the end result, especially in terms of potential transgenerational transmission. DNA methylation pattern is cell- and tissue-specific, playing a key role in determining cell fate and differentiation. Therefore, choosing the right tissue to sample is key in understanding its significance and for comparing results across different samples. As it is the case for other fields of translational research for psychiatric diseases, brain tissues would be the optimal sample to study. However, brain tissues are hardly ever available and would certainly be unpractical or indeed unethical to sample for research purposes. Indeed, brain samples are only available post-mortem, are hard to access and even when analyzed, interpreting the results requires due consideration of post-mortem changes (e.g., duration of agonal period, cause of death, delay of sampling). In contrast to brain tissue, blood sampling is more easily accessible and facilitates standardized, longitudinal studies in living individuals, making it a promising approach for identifying clinical biomarkers. Although DNA methylation patterns in blood may not fully capture the changes occurring in the brain, they can still provide valuable insights into psychiatric disorders. Alternative samples collected in the periphery (e.g., saliva) are also valuable in this setting but have been explored in a comparatively lower number of papers. Two key models have been employed in this setting of methylation profile studies: the signature model and the mirror-site model [23]. The signature model suggests that environmental factors influencing the development of psychiatric disorders may induce specific changes in blood DNA methylation. These changes can serve as biomarkers reflecting the environmental exposure associated with the disorder. The mirror-site model posits that alterations in blood DNA methylation could parallel those in the brain, offering a window into cerebral changes. Even though blood DNA methylation does not perfectly mirror brain methylation, correlations for specific CpG sites have been observed, making blood a practical surrogate for studying some aspects of brain epigenetics.

13.1.2 Histone Acetylation and Methylation

Histones are essential protein components of chromosomes that significantly influence DNA accessibility. DNA wraps around histones, forming nucleosomes, which compact and shape chromatin structure. Histones are rich in basic amino acids such as arginine and lysine, facilitating strong interactions with DNA phosphate groups. Post-translational modifications like methylation and acetylation occur primarily on lysine and arginine residues. Among the numerous histone modifications, histone methylation and acetylation are the most extensively studied. These modifications constitute the histone code, which governs chromatin accessibility and regulates gene expression. Acetylation modifies lysine residues and is controlled by enzymes known as histone acetyltransferases (HATs) and histone deacetylases (HDACs). This process of histone acetylation is associated with the activation of gene transcription.

13.1.3 Non-coding RNA

Epigenetic changes involving non-coding RNA are classified according to their sizes. Non-coding RNAs (ncRNA) are therefore distinguished into microRNAs (miRNAs) and long non-coding RNAs (lncRNAs). MicroRNAs (miRNAs) are approximately 22 nucleotides in length and function as post-transcriptional regulators. Following a multi-step maturation process, miRNAs bind to target messenger RNAs (mRNAs), either inhibiting their translation by blocking ribosome access or marking the mRNAs for degradation. Binding typically involves a perfect match between nucleotides 2 through 7 (or 2 through 8) of the miRNA, known as the seed region, and the 3′-untranslated region (3’-UTR) of the mRNA, although other binding mechanisms can occur as well. miRNAs play a critical role in regulating gene expression in the central nervous system, contributing to neuronal development, differentiation, and synaptic plasticity. Numerous studies have investigated miRNA expression profiles in both postmortem brain tissues and peripheral samples, comparing healthy controls to patients. These miRNAs target genes involved in forming networks crucial for nervous system functions and diseases. These include synaptic and transcriptional regulation, protein kinase A signaling, neuronal CREB signaling, synaptic long-term potentiation and depression, DNA methylation, as well as neuronal development, differentiation, and synaptic plasticity. MiRNAs, which are highly conserved RNA molecules that regulate over 60% of human protein-coding genes, influence cell growth and differentiation. They mature through multiple cleavage steps and integrate with RNA-induced silencing complexes (RISCs) to target mRNA transcripts, resulting in their cleavage or translational repression. In addition to differences in miRNA expression, miRNA polymorphisms have been linked to susceptibility to psychiatric disorders and as potential biomarkers for treatment response. Since miRNAs are encoded by the genome, genetic variants can influence their maturation, structure, and functions. A prominent example from these studies is miR137, which is involved in neuronal processes relevant to psychiatric disorders. In the present chapter, we will summarize some of the most relevant evidence in the field of epigenetics for psychiatric diseases, focusing on bipolar disorder (BD), major depressive disorder (MDD), and schizophrenia (SCZ).

13.2 Epigenetic in Bipolar Disorder

13.2.1 Methylation Profile in BD

Epigenetic anomalies in bipolar disorder (BD) have been the subject of past reports with a focus on DNMT1, an enzyme essential for maintaining DNA methylation patterns. Studies have observed increased DNMT1 expression in the prefrontal cortex layers I and II of BD patients. This elevation is linked to decreased expression of Reelin and GAD67, which are crucial for synaptic plasticity and GABAergic transmission. Further research confirmed that DNMT1 binding to the promoters of GAD67, RELN, and BDNF is higher in BD patients compared to controls, independent of the methylation status. This suggests a connection between primary DNA methylation defects and GABAergic pathways in BD.

13.2.2 Histone Modifications in BD

Histone modifications in BD have been less frequently studied, especially regarding peripheral HDAC gene expression. Research has reported state-dependent increases in HDAC4 mRNA levels in BD patients compared to controls and those with unipolar disorder. Additionally, HDAC6 mRNA expression decreases during both depressive episodes and remission in BD. Another study noted reduced expression of sirtuins 1, 2, and 6 in BD patients during depressive states compared to controls and patients with MDD, indicating widespread dysregulation of histone modifications in BD. Sirtuins represent a group of enzymes that depend on NAD+ and regulate cellular functions by deacetylating various proteins.

13.2.3 ncRNA in BD

Significant alterations have been observed in BD in miRNA regulation. Up to 19% of miRNAs in the prefrontal cortex were found to be altered, with a general reduction in miRNA expression levels in BD patients. Genetic studies have also identified an association between the DICER1 gene, which is involved in miRNA maturation, and suicidal behavior in individuals with BD and MDD. Despite these findings, the evidence supporting primary dysregulation of epigenetic mechanisms in BD is still preliminary and appears less compelling compared to conditions such as autism or schizophrenia [24].

13.3 Epigenetic in Major Depressive Disorder

Current studies in MDD predominantly focus on DNA methylation and non-coding RNA using peripheral blood samples, with limited exploration of histone modifications [25].

13.3.1 Methylation Profile in MDD

Current evidence shows a diverse association between DNA methylation and MDD, such as hypomethylation of the SYN2 gene linked to depression and hypermethylation of the TESC gene correlated with brain integrity changes in MDD patients. DNA methylation influences transcription factor access to gene regulatory regions, potentially altering gene expression. Variants of 5-methylcytosine, such as 5-hydroxymethylcytosine, further complicate these interactions. Non-CpG methylation, prevalent in neurons and glial cells, increases with age and synapse development, playing a significant role in gene regulation similar to CpG methylation. MeCP2 binding to non-CpG sites affects gene expression, notably impacting disorders like Rett syndrome. Environmental stress during sensitive developmental periods can lead to lasting epigenetic changes. Early life adversities appear associated with epigenetic changes in stress-related genes like the glucocorticoid receptor (GR). DNA methylation changes in MDD are observed in both brain and peripheral tissues, with some overlap in methylation patterns between tissues. However, as reported in the previous section of this chapter, tissue- and cell-type-specific analyses reveal that peripheral markers may not always reflect brain methylation accurately. Recent advances in single-cell epigenomics offer potential for more precise studies of MDD-associated methylation in specific brain cells.

13.3.2 ncRNA in MDD

Aberrant miRNA expression is linked to various disorders, including MDD. Studies have found altered miRNA levels in the peripheral blood and brain regions of MDD patients. MiRNAs such as miR-106b-5p, miR-34, and miR-218 have been implicated in psychiatric disorders by affecting neuronal differentiation, synaptic plasticity, and neurotransmitter regulation. CircRNAs, which form from precursor mRNA through direct splicing, also play a role in depression. They are involved in regulating neuronal apoptosis, neuroglial activities, and neurotransmitter function. For example, circRNAs like circPTK2 and circHIPK2 have been shown to influence neuronal survival and astrocyte function, respectively. LncRNAs, despite not encoding proteins, are crucial for neural function, affecting processes such as neurogenesis and cognitive function. Alterations in lncRNA expression have been noted in MDD, influencing synaptic function and neuroglial cell activity. Specific lncRNAs, like lncRNA TCONS 00019174, have been linked to neuronal damage and synaptic dysfunction in depression. Antidepressant treatments, including SSRIs, SNRIs, and ketamine, can alter ncRNA expression. For example, SSRIs like fluoxetine have been shown to upregulate miR-16 and miR-320a while normalizing miRNA levels. Similarly, SNRIs like duloxetine and desvenlafaxine affect miRNA levels associated with brain activity and connectivity. Ketamine modulates miRNAs related to its antidepressant effects, such as miR-206, which influences BDNF expression. Understanding ncRNA regulation in response to antidepressants provides insight into their therapeutic mechanisms and potential new targets for treatment. To advance understanding further, integrating modern technologies like next-generation sequencing with single-cell approaches will enable comprehensive exploration of the epigenetic landscape across different cell types. Techniques such as Hi-C for 3D chromosome conformation and ATAC-Seq for chromatin accessibility offer new avenues to investigate how spatial genomic interactions and chromatin structure influence gene expression in MDD. Moreover, the role of metabolites in epigenetic processes, such as acetyl-CoA and S-adenosyl-methionine, underscores the link between metabolic signaling and chromatin regulation in the brain. Understanding these metabolic pathways could reveal new therapeutic targets for manipulating epigenetic states in MDD. Emerging evidence also implicates the gut microbiome and its metabolites in influencing epigenetic modifications and contributing to MDD pathophysiology. Exploring these interactions through the brain-gut-microbiota axis could provide further insights into disease mechanisms. Finally, advancements in genome editing technologies like CRISPR-Cas9 and TALENs offer unprecedented opportunities to manipulate epigenetic modifiers precisely. These tools may allow researchers to deepen our understanding for the complex relationship between specific epigenetic changes and stress-related disorders, potentially paving the way for targeted therapeutic interventions.

13.4 Epigenetics in Schizophrenia

13.4.1 Methylation Profile in SCZ

As is the case for other psychiatric conditions, the vast majority of studies probing epigenetics of SCZ have explored DNA methylation patterns. The available reports suggest an intricate pattern across various genomic regions. Methylation patterns in SCZ exhibit significant variability. Several neurotransmitter patterns have been assessed in this context, including the dopaminergic, glutamatergic, serotonin pathways. In the dopaminergic system, differential methylation of genes like DRD2 and COMT appears correlated with SCZ susceptibility and treatment response. Similarly, hypermethylation of serotonin receptor genes (HTR1A, HTR2A) and hypomethylation of glutamatergic genes (GRM2, GRM5) may be associated with part of the derangements in neurotransmission that has been reported in SCZ. Methylome-wide association studies have identified widespread methylation differences in SCZ. Studies in peripheral blood of monozygotic twins discordant for SCZ revealed differential methylation in genes related to cellular functions and immune response, suggesting broad systemic impacts of epigenetic dysregulation in SCZ pathophysiology. However, global methylation studies in SCZ have reported inconsistent findings, with some showing hypomethylation in specific populations while others found no significant differences. Advancements in technology like MeDIP and Illumina BeadChip have enabled detailed analysis of DNA methylation profiles, identifying specific genes such as ST6GALNAC1 and PUS3 with altered methylation patterns in SCZ. Interestingly, biallelic PUS3 variants have been reported to be associated with neurodevelopmental disorders, further underscoring the potential for these findings on independent samples and different modalities. Postmortem analysis on brain tissues has also reported mixed findings, suggesting that if indeed methylation changes are related to SCZ, they are likely complex and heterogeneous. Overall, while DNA methylation studies in SCZ provide insights into potential biomarkers and therapeutic targets, challenges remain in reconciling findings across different populations and brain regions. Traditional methods to measure DNA methylation profile might have limited the detection of DNA hydroxyl-methylation (5hmC), and therefore, the interpretation of the current evidence appears unclear, especially for earlier studies. The relative impact of 5hmC on the observed methylation pattern heterogeneity may be difficult to gauge at the moment.

13.4.2 Histone Modifications in SCZ

Histone modifications are increasingly recognized for their potential role in SCZ. Increased histone methyltransferase expression in the brain and blood, acting on H3, has been identified as a significant marker for SCZ. Additionally, elevated histone deacetylase (HDAC) activity has been noted in SCZ patients. Contrarily, hypoacetylation of H4 has been observed in peripheral blood mononuclear cells and the prefrontal cortex of SCZ patients, affecting the promoter regions of various genes such as HTR2C, PPM1E, GAD67, and TOMM70A. Increased methylation at H3R17 in the prefrontal cortex correlates with decreased expression of metabolic genes like MDH, CRYM, OAT, and CYTOC/CYC1.

13.4.3 ncRNA in SCZ

ncRNAs, particularly miRNAs, are another crucial component of epigenetic regulation. These small RNAs (21–25 nucleotides) regulate gene expression post-transcriptionally, either by cleaving target mRNAs or repressing their translation. MiRNAs can impact various biological processes, including neurogenesis, synaptic plasticity, and signaling. Specific miRNAs such as miR-181b, miR-219-2-3p, and let-7 g serve as candidate biomarkers for SCZ. In SCZ, dysregulation of miRNAs like miR-185, miR-195, and miR-132 has been linked to altered neurodevelopment and gene expression. For instance, miR-185's interaction with the RhoGTPase family and spine-specific proteins impacts dendritic spine development. MiR-195 inversely correlates with BDNF protein levels, and miR-346 downregulation affects glutamate receptor expression. Thus, histone modifications and ncRNAs represent vital areas of study in understanding the complex epigenetic landscape of SCZ [26].

13.5 Conclusions

Notwithstanding current limitations for the available evidence inherent in the complexity of the studied phenomena, the field of epigenetics appears promising and may represent a valuable avenue for future translational research for the field of psychiatric diseases and beyond. Future research integrating advanced genomic technologies and larger sample sizes will be crucial for elucidating the complex and heterogeneous nature of epigenetic modifications in psychiatric disorders, paving the way for personalized treatments and targeted interventions integrating epigenetic profiles in future treatment algorithms.

References

1. Kieling C, Buchweitz C, Caye A, Silvani J, Ameis SH, Brunoni AR et al (2024) Worldwide prevalence and disability from mental disorders across childhood and adolescence: evidence from the global burden of disease study. JAMA Psychiatry 81(4):347–356
2. Arias D, Saxena S, Verguet S (2022) Quantifying the global burden of mental disorders and their economic value. EClinicalMedicine 54:101675
3. Cai N, Choi KW, Fried EI (2020) Reviewing the genetics of heterogeneity in depression: operationalizations, manifestations and etiologies. Hum Mol Genet 29(R1):R10–R18
4. Dobbs D (2018) How asylums became the crucible of genetics. Nature 558(7708):28–30
5. Gejman P, Sanders A, Duan J (2010) The role of genetics in the etiology of schizophrenia. Psychiatr Clin North Am 33(1):35–66
6. Craddock N, O'Donovan M, Owen M (2005) The genetics of schizophrenia and bipolar disorder: dissecting psychosis. J Med Genet 42(3):193–204
7. Gottesman II (1991) Schizophrenia genesis: The origins of madness, Schizophrenia genesis: The origins of madness. W H Freeman/Times Books/Henry Holt & Co, New York, NY, p xiii, 296
8. Murray GK, Lin T, Austin J, McGrath JJ, Hickie IB, Wray NR (2021) Could polygenic risk scores be useful in psychiatry?: a review. JAMA Psychiatry 78(2):210–219
9. Sada-Fuente E, Aranda S, Papiol S, Heilbronner U, Moltó MD, Aguilar EJ et al (2023) Common genetic variants contribute to heritability of age at onset of schizophrenia. Transl Psychiatry 13:201
10. Richards AL, Cardno A, Harold G, Craddock NJ, Di Florio A, Jones L et al (2022) Genetic liabilities differentiating bipolar disorder, schizophrenia, and major depressive disorder, and phenotypic heterogeneity in bipolar disorder. JAMA Psychiatry 79(10):1032–1039
11. Gordovez FJA, McMahon FJ (2020) The genetics of bipolar disorder. Mol Psychiatry 25(3):544–559
12. Lilienfeld SO, Treadway MT (2016) Clashing diagnostic approaches: DSM-ICD Versus RDoC. Annu Rev Clin Psychol 12:435–463
13. Stein DJ, Shoptaw SJ, Vigo DV, Lund C, Cuijpers P, Bantjes J et al (2022) Psychiatric diagnosis and treatment in the 21st century: paradigm shifts versus incremental integration. World Psychiatry 21(3):393–414
14. Nordgaard J, Nielsen KM, Rasmussen AR, Henriksen MG (2023) Psychiatric comorbidity: a concept in need of a theory. Psychol Med 53(13):5902–5908
15. Vanheule S, Desmet M, Meganck R, Inslegers R, Willemsen J, De Schryver M et al (2014) Reliability in psychiatric diagnosis with the DSM: old wine in new barrels. Psychother Psychosom 83(5):313–314

16. Caspi A, Houts RM, Ambler A, Danese A, Elliott ML, Hariri A et al (2020) Longitudinal assessment of mental health disorders and comorbidities across 4 decades among participants in the Dunedin birth cohort study. JAMA Netw Open 3(4):e203221
17. Bryant RA, Galatzer-Levy I, Hadzi-Pavlovic D (2023) The heterogeneity of posttraumatic stress disorder in DSM-5. JAMA Psychiatry 80(2):189–191
18. Pettersson E, Lichtenstein P, Larsson H, Song J, Attention Deficit/Hyperactivity Disorder Working Group of the iPSYCH-Broad-PGC Consortium, Autism Spectrum Disorder Working Group of the iPSYCH-Broad-PGC Consortium, Bipolar Disorder Working Group of the PGC, Eating Disorder Working Group of the PGC, Major Depressive Disorder Working Group of the PGC, Obsessive Compulsive Disorders and Tourette Syndrome Working Group of the PGC, Schizophrenia CLOZUK, Substance Use Disorder Working Group of the PGC, Agrawal A et al (2019) Genetic influences on eight psychiatric disorders based on family data of 4 408 646 full and half-siblings, and genetic data of 333 748 cases and controls. Psychol Med 49(7):1166–1173
19. Bourque VR, Poulain C, Proulx C, Moreau CA, Joober R, Forgeot d'Arc B et al (2024) Genetic and phenotypic similarity across major psychiatric disorders: a systematic review and quantitative assessment. Transl Psychiatry 14(1):1–10
20. Zhang X, Cozen AE, Liu Y, Chen Q, Lowe TM (2016) Small RNA modifications: integral to function and disease. Trends Mol Med 22(12):1025–1034
21. Czamara D, Eraslan G, Page CM, Lahti J, Lahti-Pulkkinen M, Hämäläinen E et al (2019) Integrated analysis of environmental and genetic influences on cord blood DNA methylation in new-borns. Nat Commun 10(1):2548
22. Kee MZL, Teh AL, Clappison A, Pokhvisneva I, MacIssac JL, Lin DTS et al (2022) Fetal sex-specific epigenetic associations with prenatal maternal depressive symptoms. iScience 25(9):104860
23. Aberg KA, McClay JL, Nerella S, Clark S, Kumar G, Chen W et al (2014) Methylome-wide association study of schizophrenia: identifying blood biomarker signatures of environmental insults. JAMA Psychiatry 71(3):255–264
24. Legrand A, Iftimovici A, Khayachi A, Chaumette B (2021) Epigenetics in bipolar disorder: a critical review of the literature. Psychiatr Genet 31(1):1–12
25. Yuan M, Yang B, Rothschild G, Mann JJ, Sanford LD, Tang X et al (2023) Epigenetic regulation in major depression and other stress-related disorders: molecular mechanisms, clinical relevance and therapeutic potential. Signal Transduct Target Ther 8(1):1–30
26. Srivastava A, Dada O, Qian J, Al-Chalabi N, Fatemi AB, Gerretsen P et al (2021) Epigenetics of schizophrenia. Psychiatry Res 305:114218

The Metabolomics Revolution Toward Personalized Medicine: The Case of Gestational Diabetes

14

Roberta Pintus, Angelica Dessì, Alice Bosco, and Vassilios Fanos

Contents

14.1 Introduction.. 233
14.2 What Is Gestational Diabetes Mellitus?.............................. 234
14.3 Personalized Medicine and Gestational Diabetes Mellitus............. 235
14.4 Metabolomics and Gestational Diabetes Mellitus...................... 235
14.5 Conclusions... 247
References... 247

14.1 Introduction

Gestational diabetes mellitus (GDM), one of the most common obstetric complications, is defined as any degree of glucose intolerance during pregnancy [1].

This pathology can affect the health status of both the mother and the offspring in the early and long term; therefore, early diagnosis and timely treatment are essential [2].

Women with gestational diabetes may develop type 2 diabetes, cardiovascular diseases, and obesity (Moon 2022). While children of mothers with GDM could develop metabolic disorders, cardiovascular diseases, and obesity as well [3].

To date, the diagnosis is made by glycemia measurement and sometimes it is performed in later stages of pregnancy [4].

R. Pintus (✉) · A. Dessì · A. Bosco · V. Fanos
Neonatal Intensive Care Unit, Department of Surgical Sciences, University of Cagliari, AOU, Cagliari, Italy
e-mail: angelicadessi@unica.it

L. Emmi, V. Fanos (eds.), *Epigenetics and Rare Diseases*, Rare Diseases of the Immune System, https://doi.org/10.1007/978-3-032-02902-7_14

Furthermore, it is impossible to evaluate the actual effect of therapy in each patient.

Thus, there is a need to change the approach in the treatment, taking into account a personalized medicine.

Metabolomics could be useful in the investigation of GDM due to its ability to snapshot the metabolic state of each subject that could lead to the discovery of new biomarkers to identify women at risk and the effect of the therapeutic approach [5].

14.2 What Is Gestational Diabetes Mellitus?

GDM could be considered a form of type 2 diabetes. Indeed, it is characterized by reduced insulin secretion and increased insulin resistance. Even if insulin resistance is physiological in normal pregnancies, in case of GDM this is not compensated [6].

From the diagnostic point of view, guidelines recommend the measurement of blood sugar at the first appointment in pregnancy for all women with no previous tests to single out those with diabetes prior to becoming pregnant or physiological pregnancy of women presenting specific risk factors [4].

These risk factors are the following: obesity (body mass index higher than 30), gestational diabetes in previous pregnancies, high level of glucose (100/125 mg/dL) before or at the beginning of the pregnancy, age older than 35 years, fetal macrosomia in previous pregnancy, positive family anamnesis [7].

Furthermore, women coming from South Asia, Caribbean, and Middle East should be screened as well due to the high incidence of this disease in those areas [8].

Women with fasting glycemia higher than 92 mg/dl and higher than 180 mg/dl, and 153 mg/dl after 1 and 2 hours respectively are considered to have GDM [4].

As already mentioned, GDM causes an increased risk of morbidity in both the mother and the child. The acute complications concerning the mother could be: higher risk of spontaneous abortion, polyhydramnios, hypertension in pregnancy, pyelonephritis and other infections, preterm delivery, hypoglycemia, ketoacidosis, and cesarean delivery. The late complications for the mother are type 2 diabetes or a metabolic syndrome or cardiological diseases. Thus, after pregnancy, it is important for these women to be followed up [3].

The early complications for the neonate are shoulder dystocia, macrosomy, respiratory distress, hyperbilirubinemia, polycythemia, hypocalcemia, and hypoglycemia. While the late complications are cardiovascular defects and higher risk of overweight and obesity in the adolescent age as well as the higher risk of developing type 2 diabetes [3].

The reduced action of insulin in GDM causes an excessive increase of nutrients in the blood (glucose, lipids, amino acids), which, on passing through the placenta, determines hyperinsulinism capable of favouring a consequent increase in adipose tissue with fetal organomegaly and macrosomia. Furthermore, since the excessive glucose crosses the placenta, it may lead to a condition of constant hyperglycemia in the fetus that could result in neonatal hypoglycemia at birth [9].

14.3 Personalized Medicine and Gestational Diabetes Mellitus

Today, the management of GDM focuses on lifestyle interventions (diet and exercise) and pharmacotherapy (insulin and oral hypoglycemic agents) to maintain euglycemia. However, these strategies are often one-size-fits-all, neglecting individual metabolic variations that could impact treatment response.

In general, personalized medicine aims to tailor medical treatment to individual characteristics, including genetic, environmental, and lifestyle factors. In GDM, personalized approaches could involve risk prediction, tailored intervention, and monitoring to allow therapy adjustments.

Regarding risk prediction, metabolomics could improve risk stratification, identifying women at high risk for GDM earlier in pregnancy.

Then, individual metabolomic profiles could guide dietary and pharmacologic interventions optimizing glycemia stabilization (tailored interventions).

Finally, regarding therapy adjustments, metabolomics biomarkers could allow for real-time monitoring of treatment efficacy and adjustments based on metabolomics responses.

14.4 Metabolomics and Gestational Diabetes Mellitus

Metabolomics, one of the most recent of the omics sciences, is defined as the study of the complex system of metabolites owing to its capacity of examining an individual's entire metabolic profile. This science is able of describing the chemical profile of a biological system in terms of low-molecular-weight metabolites present in cells, tissues, organs, and biological fluids [10].

Indeed, a metabolomic analysis is capable of detecting analytes of relatively low molecular weight (up to 1000 Da), among which amino acids, oligopeptides, sugars, steroids, biliary acids, simple and compound fatty acids, and intermediate compounds of many biochemical pathways [10]. The metabolomic technique analysis available to date are: nuclear magnetic resonance (NMR) and gas or liquid chromatography coupled with mass spectrometry (LC-MS and GC-MS). The most common biological samples used in metabolomics are urine, plasma, serum, feces, but other biofluids can be used as well (liquor, saliva, gastric/pancreatic juices, aqueous humor) or tissues. Each biological fluid has a unique and characteristic biochemical composition that changes in response to physiological or pathophysiological stimuli to generate a metabolic "fingerprint" [10].

Thus, it is possible to have "in real time" pictures of the influences on the organism of biochemical perturbations caused by diseases, drugs, and toxins or diet.

The aim of such characterization is that of developing "tailored" therapeutic approaches for the single individual and/or group of individuals, where for group is meant persons having a common phenotypical profile deriving from a similar gene–environment interaction.

In recent years, many investigations have been performed on animals, adults, children, and neonates using a metabolomic approach that makes it possible to study important perinatal and postnatal pathologies such as asphyxia and hypoxia, congenital diseases, the state of nutrition, kidney diseases, nephrotoxicity, cardiovascular disease, neurological disorders, and many more including GDM [11].

As can be seen in the guidelines, at present the diagnosis of GDM is based solely on the finding of high values of glucose in the blood. Furthermore, in cases not considered at risk, it is performed toward the end of the 6 months of pregnancy when, if it is found altered, the fetus is already affected by a series of metabolic influences capable of impacting on its growth [11].

For this reason, metabolomics would appear to be the most suitable method since it is capable of providing a real time picture of the metabolic state of an individual it would help to identify biomarkers to early predict if a woman is prone to GDM and the subsequent immediate treatment to prevent the development of GDM preserving the health status of the fetus as well.

In the last 5 years, there was an explosion of articles regarding the application of metabolomics in the investigation of GDM.

Especially during the Covid-19 pandemic, the interest of the scientific community for this disease increased massively.

Thus, the insertion of "gestational diabetes" and "metabolomics" as keywords on PubMed resulted in 319 results, 159 of them were experimental studies on humans.

Since a comprehensive review of all the studies falls beyond the scope of this chapter, we limited the number of studies analyzed to those published in the last 3 years, and they are summarized in Table 14.1.

In 2023, we analyzed the urinary metabolome of newborns from GDM mothers treated with insulin or diet collected at different timepoints. Only at birth there were significant variations of metabolites involved in citrate, glycine, serine, and threonine metabolism in newborns of insulin-treated mothers. This seemed to be due to the impact of early nutrition that can alter the health programming as well [12].

While Burzynska-Pediziwatr et al. analyzed plasma of GDM and controls subject. They noted higher concentration of 2-hydroxy and 3-hydroxy butyric acid, oxalic acid and glycerol and decreased level of myristic acid and oleic acid. 2-Hydroxy butyric acid is a marker of increased oxidative stress connected with beta-cell dysfunction and insulin resistance. It is also positively correlated with the OGTT test as well as 3-hydroxibutyric acid. While oxalic acid is a competitive inhibitor of lactate dehydrogenate that has been related to kidney damage in a previous study. Finally, glycerol is an indicator of gluconeogenesis insulin resistance induced. Decrease myristic acid is a protective molecule against insulin resistance as well as oleic acid and their intake could prevent GDM [13].

Furthermore, Wu et al. through their metabolomic analysis of plasma of GDM women found altered level of different types of bile acids in GDM women. It is the first study that correlates a given bile acid with the risk of GDM since they are involved in glucose metabolism at different levels. Indeed, for instance, some such as taurolic acid are positively associated with GDM [14].

Table 14.1 Metabolomics investigations of gestational diabetes mellitus in the last 3 years

Authors	Patients	Methods	Sample	Metabolite results	Biological hypothesis
Dessì [12]	21 newborns of GDM 5 newborns of non-GDM at different time points	NMR, GC-MS	Urine	Altered levels of acetate, lactate, glycyl-proline/proline, isocitrate, N-N-dimethylglycine, N-acetylglucosamine, N-carbamoyl-aspartate at T1 in insulin NDM	Changes certainly due to GDM only at this time point
Burzynska-Pedziwiatr [13]	32 GDM 32 controls	GC-Q-MS	Plasma	Myristic, oleic, arachidonic, alpha linoleic acid significant changes in GDM	Oleic acid may be protective Altered lipids contribution to pathophysiology of GDM
Wu [14]	100 GDM 545 non-GDM	UPLC-MS	Plasma	Primary unconjugated and conjugated and secondary unconjugated and conjugated bile acids distinct in GDM	Bile acid role in the pathogenesis of GDM
Huhtala [15]	300 GDM	^{1}H-NMR	Serum	↑saturated, monosaturated fatty acid ↓omega 3 polyunsaturated fatty acid	All associated with insulin resistance: Distinct lipid profile. Eventual lifestyle intervention early in pregnancy

(continued)

Table 14.1 (continued)

Authors	Patients	Methods	Sample	Metabolite results	Biological hypothesis
Zhang [16]	91 GDM 308 controls	MS	Urine	Purine metabolism, Krebs cycle related both to arsenic exposure and GDM	Arsenic exposure as metabolic disorder
Li [17]	33 GDM women 33 controls	LC-MS	Serum 1 year before conception	↑phosphatidyl ethanolamines in GDM	Predictive power 1 year before pregnancy Distinctive preconception features of GDM
Tao [18]	71 GDM 89 controls	^{1}H-NMR	Serum	↑BCCA, glutarate, AAA, 3-hydroxymandelate, benzoic acid	Role of gut microbiota in the pathogenesis of GDM and vice versa: GDM perturbation of intestinal microbiota
Sormunen Harju [19]	755 obese and non-obese GDM women	^{1}H-NMR	Blood	Different metabolomic profiles between obese and non-obese women	Biomarker of different etiopathogenesis
Lyu [20]	27 GDM 27 controls	GC-MS UPLC-MS	Stool	Butyrate, revolonate, estradiol, progesterone metabolism alterations	Gut microbiota contribution to the pathogenesis of GDM
Huhtala [21]	119 women with GDM, insulin, or metformin	^{1}H-NMR	Serum	Lipid and lipoprotein alterations	Serum metabolome in pregnancy in women with GDM is related to future glucose metabolism and insulin resistance

Wang [22]	1409 women	LC-MS	Plasma	33 lipids altered in GDM	GDM was associated with a range of lipid metabolic alterations at early post-partum. Some lipids were also associated with incident T2D and mediated the progression from GDM to T2D
Heath [23]	20 GDM 20 no GDM	UPLC/MS	Plasma samples 1° and 3° trimester	Carnitine variation ↑hypoxantine, inosine, monophosphate	Fatty acid oxidation altered
Razo-Azamar [24]	62 controls 13 GDM	MS	Serum	↑carnitine	Early prediction model, lipid and BCAA impairment
Li [25]	9 GDM 9 no GDM 9 pregnant women	LC-MS/MS	Saliva	↑L-isoleucine, glucose-6-phosphate, DHA, arachidonic acid, adenosine, monophosphate, xanthine, hypoxantine	Similar to previous studies on urine and plasma
Sovio [26]	349 GDM 2347 controls	UPLC-MS	Serum	↑mannose, 4-hydroxyglutarate, 1,5-anhydroglucitol lactosil-N-palmitol sphingosine	Screening for GDM
Gao [27]	24 GDM 24 controls	UPLC-MS	Serum	↑isobutyric, isovaleric acid, valeric, caproid acid, SCFA in GDM	Gut microbiota-derived metabolites alterations in GDM correlation with clinical biomarkers

(continued)

Table 14.1 (continued)

Authors	Patients	Methods	Sample	Metabolite results	Biological hypothesis
He [28]	20 GDM 20 controls aged 35–40 years old	GC-MS	Plasma Urine	Cysteine, malonic acid, alanine, eicosanoic acid, stearic acid, arachidonic acid 2-methyl octadecanoic acid altered in GDM plasma Urine no distinction	Ethiology of GDM in advanced maternal age
Spanou [29]	62 GDM 77 controls	^{1}HNMR	Serum	↑ketone bodies, BCAA ↓one carbon metabolism, energy production, purine metabolism	GDM shares some pathways alterations with T2D
Hu [30]	60 GDM 60 controls	UHPLC-MS MS	Serum 1° trimester	↓dehydroepianidrosterone	Downregulation of fatty acid oxidation Predictive panel of GDM phenotype
Piras [31]	25 GDM 25 no GDM women	1H-NMR	Urine	↑tryptophan, trigonelline, hippurate, threonine ↓1-methylnicotidiamide, 3-hydroxy-kurenine, glycocholate, isoleucine, Valine in GDM	Before diagnosis is present, a clear panel of metabolites
Peng [32]	59 GDM 59 controls	UPLC/MS	Serum	↑L-valine, hypoxantine, eicosapentanoic acid, 2-amino-1,3,4-octadecanotriol colin in GDM	Valine: Insulin resistance Eicosapentanoic acid: Inflammation Panel of biomarker

Furse [33]	867 obese pregnant women (GDM diagnosis at 28 weeks)	MS	Plasma	↑diglycerides, triglycerides, de novo lipogenesis	Increased de novo lipogenesis-related species in pregnant women with obesity and GDM are associated with measures of offspring adiposity and may be a target for improving lifelong health
Wang [34]	1008 women	UPLC/MS MS	Plasma before diagnosis	6 and 4 lipids positively and negatively associated with BMI in GDM	The relationship between higher BMI and GDM might be mediated by the composite effect of BMI-associated lipids
Chen [35]	272 GDM 748 non-GDM	GC-MS	Hair	Different metabolites altered	Hair metabolome as screening of internal and external stimuli
Tian [36]	50 GDM 50 controls	LC-MS	Serum	↑pyroglutamic acid, phenylacetic acid, panthotenic acid ↓anhydroglucitol, calcitrol oxoproline	Early signature of GDM
Reutrakul [37]	81 obese GDM women 42 with obstructive apnea	LC-MS	Serum	No differences	

(continued)

Table 14.1 (continued)

Authors	Patients	Methods	Sample	Metabolite results	Biological hypothesis
Wang [38]	336 GDM 672 controls	UPLC/MS	Blood	10 lipids associated with GDM	Specific plasma lipids biomarkers in early pregnancy were associated with GDM. Predictive power of metabolomics
Mansell [39]	912 women	1H-NMR	Cord blood Maternal serum	↑acetoacetate, 3-hydroxybutyrate, citrate in GDM	Marker in cord blood of GDM
Chen [40]	147 GDM mothers 271 controls with their newborns	UPLC-MS	Meconium Maternal serum	Taurine, pyrimidine, B alanine metabolism, bile acid altered in meconium of GDM babies ↑riboflavin, taurine ↓glycerophosphocholine, glychocolic acid, ramnose in maternal blood and meconium	Alterations of GDM newborns microbiota and lipid metabolism both in mothers and in newborns
Raczkowska [41]	26 abnormal glucose tolerance GDM 29 controls 24 abnormal fasting plasma glucose levels	GC-MS	Serum	Alfa-hydroxybutyric acid, beta-hydroxybutyric acid, and several fatty acids, significant in a GT-GDM but not in aFG-GDM	Select women at risk of GT-GDM in the first trimester
Wu [42]	50 GDM 50 controls	UPLC MS	Breastmilk	↑4-pyridoxic acid, riboflavin, pyridoxal, 4-hydroxy-L-phenylglycine, lysophosphocholine, pantothenic acid O-succinyl-L-homoserin ↓malic acid, L-pipelicolic acid in GDM	Feeding infants with GDM breastmilk leads to excessive BWG up to 1 year. Tailored breastfeeding

Liu [43]	2290 women	MS	Blood	Amino acids, carbohydrates, fatty acids, and lipids metabolic alterations	Metabolites associated with GDM and the underlying pathophysiology to a postpartum disorder of glucose metabolism
Sikoroski [44]	600 women	MS	Serum	Several amino acid carbohydrate lipids alterations common in SA and European with GDM	Higher prevalence in south Asian women
Lu [45]	200 GDM 200 controls 2° trimester 199 GDM 199 controls 3° trimester	UPLC-MS/MS	Serum	b-methyl-oxovaleric acid in second trimester Ketoleucine alphaketovaleric acid in 3° trimester	Specific biomarkers for specific GA
Zhang [46]	50 GDM 50 no GDM	UPLC-MS	Plasma	Abnormal amino acids and lipids metabolism in GDM	Pathogenesis and prediction
Yang [47]	32 GDM 31 controls	UPLC-HRMS	Placenta	Linolenic acid and in GDM	Good predictors
Meng [48]	34 GDM 34 controls	HP-LC MS	Serum	Eicopentaenoic acid, docoesanoic acid, docopentanoic acid, arachidonic acid, citric acid, alpha-ketoglutaric acid, genistein in GDM	Panel of biomarkers

(continued)

Table 14.1 (continued)

Authors	Patients	Methods	Sample	Metabolite results	Biological hypothesis
Herrera-van Oostdam [49]	26 GDM 24 controls women and their newborns	LC-MS	Urine	↑sperimine, hexadecenoylcarnitine in newborns Significant differences in butyric, isobutyric, and uric acid in the second and third trimesters of pregnancy	Spermine: Alteration of arginine/NO pathway - > altered endothelial function
Rahman [50]	107 GDM 214 controls	UPLC-MS	Plasma	Triglycerides, phosphocholine, cholesteryl esters	Same results in different ethnic groups
Ondrejikova [51]	44 GDM 33 no GDM At 24–28, 30–36 weeks of gestation	GC-MS	Blood Umbilical cord blood	↓testosterone, estradiol, 7-oxigenated androgens	Increasing diabetogenic effects with the approaching terms are associated with increasing progesterone levels
Liu [52]	486 pregnant women	LC-MS/MS	Serum	Glycerophospholipids, sphingolipids, primary bile acids	Early diagnosis
Huhtala [53]	113 women: 97 insulin, 98 metformin	^{1}H-NMR	Cord serum	↑alanine	The lipid metabolism is related to birthweight
Koos [54]	46 GDM 46 controls	UPCL-MS/MS	Urine	7 metabolites	Prediction in early pregnancy
Zhan [55]	49 GDM 54 normal second and third trimesters	UHPLC	Plasma	Monoacylglycerol, dihydrobiopterin, 13-S-hydroxyoctadecadienoic acid, arginine, ornithine	Metabolite changes at different stages of pregnancy related to GDM

Legend: *GDM* Gestational Diabetes Mellitus, *NMR* Nuclear Magnetic Resonance, *GC-MS* Gas Chromatography-Mass Spectrometry, *GC-Q-MS* Gas Chromatography-Quadrupole-Mass Spectrometry, *UPLC* Ultra Performance Liquid Chromatography and Tandem Mass Spectrometry, *^{1}H-NMR* Proton Nuclear Magnetic Resonance, *LC-MS* Liquid Chromatography-Mass Spectrometry, *MS* Mass Spectrometry, *BCAA* Branched-Chain Amino Acids, *AAA* Aromatic Amino Acids, *SCFA* Short Chain Fatty Acids, *UHPLC-MS* Ultra-high Performance Liquid Chromatography-Mass Spectrometry, *BMI* Body Mass Index, *aFG-GDM* abnormal fasting glucose gestational diabetes mellitus, *GT-GDM* glucose tolerant gestational diabetes mellitus, *HRMS* high Resolution Mass Spectroscopy, *NO* Nitric Oxide

While Wang et al. through the analysis of plasma found lipid metabolic alterations early at post-partum and some were associated with the incidence of type 2 diabetes [22].

Even Heat et al. analyzed plasma of GDM women and detected several alterations in lipids metabolism as well which seems to be a peculiar feature of GDM due to insulin resistance [23].

Other alterations were highlighted in plasma of GDM women supporting the hypothesis that could be a different etiology of GDM in women of advanced age [28].

The same research group analyzed also the urine metabolome finding no difference. On the other hand, Piras et al., through urine analysis found a clear panel of metabolites predictive of GDM [31].

While Furse et al., in plasma metabolome of obese pregnant women with GDM highlighted several altered species of diglycerides and triglycerides that may indicate a "de novo" lipogenesis [33]. This phenomenon seems to be correlated with measures of offspring's adiposity and may be a target for improving the future health of the offspring themselves [33].

Additionally, Sormunen et al. through blood metabolomics analysis of obese and non-obese women with and without GDM supported the hypothesis that in obese women the etiopathogenesis of GDM could be different [19].

Moreover, Wang with the metabolomic analysis of this fluid before diagnosis detected different species of lipids positively and negatively associated with BMI in GDM. This could be due to the composite effect of BMI-associated lipids [34].

However, Hutala et al. in 2023 performed a metabolomics analysis of serum of GDM women and found a different and distinct lipid profile of each group. These findings may help the development of early lifestyle intervention tailored upon each pregnant woman [15].

In another study of serum of GDM women treated with insulin or metformin, the authors detected different metabolomic profile between the 2 groups during pregnancy, predictive of future insulin resistance.

On the other hand, Raczkowaska et al. analyzed serum of 2 groups of GDM women characterized by abnormal glucose tolerance and abnormal fasting plasma glucose highlighting several altered fatty acids in abnormal glucose tolerance group. These findings could be useful for an early diagnosis and a further screening of at-risk women [41].

Furthermore, Peng et al. detected different altered metabolites such as valine (related to insulin resistance) and eicosapentanoic acid (related to inflammation) as a part of a panel of biomarkers for GDM [32].

While, Li analyzed serum of GDM mothers one year before conception supporting the predictive power of metabolomics and indicating peculiar preconception features of women that will develop GDM [17]. Moreover, several other authors supported the hypothesis of predictive power of metabolomics [26, 38]. Indeed, even starting the collection of samples at 1° trimester it is possible to have a predictive panel of GDM and GDM phenotype characterized by a downregulation of fatty acid oxidation [30]. In addition, Razor-Amar et al. with serum metabolomics analysis found an early valid prediction model of GDM, based on lipids and branched chain amino acids alteration [24].

Even Koos et al. through the metabolomics analysis of urine supported the predictive power of metabolomics [54].

While, Herrera-van Dostam et al. analyzing the urine metabolome of GDM women found some metabolites (spermine) that reflect the alterations of endothelial function in GDM, providing further insights of its physiopathogenesis [49].

Furthermore, Zhang et al. performed the metabolomics analysis of urine of pregnant women exposed to arsenic stating that arsenic poisoning or exposure could mimic metabolic disorders such as GDM [16].

Moreover, findings on other biological fluids were confirmed on the metabolomics analysis of saliva of GDM mothers performed by Li et al., corroborating the role of metabolomics in the investigation of the pathophysiology of GDM [55].

Another author analyzed the hair metabolome of GDM women finding several altered metabolites [35]. Thus, even hair could be used as a screening for GDM and to monitor individual answers to internal and external stimuli.

From the fetal and neonatal point of view even through the analysis of the placenta it was possible to detect good predictor of GDM such as linolenic acid and different fatty acids [47].

While, cord serum of fetus of GDM mothers showed high levels of alanine that is directly related to insulin resistance and altered lipid metabolism that correlated with birth weight [53].

Additionally, Ondrejikova et al. with the analysis of umbilical cord blood highlighted the increasing diabetogenic effect of progesterone at the end of pregnancy in GDM women [51]. Finally, Mansell et al. found in cord blood of GDM mothers reliable biomarkers related to lipid alterations [39].

Even the metabolome of breastmilk of GDM mothers is altered. It seems to be obesogenic given the excessive birth gain one year after birth of these infants [42].

Regarding the prevalence of GDM in different ethnic groups, Sikboroski et al., through the analysis of GDM serum found out several amino acids, carbohydrate, and lipid metabolisms alterations shared among South Asian and European women with a higher prevalence in South Asian women [44]. Moreover, Rahaman et al. with plasma metabolomics found similar results in different ethnic groups [50].

The role of gut microbiota in this context goes beyond the aim of this paragraph but several studies proved the presence of dysbiosis in GDM women. Indeed, Tao et al. through metabolomics analysis of serum of these women supported the hypothesis of a role of microbiota in the pathogenesis of GDM and vice versa: namely, GDM itself could be the cause of intestinal dysbiosis in these women [18]. These hypotheses are possible since metabolomics is able to distinguish between metabolites of human and microbial origins. Again, Gao through the analysis of serum of GDM mothers found altered gut microbiota metabolites that correlate with clinical features of GDM [27]. Also, Lyu through the metabolomics analysis of stool highlighted the contribution of microbiota to GDM pathogenesis [20].

Finally, Chen et al. performed a metabolomics analysis on meconium and maternal serum detecting altered metabolites related to GDM newborns microbiota and lipid metabolism in both mothers and newborns [40].

14.5 Conclusions

The integration of metabolomics and personalized medicine in GDM faces several challenges. Among others, standardization, cost and accessibility, and clinical validation. Indeed, developing standardized protocols for metabolomic analyses to ensure reproducibility and comparability across studies it is not easy. While, the application of metabolomics is still very expensive, the cost must be reduced, and the accessibility of the technologies for widespread clinical use should be improved. Moreover, large-scale multicentered studies should be performed in order to validate metabolomic biomarkers and personalized treatment algorithms.

Despite these challenges, the potential benefits of this integrated approach are substantial. Personalized medicine, informed by metabolomic insights, could revolutionize GDM management, reducing complications and improving outcomes for both mother and child.

References

1. Lapolla A, Dalfrà MG, Lencioni C et al (2004) Epidemiology of diabetes in pregnancy: a review of Italian data. Diabetes Nutr Metab 17(6):358–367
2. Bloomgarden ZT (2000) American Diabetes Association 60th scientific sessions, 2000: diabetes and pregnancy. Diabetes Care 23(11):1699–1702
3. Moon JH, Jang HC (2022) Gestational diabetes mellitus: diagnostic approaches and maternal-offspring complications. Diabetes Metab J 46(1):3–14
4. Lende M, Rijhsinghani A (2020) Gestational diabetes: overview with emphasis on medical management. Int J Environ Res Public Health 17(24):9573
5. Lu W, Hu C (2022) Molecular biomarkers for gestational diabetes mellitus and postpartum diabetes. Chin Med J 135(16):1940–1951
6. Di Cianni G, Miccoli R, Volpe L et al (2003) Intermediate metabolism in normal pregnancy and in gestational diabetes. Diabetes Metab Res Rev 19(4):259–270
7. Zhang C, Rawal S, Chong YS (2016) Risk factors for gestational diabetes: is prevention possible? Diabetologia 59(7):1385–1390
8. Wang H, Li N, Chivese T, Werfalli M et al (2022) IDF diabetes atlas: estimation of global and regional gestational diabetes mellitus prevalence for 2021 by International Association of Diabetes in pregnancy study group's criteria. Diabetes Res Clin Pract 183:109050
9. Johns EC, Denison FC, Norman JE et al (2018) Gestational diabetes mellitus: mechanisms, treatment, and complications. Trends Endocrinol Metab 29(11):743–754
10. Noto A, Fanos V, Dessì A (2016) Metabolomics in newborns. Adv Clin Chem 74:35–61. https://doi.org/10.1016/bs.acc.2015.12.006. Epub 2016 Jan 19. PMID: 27117660
11. Fanos V, Pintus R, Dessì A (2018) Clinical metabolomics in neonatology: from metabolites to diseases. Neonatology 113(4):406–413
12. Dessì A, Bosco A, Cesare Marincola F et al (2023) Sardinian infants of diabetic mothers: a metabolomics observational study. Int J Mol Sci 24(18):13724. https://doi.org/10.3390/ijms241813724. PMID: 37762025; PMCID: PMC10530546

13. Burzynska-Pedziwiatr I, Dudzik D, Sansone A et al (2023) Targeted and untargeted metabolomic approach for GDM diagnosis. Front Mol Biosci 9:997436
14. Wu Y, Wang Z, Zhao Z et al (2023) Bile acid metabolites in early pregnancy and risk of gestational diabetes mellitus: results from a prospective cohort study. Diabetes Obes Metab 25(8):2255–2267
15. Huhtala M, Rönnemaa T, Tertti K (2023) Insulin resistance is associated with an unfavorable serum lipoprotein lipid profile in women with newly diagnosed gestational diabetes. Biomolecules. 13(3):470
16. Zhang Q, Tian M, Zhang X et al (2023) Metabolic biomarkers linking urinary arsenic species to gestational diabetes mellitus: a cross-sectional study in Chinese pregnant women. Sci Total Environ 892:164761
17. Li LJ, Wang X, Chong YS et al (2023) Exploring preconception signatures of metabolites in mothers with gestational diabetes mellitus using a non-targeted approach. BMC Med 21(1):99
18. Tao M, Lu G, Zhang S, Li P (2023) Root causes of long-term complications of gestational diabetes mellitus: metabolic disturbances of the host and gut microbiota. Clin Chim Acta 548:117490
19. Sormunen-Harju H, Huvinen E, Girchenko PV et al (2023) Metabolomic profiles of nonobese and obese women with gestational diabetes. J Clin Endocrinol Metab 108(11):2862–2870
20. Lyu X, Wang S, Zhong J et al (2023) Gut microbiome interacts with pregnancy hormone metabolites in gestational diabetes mellitus. Front Microbiol 14:1175065
21. Huhtala MS, Rönnemaa T, Paavilainen E et al (2023) Prediction of pre-diabetes and type 2 diabetes nine years postpartum using serum metabolome in pregnant women with gestational diabetes requiring pharmacological treatment. J Diabetes Complicat 37(7):108513
22. Wang G, Buckley JP, Bartell TR et al (2023) Gestational diabetes mellitus, postpartum Lipidomic signatures, and subsequent risk of type 2 diabetes: a Lipidome-wide association study. Diabetes Care 46(6):1223–1230
23. Heath H, Rosario R, McMichael LE et al (2023) Gestational diabetes is characterized by decreased medium-chain Acylcarnitines and elevated purine degradation metabolites across pregnancy: a case-control time-course analysis. J Proteome Res 22(6):1603–1613
24. Razo-Azamar M, Nambo-Venegas R et al (2023) An early prediction model for gestational diabetes mellitus based on metabolomic biomarkers. Diabetol Metab Syndr 15(1):116
25. Li Y, Feng Y, Yang Z, Zhou Z et al (2023) Untargeted metabolomics of saliva in pregnant women with and without gestational diabetes mellitus and healthy non-pregnant women. Front Cell Infect Microbiol 13:1206462
26. Sovio U, Clayton GL, Cook E et al (2022) Metabolomic identification of a novel, externally validated predictive test for gestational diabetes mellitus. J Clin Endocrinol Metab 107(8):e3479–e3486
27. Gao Y, Chen H, Li J, Ren S et al (2022) Alterations of gut microbiota-derived metabolites in gestational diabetes mellitus and clinical significance. J Clin Lab Anal 36(4):e24333
28. He XL, Hu XJ, Luo BY et al (2022) The effects of gestational diabetes mellitus with maternal age between 35 and 40 years on the metabolite profiles of plasma and urine. BMC Pregnancy Childbirth 22(1):174
29. Spanou L, Dimou A, Kostara CE et al (2022) A study of the metabolic pathways affected by gestational diabetes mellitus: comparison with type 2 diabetes. Diagnostics (Basel) 12(11):2881
30. Hu T, An Z, Li H et al (2022) UHPLC-MS/MS-based metabolomics and clinical phenotypes analysis reveal broad-scale perturbations in early pregnancy related to gestational diabetes mellitus. Dis Markers 2022:4231031
31. Piras C, Neri I, Pintus R et al (2022) First trimester metabolomics 1H-NMR study of the urinary profile predicts gestational diabetes mellitus development in obese women. J Matern Fetal Neonatal Med 35(25):8275–8283

32. Peng ML, Zhang Z, Zhou M et al (2022) Identification of differential metabolites using untargeted metabolomics between gestational diabetes and normal pregnant women. Int J Gynaecol Obstet 159(3):903–911
33. Furse S, Koulman A, Ozanne SE et al (2022) Altered lipid metabolism in obese women with gestational diabetes and associations with offspring adiposity. J Clin Endocrinol Metab 107(7):e2825–e2832
34. Wang Y, Wu P, Huang Y et al (2022) BMI and lipidomic biomarkers with risk of gestational diabetes in pregnant women. Obesity (Silver Spring) 30(10):2044–2054
35. Chen X, Zhao X, Jones MB et al (2022) The relationship between hair metabolites, air pollution exposure and gestational diabetes mellitus: a longitudinal study from pre-conception to third trimester. Front Endocrinol (Lausanne) 13:1060309
36. Tian M, Ma S, You Y et al (2021) Serum metabolites as an indicator of developing gestational diabetes mellitus later in the pregnancy: a prospective cohort of a Chinese population. J Diabetes Res 2021:8885954
37. Reutrakul S, Chen H, Chirakalwasan N et al (2021) Metabolomic profile associated with obstructive sleep apnoea severity in obese pregnant women with gestational diabetes mellitus: a pilot study. J Sleep Res 30(5):e13327
38. Wang Y, Huang Y, Wu P et al (2021) Plasma lipidomics in early pregnancy and risk of gestational diabetes mellitus: a prospective nested case-control study in Chinese women. Am J Clin Nutr 114(5):1763–1773
39. Mansell T, Vlahos A, Collier F et al (2022) The newborn metabolome: associations with gestational diabetes, sex, gestation, birth mode, and birth weight. Pediatr Res 91(7):1864–1873
40. Chen T, Qin Y, Chen M, Zhang Y et al (2021) Gestational diabetes mellitus is associated with the neonatal gut microbiota and metabolome. BMC Med 19(1):120
41. Raczkowska BA, Mojsak P, Rojo D et al (2021) Gas chromatography-mass spectroscopy-based metabolomics analysis reveals potential biochemical markers for diagnosis of gestational diabetes mellitus. Front Pharmacol 12:770240
42. Wu Y, Yu J, Liu X, Wang W, Chen Z, Qiao J, Liu X, Jin H, Li X, Wen L, Tian J, Saffery R, Kilby MD, Qi H, Tong C, Baker PN et al (2021) Gestational diabetes mellitus-associated changes in the breast milk metabolome alters the neonatal growth trajectory. Clin Nutr 40(6):4043–4054
43. Liu Y, Kuang A, Bain JR et al (2021) Maternal metabolites associated with gestational diabetes mellitus and a postpartum disorder of glucose metabolism. J Clin Endocrinol Metab 106(11):3283–3294
44. Sikorski C, Azab S, de Souza RJ et al (2022) Serum metabolomic signatures of gestational diabetes in south Asian and white European women. BMJ Open Diabetes Res Care 10(2):e002733
45. Lu W, Luo M, Fang X et al (2021) Discovery of metabolic biomarkers for gestational diabetes mellitus in a Chinese population. Nutr Metab (Lond) 18(1):79
46. Zhang H, Zhao Y, Zhao D et al (2021) Potential biomarkers identified in plasma of patients with gestational diabetes mellitus. Metabolomics 17(11):99
47. Yang Y, Pan Z, Guo F et al (2021) Placental metabolic profiling in gestational diabetes mellitus: An important role of fatty acids. J Clin Lab Anal 35(12):e24096
48. Meng X, Zhu B, Liu Y et al (2021) Unique biomarker characteristics in gestational diabetes mellitus identified by LC-MS-based metabolic profiling. J Diabetes Res 2021:6689414
49. Herrera-Van Oostdam AS, Salgado-Bustamante M, Lima-Rogel V et al (2021) Urinary Metabolomic profile of neonates born to women with gestational diabetes mellitus. Meta 11(11):723
50. Rahman ML, Feng YA, Fiehn O, Albert PS et al (2021) Plasma lipidomics profile in pregnancy and gestational diabetes risk: a prospective study in a multiracial/ethnic cohort. BMJ Open Diabetes Res Care 9(1):e001551

51. Ondřejíková L, Pařízek A, Šimják P et al (2021) Altered Steroidome in women with gestational diabetes mellitus: focus on neuroactive and immunomodulatory steroids from the 24th week of pregnancy to labor. Biomolecules 11(12):1746
52. Liu L, Liu L, Wang J et al (2021) Differentiation of gestational diabetes mellitus by nuclear magnetic resonance-based metabolic plasma analysis. J Biomed Res 35(5):351–360
53. Huhtala MS, Rönnemaa T, Pellonperä O et al (2021) Cord serum metabolome and birth weight in patients with gestational diabetes treated with metformin, insulin, or diet alone. BMJ Open Diabetes Res Care 9(1):e002022
54. Koos BJ, Gornbein JA (2021) Early pregnancy metabolites predict gestational diabetes mellitus: implications for fetal programming. Am J Obstet Gynecol 224(2):215.e1–215.e7
55. Zhan Y, Wang J, He X, Huang M et al (2021) Plasma metabolites, especially lipid metabolites, are altered in pregnant women with gestational diabetes mellitus. Clin Chim Acta 517:139–148

Epigenetics and Microbiome

15

Niccolò Meriggi, Mariela Mejia Monroy, Sofia Chioccioli, and Carlotta De Filippo

Contents

15.1 Human Microbiome and Its Role in Health 252
15.2 Gut Microbiota: Metabolites and Intestinal Homeostasis 253
15.3 Gut Microbiota and Intestinal Epigenetics Control the Host's Metabolic and Physiological State 254
15.4 Main Epigenetic Mechanisms Regulated by Gut Microbial Communities 256
15.4.1 Methylation and Demethylation 256
15.4.2 Histone Modifications: Acetylation, Deacetylation, and Crotonylation 258
15.4.3 Biotinylation, Phosphorylation, and RNA Interference in Epigenetic Regulation 260
15.5 Microbial Influence on lncRNAs, circRNAs, and miRNAs 261
15.6 Epigenetic Regulation of Immune Activity by the Microbiome 261
15.7 Role of Microbiome in Epigenetic Pathogen-Mediated Reprogramming: Interdependence and Behavioral Shifts 263
15.8 Conclusions 264
References 265

Authors Niccolò Meriggi and Mariela Mejia Monroy have equally contributed to this chapter.

N. Meriggi · C. De Filippo (✉)
Institute of Agricultural Biology and Biotechnology, National Research Council (CNR), Pisa, Italy
e-mail: carlotta.defilippo@cnr.it

M. M. Monroy
NEUROFARBA Department, Pharmacology and Toxicology Section, University of Florence, Florence, Italy

Department of Molecular and Developmental Medicine, University of Siena, Siena, Italy

S. Chioccioli
NEUROFARBA Department, Pharmacology and Toxicology Section, University of Florence, Florence, Italy

L. Emmi, V. Fanos (eds.), *Epigenetics and Rare Diseases*, Rare Diseases of the Immune System, https://doi.org/10.1007/978-3-032-02902-7_15

15.1 Human Microbiome and Its Role in Health

The term "human microbiome" was first introduced by Joshua Lederberg in 2001, who defined it as "the ecological community of commensal, symbiotic, and pathogenic microorganisms that literally share our body space" [1]. This community includes a diverse array of microorganisms, such as eukaryotes, archaea, bacteria, and viruses, with bacteria being the most numerous and extensively studied members of the human microbiome. These microorganisms typically coexist without harm in various parts of the body, and when in balance, commensal bacteria play vital roles in maintaining human health. Microbial communities are dynamic and can change in response to environmental factors, including diet, medications, and other exposures. By shifting the microbiome's taxonomic composition, these factors can lead to significant changes in the production of their related metabolites, consequently affecting the functionality of the microbiome and the impact on human health.

For example, gut microbiota can synthesize essential vitamins that the human genome does not encode, providing a metabolic complement. They also aid in digesting complex food sources, releasing nutrients, supporting immune function, and producing anti-inflammatory compounds that help inhibit the activity of pathogenic microbes. However, disruptions to microbiome composition or imbalances (known as dysbiosis) have been associated with a wide range of diseases, including asthma, atopic dermatitis, periodontal disease, obesity, cancer, Crohn's disease, and other gastrointestinal disorders [2]. While much of the medical literature has focused on the microbiome's role in disease, fewer studies have investigated its physiological functions. The 2007 Human Microbiome Project (HMP), led by the NIH, characterized the microbiome across body sites in healthy individuals, revealing the microbiome's key roles in health. This project provided groundbreaking insights into host-microbiome interactions and laid the groundwork for strategies to monitor and manipulate the microbiome to improve health [1]. Concluding in 2013, the HMP was succeeded by the Integrated Human Microbiome Project (iHMP), which uses a longitudinal, multi-omics approach to further expand the understanding of human microbial communities (https://hmpdacc.org/ihmp/).

According to all the evidence described, the microorganisms inhabiting the gastrointestinal tract appear as a dynamic ecosystem that has co-evolved with the host becoming directly responsible for the host's health. This relationship is mutually beneficial, with the microbiota compensating for metabolic functions the human body cannot perform [3]. Thus, the gut microbiome can be considered an organ due to its complexity and vital roles. While it lacks a defined structure like other organs, the microbiota weighs approximately 1.5–2 kg and functions as an organized system of cells, like a complex ecosystem. It is dominated by five major bacterial groups (phyla): Bacteroidetes, Firmicutes, Actinobacteria, Proteobacteria, and Fusobacteria. Most bacteria in the gut are strict anaerobes, outnumbering facultative anaerobes and aerobes. These bacteria perform essential functions, such as protecting the gut against pathogens, aiding in digestion, and producing vitamins and short-chain fatty acids (SCFAs). The gut microbiota's influence on immunity and metabolism highlights its role as an "organ" crucial for health. The selection of specific SCFA-producing bacteria is important for maintaining the integrity of the intestinal epithelium, harvesting energy, regulating immunity, and preventing pathogen overgrowth. In recent years, the focus on gut health has grown, particularly in preventive medicine. However, defining gut health remains a challenge and is often interpreted differently across scientific and clinical contexts [4]. Some experts define a healthy gut as the absence of diagnosed gastrointestinal diseases or disorders such as intestinal bowel diseases (Crohn's disease, ulcerative colitis), intestinal bowel syndrome (IBS), or celiac disease. While this clinical approach provides a clear criterion—absence of identifiable medical conditions—it may overlook subtler health issues not correlated with diagnostic criteria.

15.2 Gut Microbiota: Metabolites and Intestinal Homeostasis

The gut microbiota plays an essential role in maintaining intestinal homeostasis through an intricate balance of metabolic and defensive functions. One key mechanism is "colonization resistance", where commensal bacteria inhibit the establishment of pathogenic microorganisms by competing for essential nutrients and adhesion sites on the intestinal mucosa [5]. This defensive strategy is further bolstered by the production of antimicrobial compounds and metabolites that suppress pathogen growth, maintaining a stable and protective microbial community. They also produce SCFAs such as acetate, butyrate, and propionate, contributing to the reduction of intestinal pH, creating an unfavorable environment for pathogens [6, 7]. Beyond defense, SCFAs, derived from carbohydrate fermentation, play essential roles in mucosal immunity, gut motility, colonic water absorption, and energy provision for colonic cells [8, 9].

SCFAs also interact with G-protein-coupled receptors (GPR41 and GPR43), stimulating the release of gut hormones like peptide YY (PYY) and glucagon-like peptide 1 (GLP-1), which enhance intestinal barrier integrity, regulate gastric emptying, and promote satiety. Additionally, SCFAs, along with polyphenols, upregulate tight junction proteins such as claudin and zonula occludens-1 (ZO-1), strengthening the intestinal barrier and reducing permeability.

Gut bacteria further contribute to host metabolism by synthesizing essential vitamins such as K and B vitamins [10], transforming bile acids, and metabolizing xenobiotics and toxins [11]. Secondary bile acids produced by gut microorganisms activate nuclear receptors, profoundly influencing host metabolic pathways [12]. Moreover, bacterial enzymes transform dietary compounds such as lignans and isoflavones and modulate host detoxification pathways by altering gene expression and enzyme activity [13].

Specific gut-derived metabolites also regulate immune responses and inflammation. For example, tryptophan metabolites activate the aryl hydrocarbon receptor (AHR) in intestinal immune cells, promoting IL-22 production and supporting innate and adaptive immunity [14]. Indole derivatives modulate mucosal homeostasis through pregnane X receptor (PXR) signaling, reducing inflammation and improving intestinal permeability [15]. Polyphenols and SCFAs downregulate pro-inflammatory cytokines like IL-1β, IL-6, and TNF-α, further mitigating inflammation and supporting homeostasis.

All the evidence described above highlights that the microbiome represents a central modulatory agent in the host physiology. Collectively, all the functions described underscore the centrality of gut microbiota and its metabolites in preserving intestinal health, mediating host nutrition, and regulating systemic metabolic and immune processes. Although attention has mainly focused on the direct action of metabolites and/or how the host metabolizes them, there are also deeper levels of modulation such as epigenetic activity which needs to be further considered and studied.

15.3 Gut Microbiota and Intestinal Epigenetics Control the Host's Metabolic and Physiological State

The relationship between the gut microbiota and epigenetic mechanisms is connected to the microbial signals transmitted through bacterial sensing pathways. Gut microbes influence key epigenetic modifications, including the involvement in addition/removal of methyl, acetyl, and crotonyl groups on histone lysine residues, histone phosphorylation, DNA methylation, and likely hydroxylation, and the regulation of noncoding RNA. These epigenetic changes form a dynamic interface that allows the microbiota to interact with the host genome, shaping health and disease outcomes. Furthermore, the intimate connection between the microbiota and the single layer of intestinal epithelial cells is crucial for regulating essential biological processes, such as metabolism, nutrient absorption, neuronal development, and angiogenesis. This complex interplay underscores the microbiota's central role in maintaining host homeostasis and its significant influence on disease pathogenesis. The main mechanisms by which the microbiota influences host physiology through epigenetic regulation can be resumed as follows (Fig. 15.1):

1. **Impact of the gut microbiota biosynthesis or metabolism in the availability of host endogenous epigenetically relevant metabolite precursors.** Epigenetic regulation involves modifications to DNA or histones that affect gene expression without altering the underlying genetic code and this effect can occur by microbial metabolites. Metabolite precursors involved in epigenetic modifications include (i) SCFAs such as butyrate, propionate, and acetate, (ii) amino acids (e.g., tryptophan, methionine), (iii) one-carbon metabolism intermediates, which are involved in methylation processes (e.g., vitamins B2, B9, and B12) through S-adenosylmethionine (SAM), the main substrate for DNA and histone methylation, (iv) nucleotides involved in DNA synthesis and repair.
2. **Regulation of the activity or expression of epigenetic-modifying enzymes.** Dysbiosis and a decrease in microbiota diversity can influence the levels of nutrients and metabolites that may regulate DNA methylation and histone modifications altering host gene expression. This can occur by directly inhibiting the

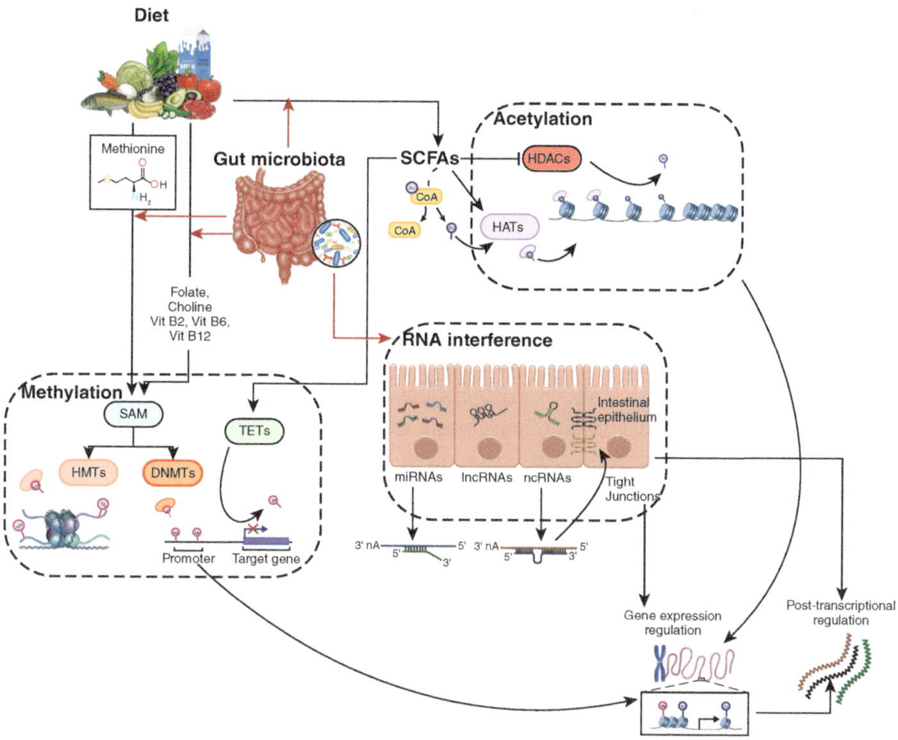

Fig. 15.1 Main interactions of the gut microbiota with the host epigenome. Commensal microbes produce essential metabolites that act as substrates, cofactors, or regulators of epigenetic enzymes. *HDACs* histone deacetylases, *HATs* histone acetyltransferases, *SCFAs* short-chain fatty acids, *SAM* S-adenosylmethionine, *HMTs* histone methyltransferases, *DNMTs* DNA methyltransferases, *TETs* ten-eleven translocation enzymes, *miRNAs* microRNAs, *lncRNAs* long noncoding RNAs, *ncRNAs* Noncoding RNAs

enzymes involved in these processes. For instance, SCFAs directly inhibit host histone deacetylases (HDACs) deacetylase activity [16] resulting in chromatin changes generally associated with increased expression of target genes [17]. SCFAs also affect histone decrotonylation and acylation [18].

The intricate relationship between the gut microbiota and host immune cells maturation has been increasingly recognized as a key element in the regulation of epigenetic mechanisms. One of the proposed mechanisms underlying this interaction is the modulation of chromatin states. Seeley et al. [19] provided compelling evidence supporting this hypothesis, demonstrating that prolonged exposure of bone-marrow-derived macrophages to lipopolysaccharides (LPS) led to the transcriptional silencing of LPS-induced genes through the restriction of chromatin remodeling.

Epigenetic modifications, particularly DNA methylation, play a pivotal role in mediating the effects of microbial signals on host physiology. Thaiss et al. explored the influence of microbiota-derived metabolites on the host's circadian transcriptional and epigenetic rhythms, revealing that these environmental cues contribute to the synchronization of metabolic pathways in peripheral organs such as the liver. They observed that bacterial populations within the colon exhibit rhythmic fluctuations in localization and metabolite secretion throughout the day. Disrupting this rhythmicity through antibiotic treatment resulted in a reorganization of metabolic pathways in both the intestine and the liver, subsequently altering diurnal gene expression patterns [20].

Further insights into the interplay between microbiota composition and epigenetic modifications have emerged from studies employing the next-generation sequencing of DNA methylomes. A study conducted on pregnant women revealed a correlation between dominant gut microbiota phyla—Bacteroidetes, Proteobacteria, and Firmicutes (recently reclassified as Bacteroidota, Pseudomonadota, and Bacillota, respectively)—and specific epigenetic profiles. Notably, individuals in whom Firmicutes predominated exhibited methylation patterns associated with genes linked to cardiovascular diseases, lipid metabolism, obesity, and inflammatory responses [21].

In addition to these findings, alterations in DNA methylation have been associated with increased expression and activity of DNA demethylase enzymes such as Tet3 and Dnmt1 in conventionally housed (CNV) mice, suggesting that microbial environments influence epigenetic regulation at a systemic level [22]. Collectively, these studies highlight the profound impact of gut microbiota on the host epigenome, emphasizing its role in shaping immune function, metabolic homeostasis, and disease susceptibility. Furthermore, the microbiota interacts with host chromatin through histone modifications, both under steady-state conditions and in the context of inflammation and disease [23].

The influence of gut microbiota on host epigenetics extends beyond individual histone modifications. In a study using post-colonized germ-free (GF) mice, Krautkramer et al. demonstrated that microbial colonization globally regulates histone acetylation and methylation across various host tissues in a manner dependent on dietary composition. Their findings highlighted that a "Western type" diet, characterized by high fat and sucrose content and a low polysaccharide intake, prevented host histone modifications due to a reduced production of SCFAs. Strikingly, they also showed that the epigenetic phenotypes observed in GF mice colonized with gut microbiota could be recapitulated through SCFA supplementation alone, emphasizing the critical role of microbial-derived metabolites in shaping host epigenetic landscapes [24].

15.4 Main Epigenetic Mechanisms Regulated by Gut Microbial Communities

15.4.1 Methylation and Demethylation

Epigenetic-modifying enzymes, such as DNA methyltransferases (DNMTs), histone methyltransferases (HMTs), and histone acetyltransferases (HATs), rely on methyl and acetyl donors for their catalytic activity. The gut microbiota influences

DNA methylation by modulating the availability of key methyl donors, including microbial-produced folate, dietary methionine, and other components of one-carbon metabolism. This pathway is essential for synthesizing S-adenosylmethionine (SAM), the primary methyl donor for DNA methylation, which gut commensals can produce from dietary methionine and folate. Key elements of one-carbon metabolism—folate, choline, methionine, and vitamins B2, B6, and B12—are indispensable for the host, which depends on dietary intake and microbial contributions [25]. Folate facilitates the remethylation of homocysteine to methionine, ensuring a steady SAM supply. Microbiota dysbiosis can disrupt SAM levels, thereby altering DNA and histone methylation.

Variations in DNA methylation, influenced by these metabolic factors, regulate gene activation and silencing across tissues, underscoring the microbiota's crucial role in host epigenetic processes. While hosts possess intrinsic metabolic pathways for synthesizing some of these molecules, the gut microbiota serves as a significant additional source, producing essential metabolites that act as substrates, cofactors, or regulators of epigenetic enzymes.

Vitamin B2, also known as riboflavin, is a precursor for synthesis of flavin-adenine dinucleotide (FAD), a cofactor for the synthesis of 5-methyltetrahydrofolate [26] while vitamin B6 acts as coenzyme through a different pathway. Vitamin B12, on the other hand, acts as coenzyme in the production of SAM [25]. Evidence that the microbiota influences the host epigenome comes from observed changes in DNA methylation. In GF mice, DNA methylation at the 5′ CpG island of the toll-like receptor 4 (*Tlr4*) gene in large intestine epithelial cells (IECs) was reduced compared to CNV mice, leading to lower *Tlr4* expression and decreased responsiveness to LPS. Similarly, demethylation was found at the *Cxcl16* gene in the colon and lungs of GF mice, correlating with reduced mucosal invariant natural killer T cell accumulation in the absence of microbiota. The microbiota also regulates DNA methylation in regulatory T cells (Tregs), enhancing the expression of *Uhrf1*, a DNA methylation adaptor protein that interacts with DnECs mt1 and HDAC1 [27]. Deletion of *Uhrf1* in T cells disrupted normal cell cycle gene expression in Tregs and resulted in spontaneous colitis.

Studies have highlighted that the microbiota plays a key role in shaping the mammalian DNA methylome across various cell types and tissues, influencing both normal function and disease. The microbiota-derived SCFA acetate is an acetyl group donor that affects alpha-ketoglutarate, an intermediate of the tricarboxylic acid (TCA) and co-substrate of ten-eleven translocation (TET) enzymes, a family of DNA and histone demethylases [28]. Ansari et al. through whole-genome bisulfite sequencing of conventionally raised and germ-free mice, identified localized, TET2/3-dependent DNA methylation changes at regulatory elements driven by commensal microbiota exposure. This process activates "early sentinel" response genes critical for maintaining intestinal homeostasis. Moreover, during dextran sodium sulfate-induced inflammation, microbiota exposure leads to substantial alterations in DNA methylation and chromatin accessibility, reshaping gene expression patterns associated with colitis and colon cancer [22]. In humans, DNA methylation in colonic biopsies was linked to microbiota composition, inflammation, and disease status, particularly in ulcerative colitis, which is often a precursor to colorectal cancer. Higher levels of *Fusobacterium* were tied to increased DNA methylation at colorectal cancer-related genes in ulcerative colitis patients, and microbial dysbiosis was shown to promote DNA methylation changes and epithelial hyperplasia in GF mice [29]. Similar microbiota-related DNA methylation changes were also seen in obese humans, especially in genes that regulate energy metabolism. These findings suggest that alterations in the microbiota-driven DNA methylation axis could be predictive of inflammation-related cancer development.

15.4.2 Histone Modifications: Acetylation, Deacetylation, and Crotonylation

Human HATs are classified into four families: GNAT (including GNC5 and PCAF), MYST (such as MOZ and TIP60), p300/CBP, and SRC (including SRC-1 and ACTR). Histone deacetylases (HDACs), categorized into four groups, either require zinc (Groups I, II, IV) or NAD+ (Group III) for their activity. HATs promote transcription by acetylating histones, while HDACs remove acetyl groups, leading to gene repression.

Host-microbiota interactions are partly mediated by small-molecule metabolites, which can act as receptor ligands, enzyme inhibitors, or metabolic precursors. The microbial metabolome is shaped by the composition of the microbiota and the host's diet. Microbial metabolites, such as SCFAs, can regulate histone acetylation, methylation, and crotonylation. Butyrate and acetate contribute to histone acetylation by providing intermediate molecules essential for HATs. Butyrate is a particular energy source for IECs acting as a readily available carbon source that undergoes β-oxidation to produce acetyl-coenzyme A (CoA), which fuels the TCA cycle and supports ATP production. Acetyl-CoA is the primary acetyl donor for HATs [30]. Microbiota and SCFAs play a crucial regulatory role in the nervous system mediating microglia development and homeostasis in the central nervous system [31]. Microglia in GF mice have reduced activation and interferon (IFN) type I signaling, along with altered expression of genes related to histone acetylation and methylation (e.g., *HDAC1*, *Kdm6b*) and SCFAs administration partially restores microglial function. Epicatechin and epigallocatechin-3-gallate, bioactive compounds commonly found in tea, can function as HAT inhibitors, an enzyme involved in regulating gene expression through histone modification. However, these compounds can also be degraded and metabolized by the intestinal microbiota, which may influence their bioavailability and biological effects.

HDAC activity is regulated by microbial metabolites, dietary factors, and hormones. Among these, SCFAs such as butyrate and propionate, play a key role by inhibiting HDACs, thereby modulating gene expression. HDACs are transcriptional repressors that remove acetyl groups from histone lysine residues, leading to chromatin condensation and transcriptional silencing. By inhibiting HDAC activity, SCFAs prevent chromatin compaction, maintaining a more open structure that facilitates gene transcription and enhances chromatin accessibility.

Other bacterial byproducts, such as sulforaphane cysteine from cruciferous vegetables and allyl compounds from garlic, also function as HDAC inhibitors. Butyrate, known for its anticancer properties, modulates gene expression in colon cancer cells by inhibiting HDACs and promoting apoptosis. It acetylates the transcription factor SP1, reducing its ability to repress tumor-suppressor genes like *p21* and *BAK*, thereby inducing cell cycle arrest and apoptosis [25]. Additionally, butyrate has been shown to promote cell differentiation, trigger apoptosis, and inhibit tumor cell proliferation, primarily through extensive histone hyperacetylation. The reciprocal relationship between host epigenetics and the gut microbiota (the host-microbiome axis of epigenetic interactions) has also been highlighted. HDACs play a crucial role in regulating interactions between the gut microbiota and various intestinal cell types. Disrupting these pathways can alter intestinal homeostasis, increase susceptibility to inflammation, and reshape the gut microbiota composition.

Alenghat et al. demonstrated that mice with an IEC-specific deletion of histone deacetylase 3 (HDAC3ΔIEC) exhibit widespread dysregulation of IEC gene expression, including reduced antimicrobial gene activity. These mice experience Paneth cell loss, impaired IEC function, altered gut microbiota, and heightened susceptibility to intestinal damage and inflammation, highlighting HDAC3's essential role in maintaining intestinal homeostasis. Notably, under germ-free conditions, IEC gene expression, Paneth cell homeostasis, and barrier function are largely restored, suggesting that HDAC3 integrates commensal-derived signals to regulate epithelial responses necessary for host-microbiota balance [32]. Additionally, the microbiota helps maintain an anti-inflammatory state in macrophages by promoting HDAC3-mediated deacetylation of the pro-inflammatory cytokine IL-12β [33]. The colonization of *Clostridium scindens* in mice enhanced the expression of H3K27-targeting demethylase in bone marrow-derived granulocyte-monocyte progenitors (GMPs), leading to an increase in activating H3K4me3 marks and a decrease in repressive H3K27Ac marks at the promoter regions of the granulopoiesis genes *C/EBPA* and *C/EBPB*. These epigenetic modifications resulted in increased neutrophil production and enhanced protection against *Entamoeba histolytica* infection [34].

Moreover, in mice, daily fluctuations in microbiota composition and abundance were mirrored by rhythmic changes in histone modifications associated with active promoters (H3K27Ac and H3K4me3), enhancer activity (H3K27Ac and H3K4me2), and the expression of circadian genes in colonic IECs. However, these chromatin and transcriptional oscillations were significantly disrupted when the microbiota was depleted through antibiotics or in GF mice, suggesting that the microbiota's natural daily variations are crucial for regulating the circadian transcriptome and maintaining normal circadian rhythms [20].

Beyond their role in HDAC inhibition, SCFAs also influence other histone-modifying enzymes. Both butyrate and propionate have been identified as inhibitors of sirtuin deacetylase enzymes, further reinforcing their impact on chromatin dynamics. Interestingly, Sirtuin 1 (SIRT1), the most evolutionarily conserved NAD + -dependent protein deacetylase, might be an important mediator of host-microbe interactions as found by Wellman et al. Using mice with localized intestinal deletion of SIRT1, a conserved protein deacetylase, it was found that SIRT1 prevented intestinal inflammation by regulating intestinal microbiota [35]. This SIRT1-deficient mice model also has more goblet and Paneth cells with a consequent modulation of the intestinal microbiota, for instance, due to protection from pathogens. Evidence suggests the importance of understanding how epigenetic elements, like the activity of HDAC3 and SIRT1, influence the composition of the intestinal microbiota.

Lysine crotonylation is a relatively unexplored histone modification regulated by specialized histone crotonyltransferases and decrotonylases.

Studies have shown that crotonylation levels on histones H3 and H4 (H3K18Cr, H3K8Cr, and H4K8Cr) are significantly reduced in the colon of antibiotic-treated mice. However, treatment with the HDAC inhibitor butyrate restores crotonylation, highlighting a direct link between this epigenetic modification and the gut microbiota [18]. Non-mucosal mononuclear phagocytes, such as macrophages and dendritic cells, depend on microbial signals to epigenetically and transcriptionally activate genes involved in interferon (IFN) signaling and T cell responses. In GF or antibiotic-treated mice, a reduction in the activating histone mark *H3K4me3* is observed in genes related to pro-inflammatory responses, such as type I IFNs. Conversely, there is an increase in the repressive mark *H3K27me3* in genes involved in metabolic processes, compared to cells exposed to microbiota. Moreover, butyrate enhances histone crotonylation by inhibiting HDAC-mediated decrotonylase activity in IECs. This effect leads to a widespread increase in crotonylation across the genome, including within repetitive DNA regions, rather than merely increasing crotonylation at specific sites. Importantly, microbiota depletion through antibiotic treatment significantly reduces histone crotonylation, emphasizing its strong dependence on microbial metabolites.

15.4.3 Biotinylation, Phosphorylation, and RNA Interference in Epigenetic Regulation

Biotinylation, phosphorylation, and RNA interference are key epigenetic mechanisms influenced by diet, microbial interactions, and cellular signaling.

Biotinylation involves the covalent attachment of biotin to histones and plays a crucial role in gene repression and genome stability. Biotin deficiency has been associated with increased chromosomal instability and transposable element activity. This process is catalyzed by holocarboxylase synthetase, biotinidase, and microbial BirA ligase, which regulate gene expression in both prokaryotes and eukaryotes.

Phosphorylation of histones and non-histone proteins modulates gene expression through the activity of kinases and phosphatases. By altering chromatin structure and protein function, phosphorylation serves as a dynamic regulatory mechanism in response to cellular and environmental signals.

RNA interference (RNAi) is another key regulatory pathway that influences gene expression at both the transcriptional and post-transcriptional levels. Noncoding

RNAs (ncRNAs) are differentially expressed in response to variations in the gut microbiota's composition and metabolism. Based on their nucleotide length, they are classified into microRNAs (miRNAs), circular RNAs (circRNAs), and long noncoding RNAs (lncRNAs). Studies in germ-free, mono-colonized, and CNV mice have shown that ncRNAs regulate gene expression at both the gene and chromosomal levels [23].

Although the precise regulatory mechanisms of ncRNAs are not fully understood, they play significant roles in host responses to microbiome-related conditions such as obesity and cancer. Additionally, bacterial ncRNAs carried in outer-membrane vesicles can influence the host epigenome, while dietary ncRNAs impact both the microbiota and host gene expression, further highlighting the complexity of host-microbiota interactions.

15.5 Microbial Influence on lncRNAs, circRNAs, and miRNAs

The expression patterns of lncRNAs can distinguish different microbial populations in the intestine [36]. Additionally, colonization of germ-free (GF) mice with specific pathogen-free microbiota alters the expression of circRNAs in the blood and within the lung tumor microenvironment [37]. miRNAs are particularly sensitive to gut microbiota composition, significantly impacting host physiology [38]. Commensal microbes regulate host miRNA expression, affecting genes involved in drug resistance, inflammation, and epithelial barrier integrity. Probiotics also modulate miRNA levels, further emphasizing the complex relationship between microbial communities, epigenetics, and host health. For example, the probiotic strains *Lactobacillus fermentum* and *Lactobacillus salivarius*, isolated from human breast milk, were shown to enhance *miR-155* and *miR-223*, while restoring *miR-150* and *miR-143* expression in a dextran sodium sulfate (DSS)-induced colitis mouse model. These effects correlated with the improvement of microbiota dysbiosis and the restoration of short-chain fatty acid- and lactic acid-producing bacterial populations, promoting anti-inflammatory responses and strengthening intestinal barrier function [38]. Although miRNAs regulate approximately 30% of eukaryotic genes, their precise mechanisms and broader roles in health and disease remain active areas of research.

15.6 Epigenetic Regulation of Immune Activity by the Microbiome

The microbiome plays a significant role in regulating immune responses through epigenetic mechanisms. For instance, Takahashi et al. demonstrated that gut bacteria help maintain the methylation of the Tlr4 gene in mouse intestinal cells [39].

In the presence of bacteria, the CpG motifs of *Tlr4* showed higher methylation in intestinal cells compared to CD45+ splenic cells, thereby reducing excessive inflammatory responses. In vitro studies with human cell lines further supported these findings, revealing that gut microbes regulate inflammation via *Tlr4* promoter methylation. Interestingly, GF mice had lower methylation in the 5′ region of the *Tlr4* gene in large intestinal cells compared to CNV mice, suggesting a more significant role of microbiota-driven methylation in the large intestine. Beyond regulating inflammation, gut microbiota also enhances immune function by promoting transcriptionally active histone methylation at inflammatory gene loci. Microbiota-induced histone modifications are linked to changes in chromatin accessibility across different tissues. For example, microbiota can affect histone acetylation and methylation in the colon, liver, and adipose tissue, depending on diet. Supplementing GF mice with SCFAs partially reversed these microbiota-sensitive histone modifications and altered gene expression patterns [24]. In mice, commensal bacteria induced H3 trimethylation at the promoters of inflammatory genes like *IL-6* and *IFNb1*, which primed NK cells for a faster response to intracellular pathogens. These findings reinforce the role of gut microbiota in preparing the immune system for rapid defense.

In intestinal or bone marrow-derived macrophages treated with butyrate, HDAC activity was inhibited, leading to increased histone acetylation at secondary response genes like *IL-6*, *IL-12*, and STAT6 signaling mediators. However, butyrate also reduced the expression of these genes, promoting an anti-inflammatory M2 macrophage phenotype that alleviates inflammation in DSS-induced colitis [40]. SCFAs like butyrate and propionate enhance immune responses against enteric pathogens by promoting monocyte differentiation into macrophages and activating an antimicrobial transcriptional program via HDAC3 inhibition. Additionally, SCFAs facilitated histone acetylation and IL-22 production by ILCs and CD4+ T cells through GPR41 activation and HDAC inhibition, providing better protection against intestinal inflammation during bacterial infections [41].

Butyrate also inhibited HDACs and activated GPR109A in colonic macrophages and dendritic cells, promoting an anti-inflammatory Treg response [42]. Clostridia-derived butyrate increased circulating Treg frequencies by inducing histone H3 acetylation in the Foxp3 regulatory regions [17]. Additionally, post-transcriptional acetylation of Foxp3 proteins in peripheral CD4+ T cells enhanced their expression and stability. Similarly, propionate treatment or deletion of HDAC3, HDAC6, or HDAC9 in mice led to increased H3K9 acetylation at the Foxp3 locus and stabilized Foxp3 protein [43]. Acetate produced by gut microbiota inhibited HDAC activity and increased histone acetylation in CD3/CD28-activated human CD4+ T cells, though the exact effects of acetate on HDAC inhibition are still debated [44]. These findings highlight how SCFAs regulate epigenetic and transcriptional responses in immune cells, supporting an anti-inflammatory phenotype during colitis and pathogenic infections (Fig. 15.2).

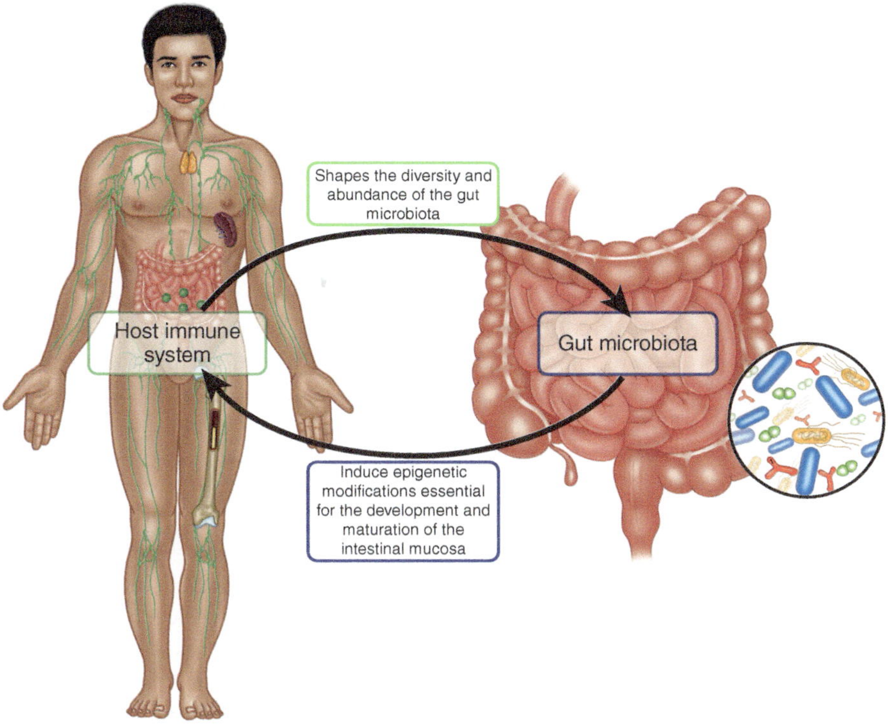

Fig. 15.2 Immune system and gut microbiota interaction. The immune system is implicated in the regulation of intestinal microbial communities, while the metabolites produced by intestinal microorganisms can modulate immune activity, producing a bidirectional regulatory system

15.7 Role of Microbiome in Epigenetic Pathogen-Mediated Reprogramming: Interdependence and Behavioral Shifts

The gut microbiota and epigenetic machinery have a bidirectional relationship that influences host health, behavior, and disease susceptibility. Intestinal HDACs regulate the composition of the microbiota, as demonstrated in HDAC3ΔIEC mice, where the deletion of HDAC3 disrupted gut microbial diversity, impaired epithelial function, and increased inflammation. This condition was reversible upon restoring the GF condition. SCFAs produced by commensal bacteria modulate immune responses by inhibiting HDACs, while HDACs themselves regulate inflammation, epithelial proliferation, and microbial interactions. Microbial metabolites, such as spermidine, butyrate, and polyphenols, serve as neuromodulators, influencing histone acetylation and gene expression related to learning and memory. Spermidine, produced by *Bacteroides thetaiotaomicron* and *Fusobacterium varium*, has been linked to improved memory and enhanced resilience against aging. Sodium butyrate, primarily produced by bacterial species belonging to *Clostridium*, *Eubacterium*, and *Butyrivibrio* genera, has been shown to enhance cognitive function in Alzheimer's models and protect neurons from degeneration in Parkinson's and Huntington's diseases by restoring histone acetylation [45].

Polyphenols, another group of microbial metabolites, act as HDAC inhibitors, potentially mitigating neurodegenerative disorders through epigenetic modulation. These findings highlight how changes in gut microbiota can lead to significant neuro-epigenetic effects, suggesting that probiotics could play a role in maintaining mental and cognitive health. Furthermore, bacterial endotoxins like LPS modulate epigenetic responses, particularly in inflammation. LPS treatment of bovine peripheral blood mononuclear cells increases the expression of genes of inflammatory cytokines such as *TNF, IL-2, IFN-γ, IL-1β, IL-4*, and *IL-8* while downregulating *HDAC6, HDAC7, DNMT3A*, and *DNMT3B*. HDAC inhibitors like trichostatin A suppress most of these inflammatory markers, except for IL-1β, which remains unchanged [46].

Similar mechanisms are seen in LPS-induced lung inflammation, where acetylation of histones H3 and H4 increases IL-6 expression, exacerbating inflammation. However, this inflammatory response can be reversed by anacardic acid, an HDAC inhibitor, demonstrating the potential therapeutic benefits of epigenetic modulation in controlling inflammation. Together, these findings underscore the profound impact of microbial interactions on epigenetic regulation and highlight the need for further research into microbiota-targeted therapies for disease prevention and treatment.

15.8 Conclusions

In conclusion, the intricate interaction between the gut microbiota and intestinal epigenetics reveals a dynamic and profound relationship that significantly impacts the host's metabolic and physiological state. This relationship is mediated through metabolites derived from the microbiota, such as SCFAs, amino acids, and vitamins, which serve as critical precursors for epigenetic modifications, including DNA methylation, histone acetylation, and phosphorylation. These metabolites not only influence the activity of enzymes responsible for epigenetic regulation but also modulate the expression of noncoding RNAs, further complicating the regulation of gene expression. In this way, the microbiota can shape the host's gene expression profile, influencing key biological processes such as metabolism, immune function, neuronal development, and angiogenesis. The ability of gut microbes to directly influence chromatin remodeling, histone modifications, and DNA methylation highlights the microbiota's central role in maintaining homeostasis and controlling disease pathways. Dysbiosis, or an imbalance in microbial diversity, can disrupt these epigenetic processes, leading to alterations in gene expression that contribute to conditions such as obesity, cardiovascular disease, and inflammation. These disruptions are not only due to microbial metabolites but can also reflect more systemic, long-term epigenetic changes, including the modulation of circadian rhythms and immune maturation, processes that are also influenced by microbial activity.

What complicates this relationship further is the emerging evidence suggesting that microbial composition is not only influenced by environmental factors such as diet and antibiotics but also has the potential to influence epigenetic modifications in a host-specific manner. The impact of a "Western type" diet, for example, illustrates the importance of dietary habits in shaping microbial influence on the host's epigenetic landscape, with the potential to reverse or exacerbate disease states. The fact that SCFAs supplementation alone can replicate epigenetic changes observed in GF mice underscores the crucial role of microbial-derived metabolites in shaping the epigenome, independently of other environmental factors.

A particularly intriguing aspect of this relationship is the bidirectional influence between the host's epigenome and the microbiota. The host's epigenetic modifications may not only respond to microbial signals but also influence microbial behavior and composition in return, creating a feedback loop. Changes in the microbiota can alter epigenetic regulation, which, in turn, may reshape the microbiota itself, influencing disease outcomes. This reciprocal regulation opens exciting possibilities for therapeutic interventions aimed at restoring microbial balance and improving host health. Whether through dietary adjustments, probiotics, or microbiota-targeted therapies, manipulating the microbial ecosystem to promote beneficial epigenetic outcomes offers a promising path for personalized medicine and disease prevention. Thus, the gut microbiota is far more than a passive participant in digestion and immunity—it acts as a crucial regulator of gene expression and overall health. While the full extent of its role in shaping the host's epigenome is still being explored, it is evident that a deeper understanding of this relationship will be pivotal in developing therapeutic strategies to address the rising prevalence of metabolic and inflammatory diseases.

References

1. NIH HMP Working Group et al (2009) The NIH human microbiome project. Genome Res 19(12):2317–2323. https://doi.org/10.1101/gr.096651.109
2. Dietert RR, Dietert JM (2015) The microbiome and sustainable healthcare. Healthcare (Basel, Switzerland) 3(1):100–129. https://doi.org/10.3390/healthcare3010100
3. Thursby E, Juge N (2017) Introduction to the human gut microbiota. Biochem J 474(11):1823–1836. https://doi.org/10.1042/BCJ20160510
4. Van Hul M, Cani PD, Petitfils C, De Vos WM, Tilg H, El-Omar EM (2024) What defines a healthy gut microbiome? Gut 73(11):1893–1908. https://doi.org/10.1136/gutjnl-2024-333378
5. Caballero-Flores G, Pickard JM, Núñez G (2023) Microbiota-mediated colonization resistance: mechanisms and regulation. Nature Rev Microbiol 21(6):347–360. https://doi.org/10.1038/s41579-022-00833-7
6. Guarner F, Malagelada JR (2003) Gut flora in health and disease. Lancet 361:512–519. https://doi.org/10.1016/S0140-6736(03)12489-0
7. Stecher B, Hardt WD (2008) The role of microbiota in infectious disease. Trends Microbiol 16:107–114. https://doi.org/10.1016/j.tim.2007.12.008

8. Zhang YJ, Li S, Gan RY, Zhou T, Xu DP, Li HB (2015) Impacts of gut bacteria on human health and diseases. Int J Mol Sci 16:7493–7519. https://doi.org/10.3390/ijms16047493
9. Rooks MG, Garrett WS (2016) Gut microbiota, metabolites and host immunity. Nature Rev Immunol 16:341–352. https://doi.org/10.1038/nri.2016.42
10. Morowitz MJ, Carlisle EM, Alverdy JC (2011) Contributions of intestinal bacteria to nutrition and metabolism in the critically ill. Sur Clin North America 91:771–viii. https://doi.org/10.1016/j.suc.2011.05.001
11. Macfarlane S, Steed H, Macfarlane GT (2009) Intestinal bacteria and inflammatory bowel disease. Crit Rev Clin Lab Sci 46:25–54. https://doi.org/10.1080/10408360802485792
12. Winston JA, Theriot CM (2020) Diversification of host bile acids by members of the gut microbiota. Gut Microbes 11(2):158–171. https://doi.org/10.1080/19490976.2019.1674124
13. Koppel N, Maini Rekdal V, Balskus EP (2017) Chemical transformation of xenobiotics by the human gut microbiota. Science 356(6344):eaag2770. https://doi.org/10.1126/science.aag2770
14. Dong F, Perdew GH (2020) The aryl hydrocarbon receptor as a mediator of host-microbiota interplay. Gut Microbes 12(1):1859812. https://doi.org/10.1080/19490976.2020.1859812
15. Ma X, Chen J, Tian Y (2015) Pregnane X receptor as the "sensor and effector" in regulating epigenome. J Cell Physiol 230(4):752–757. https://doi.org/10.1002/jcp.24838
16. Cousens LS, Gallwitz D, Alberts BM (1979) Different accessibilities in chromatin to histone acetylase. J Biol Chem 254(5):1716–1723. https://doi.org/10.1016/S0021-9258(17)37831-6
17. Furusawa Y, Obata Y, Fukuda S, Endo TA, Nakato G, Takahashi D, Nakanishi Y, Uetake C, Kato K, Kato T, Takahashi M, Fukuda NN, Murakami S, Miyauchi E, Hino S, Atarashi K, Onawa S, Fujimura Y, Lockett T, Clarke JM, Topping DL, Tomita M, Hori S, Ohara O, Morita T, Koseki H, Kikuchi J, Honda K, Hase K, Ohno H (2013) Commensal microbe-derived butyrate induces the differentiation of colonic regulatory T cells. Nature 504:446–450. https://doi.org/10.1038/nature12721
18. Fellows R, Denizot J, Stellato C, Cuomo A, Jain P, Stoyanova E, Balázsi S, Hajnády Z, Liebert A, Kazakevych J, Blackburn H, Corrêa RO, Fachi JL, Sato FT, Ribeiro WR, Ferreira CM, Perée H, Spagnuolo M, Mattiuz R, Matolcsi C, Guedes J, Clark J, Veldhoen M, Bonaldi T, Vinolo MAR, Varga-Weisz P (2018) Microbiota derived short chain fatty acids promote histone crotonylation in the colon through histone deacetylases. Nat Commun 9:105
19. Seeley JJ, Baker RG, Mohamed G, Bruns T, Hayden MS, Deshmukh SD, Freedberg DE, Ghosh S (2018) Induction of innate immune memory via microRNA targeting of chromatin remodelling factors. Nature 559:114–119. https://doi.org/10.1038/s41586-018-0253-5
20. Thaiss CA, Levy M, Korem T, Dohnalová L, Shapiro H, Jaitin DA, David E, Winter DR, Gury-BenAri M, Tatirovsky E, Tuganbaev T, Federici S, Zmora N, Zeevi D, Dori-Bachash M, Pevsner-Fischer M, Kartvelishvily E, Brandis A, Harmelin A, Shibolet O, Halpern Z, Honda K, Amit I, Segal E, Elinav E (2016) Microbiota diurnal rhythmicity programs host transcriptome oscillations. Cell 167(6):1495–1510.e12. https://doi.org/10.1016/j.cell.2016.11.003
21. Kumar H, Lund R, Laiho A, Lundelin K, Ley RE, Isolauri E, Salminen S (2014) Gut microbiota as an epigenetic regulator: pilot study based on whole-genome methylation analysis. MBio 5:e02113–e02114
22. Ansari I, Raddatz G, Gutekunst J, Ridnik M, Cohen D, Abu-Remaileh M, Tuganbaev T, Shapiro H, Pikarsky E, Elinav E, Lyko F, Bergman Y (2020) The microbiota programs DNA methylation to control intestinal homeostasis and inflammation. Nat Microbiol 5(4):610–619. https://doi.org/10.1038/s41564-019-0659-3
23. Woo V, Alenghat T (2022) Epigenetic regulation by gut microbiota. Gut Microbes 14(1):2022407. https://doi.org/10.1080/19490976.2021.2022407
24. Krautkramer KA, Kreznar JH, Romano KA, Vivas EI, Barrett-Wilt GA, Rabaglia ME, Keller MP, Attie AD, Rey FE, Denu JM (2016) Diet-microbiota interactions mediate global epigenetic programming in multiple host tissues. 2016. Mol Cell 64:982–992. https://doi.org/10.1016/j.molcel.2016.10.025
25. Bhat MI, Kapila R (2017) Dietary metabolites derived from gut microbiota: critical modulators of epigenetic changes in mammals. Nutr Rev 75(5):374–389. https://doi.org/10.1093/nutrit/nux001

26. Anderson OS, Sant KE, Dolinoy DC (2012) Nutrition and epigenetics: an interplay of dietary methyl donors, one-carbon metabolism and DNA methylation. J Nutr Biochem 23:853–859
27. Obata Y, Furusawa Y, Endo TA, Sharif J, Takahashi D, Atarashi K, Nakayama M, Onawa S, Fujimura Y, Takahashi M, Ikawa T, Otsubo T, Kawamura YI, Dohi T, Tajima S, Masumoto H, Ohara O, Honda K, Hori S, Ohno H, Koseki H, Hase K (2014) The epigenetic regulator Uhrf1 facilitates the proliferation and maturation of colonic regulatory T cells. Nat Immunol 15(6):571–579. https://doi.org/10.1038/ni.2886
28. An J, Rao A, Ko M (2017) TET family dioxygenases and DNA demethylation in stem cells and cancers. Exp Mol Med 49:e323. https://doi.org/10.1038/emm.2017.5
29. Tahara T, Hirata I, Nakano N, Tahara S, Horiguchi N, Kawamura T, Okubo M, Ishizuka T, Yamada H, Yoshida D, Ohmori T, Maeda K, Komura N, Ikuno H, Jodai Y, Kamano T, Nagasaka M, Nakagawa Y, Tuskamoto T, Urano M, Shibata T, Kuroda M, Ohmiya N (2017) Potential link between Fusobacterium enrichment and DNA methylation accumulation in the inflammatory colonic mucosa in ulcerative colitis. Oncotarget 8(37):61917–61926. https://doi.org/10.18632/oncotarget.18716
30. Lund PJ, Gates LA, Leboeuf M, Smith SA, Chau L, Lopes M, Friedman ES, Saiman Y, Kim MS, Shoffler CA, Petucci C, Allis CD, Wu GD, Garcia BA (2022) Stable isotope tracing in vivo reveals a metabolic bridge linking the microbiota to host histone acetylation. Cell Rep 41(11):111809. https://doi.org/10.1016/j.celrep.2022.111809
31. Erny D, Hrabě de Angelis AL, Jaitin D, Wieghofer P, Staszewski O, David E, Keren-Shaul H, Mahlakoiv T, Jakobshagen K, Buch T, Schwierzeck V, Utermöhlen O, Chun E, Garrett WS, McCoy KD, Diefenbach A, Staeheli P, Stecher B, Amit I, Prinz M (2015) Host microbiota constantly control maturation and function of microglia in the CNS. Nat Neurosci 18(7):965–977. https://doi.org/10.1038/nn.4030
32. Alenghat T, Osborne LC, Saenz SA, Kobuley D, Ziegler CG, Mullican SE, Choi I, Grunberg S, Sinha R, Wynosky-Dolfi M, Snyder A, Giacomin PR, Joyce KL, Hoang TB, Bewtra M, Brodsky IE, Sonnenberg GF, Bushman FD, Won KJ, Lazar MA, Artis D (2013) Histone deacetylase 3 coordinates commensal-bacteria-dependent intestinal homeostasis. Nature 504:153–157. https://doi.org/10.1038/nature1268
33. Kobayashi T, Matsuoka K, Sheikh SZ, Russo SM, Mishima Y, Collins C, deZoeten EF, Karp CL, Ting JP, Sartor RB, Plevy SE (2012) IL-10 regulates Il12b expression via histone deacetylation: implications for intestinal macrophage homeostasis. J Immunol 189(4):1792. https://doi.org/10.4049/JIMMUNOL.1200042
34. Burgess SL, Leslie JL, Uddin J, Oakland DN, Gilchrist C, Moreau GB, Watanabe K, Saleh M, Simpson M, Thompson BA, Auble DT, Turner SD, Giallourou N, Swann J, Pu Z, Ma JZ, Haque R, Petri WA Jr (2020) Gut microbiome communication with bone marrow regulates susceptibility to amebiasis. J Clin Invest 130(8):4019–4024. https://doi.org/10.1172/JCI133605
35. Wellman AS, Metukuri MR, Kazgan N, Xu X, Xu Q, Ren NSX, Czopik A, Shanahan MT, Kang A, Chen W, Azcarate-Peril MA, Gulati AS, Fargo DC, Guarente L, Li X (2017) Intestinal epithelial Sirtuin 1 regulates intestinal inflammation during aging in mice by altering the intestinal microbiota. Gastroenterology 153(3):772–786. https://doi.org/10.1053/j.gastro.2017.05.022
36. Liang L, Ai L, Qian J, Fang JY, Xu J (2015) Long noncoding RNA expression profiles in gut tissues constitute molecular signatures that reflect the types of microbes. Sci Rep 5:11763. https://doi.org/10.1038/srep11763
37. Zhu Z, Huang J, Li X, Xing J, Chen Q, Liu R, Hua F, Qiu Z, Song Y, Bai C, Mo YY, Zhang Z (2020) Gut microbiota regulate tumor metastasis via circRNA/miRNA networks. Gut Microbes 12:1788891. https://doi.org/10.1080/19490976.2020.1788891
38. Rodríguez-Nogales A, Algieri F, Garrido-Mesa J, Vezza T, Utrilla MP, Chueca N, Garcia F, Olivares M, Rodríguez-Cabezas ME, Gálvez J (2017) Differential intestinal anti-inflammatory effects of lactobacillus fermentum and lactobacillus salivarius in DSS mouse colitis: impact on microRNAs expression and microbiota composition. Mol Nutr Food Res 61:1700144. https://doi.org/10.1002/mnfr.201700144

39. Takahashi K, Sugi Y, Hosono A, Kaminogawa S (2009) Epigenetic regulation of TLR4 gene expression in intestinal epithelial cells for the maintenance of intestinal homeostasis. J Immunol 183(10):6522–6529. https://doi.org/10.4049/jimmunol.0901271
40. Chang PV, Hao L, Offermanns S, Medzhitov R (2014) The microbial metabolite butyrate regulates intestinal macrophage function via histone deacetylase inhibition. Proc National Acad Sci USA 111(6):2247–2252. https://doi.org/10.1073/pnas.1322269111
41. Yang W, Yu T, Huang X, Bilotta AJ, Xu L, Lu Y, Sun J, Pan F, Zhou J, Zhang W, Yao S, Maynard CL, Singh N, Dann SM, Liu Z, Cong Y (2020) Intestinal microbiota-derived short-chain fatty acids regulation of immune cell IL-22 production and gut immunity. Nat Commun 11(1):4457. https://doi.org/10.1038/s41467-020-18262-6
42. Kaisar MMM, Pelgrom LR, van der Ham AJ, Yazdanbakhsh M, Everts B (2017) Butyrate conditions human dendritic cells to prime type 1 regulatory T cells via both histone deacetylase inhibition and G protein-coupled receptor 109A signaling. Front Immunol 8:1429. https://doi.org/10.3389/fimmu.2017.01429
43. Smith PM, Howitt MR, Panikov N, Michaud M, Gallini CA, Bohlooly-Y M, Glickman JN, Garrett WS (2013) The microbial metabolites, short-chain fatty acids, regulate colonic T reg cell homeostasis. Science (80-) 341(6145):569–573. https://doi.org/10.1126/science.1241165
44. Bolduc J-F, Hany L, Barat C, Ouellet M, Tremblay MJ (2017) Epigenetic metabolite acetate inhibits class I/II histone deacetylases, promotes histone acetylation, and increases HIV-1 integration in CD4 + T cells. J Virol 91(16). https://doi.org/10.1128/jvi.01943-16
45. Stilling RM, Dinan TG, Cryan JF (2014) Microbial genes, brain behaviour—epigenetic regulation of the gut–brain axis. Genes Brain Behav 13:69–86
46. Doherty R, O'Farrelly C, Meade K (2013) Epigenetic regulation of the innate immune response to LPS in bovine peripheral blood mononuclear cells (PBMC). Vet Immunol Immunopathol 154:102–110

Index

A

Acetylation, 21
Activated macrophages, 171
Allergic disease, 146
  future of epigenetics in, 157, 158
  genome-wide association studies (GWAS), 146, 147
  omics
    biomarkers and epigenomic aging, 149, 150
    pathogenesis and natural history, 147–149
    pharmacogenomics, 150, 151
  pediatric, 158
    atopic march, 151
    comprehensive meta-analysis, 152
    CpGs in cord blood, 152
    eczema, 155
    eosinophil miRNAs identification, 152
    food allergy, 156, 157
    respiratory allergies from AR and asthma, 153–155
Allergic rhinitis (AR), 146, 153–155
α-iduronidase, 202
Anti-neutrophil cytoplasmic antibody (ANCA)-associated vasculitis (AAV), 113
  characteristics, 113
  GPA and MPA
    charachteristics, 113
    DNA methylation and histone modifications in, 114, 115
    miRNAs in, 115–117
    pathophysiology of, 113
  subtypes, 113
Apabetalone, 213
Apical radial glial cells, 49
Argonauta protein (AGO2), 23, 24
Assessment of Lifestyle and Allergic Disease During Infancy (ALADDIN) birth cohort, 148
Asthma, 146, 153–155
Atopic dermatitis (AD), 146, 155
Atopic march, 151
ATP-dependent chromatin remodelling complexes, 84
Autoimmune diseases, 179
Autoinflammatory diseases, 164

B

Baby bottles, 34
Behçet's disease
  DNA methylation and chromatin accessibility, 130–132
  environmental factors, 129
  incidence, 129
  lncRNAs and circRNAs in, 135, 136
  miRNAs in, 132–134
  pathophysiology, 129
  symptoms, 129
  treatment, 129
Behçet syndrome, 173
Bioinformatic analysis, 2
Biomarkers, 16, 149
Biotin deficiency, 260
Bipolar disorder (BD), 226
Bisphenol A (BPA), 69
Bisphenols, 31–34
Body mass index (BMI), 64

C

C-alpha-formylglycine (FGly), 204
Cancer, 26
Cancer stem cells (CSCs), 85

L. Emmi, V. Fanos (eds.), *Epigenetics and Rare Diseases*, Rare Diseases of the Immune System, https://doi.org/10.1007/978-3-032-02902-7

Candidate gene studies, 146
Canned foods, 33, 34, 44
CAPS-FCAS, 168
CAPS-MWS, 168
CD4+ T cell, 155
CD8+ T cells, 184
Cell surface markers expression, 85
Cellular apoptosis, 182
Cellular danger response (CDR), 70
Cerebral X-linked adrenoleukodystrophy (CALD), 213
Chromatin, 7, 8
  modifying enzymes, 82
  remodeling, 22, 23
  states, 82
  structure, 20
Chromatin target gene PRMT1 (CHTOP), 185
Chromodomain helicase DNA-binding (CHD), 22
Chronic non-bacterial osteomyelitis (CNO), 172
CINCA-NOMID, 168
Circular RNAs (circRNAs), 181, 182
Circulating microRNA (ci-miRNA), 134, 174
Clonal evolution, 86
Clostridium scindens, 259
Copper chelator D-penicillamine, 213
Copy number variants (CNVs), 147, 195
Coronary artery aneurysms (CAAs), 120
Cortical neurogenesis, 50
CpG island methylator phenotype (CIMP), 80
Cryopyrin-associated periodic syndromes, 167
Cytotrophoblasts, 58

D

Demethylation, 81
Developmental Origins of Health and Disease (DOHaD) hypothesis, 16
Diabetes, 36, 37, 40–43
Diagnosis, 106
Dichlorodiphenyltrichloroethane (DDT), 32, 35, 36
Diet and epigenetics, 33
Diethylstilbestrol (DES), 34
Dioxin-like compounds (DLCs), 36
Dioxins, 32, 36, 41, 42
DNA demethylation, 19
DNA hypomethylation, 203
DNA methylation, 4, 6, 7, 9–12, 16–18, 20, 24, 60–62, 64–67, 80, 101, 148, 152, 155, 157, 166, 174, 180, 182–184, 186, 195–201, 203, 205–208, 213, 214, 227, 254, 255, 257, 264
DNA methyltransferases (DNMTs), 7, 88, 256
DNA methyltransferase 3a (DMNT3A), 205
DNA repair, 20
DNA sequence, 3, 4, 11
Driver variants, 78
Drosophila melanogaster, 2

E

EDCs, *see* Endocrine-disrupting chemicals
Encyclopedia of DNA Elements (ENCODE), 10
Endocrine-disrupting chemicals (EDCs), 33, 44, 67
Endocrine disruptors (EDs), 30–42, 44
Endogenous RNAs (ceRNAs), 182
Enquiring about tolerance (EAT), 156
Environmental factors, 107
Environmental stimulus, 16
Environmental toxins, 60
Eosinophilic granulomatosis with polyangiitis (EGPA), 113
Epi-cblC, 197, 208–210, 212
Epidrugs, 88
Epigenetic mechanisms in children with cow's Milk Allergy (EPICMA), 156
Epigenetics, 30, 179
  aberrations, 89–91
  aging, 159
  biological phenomena, 6
  changes, 222
  definition, 4, 16, 185
  effects, 31
  factors, 164
  function, 6–10
  future perspectives, 11, 12
  markers, 137
  mechanisms, 16–18, 20–27, 51, 79, 101, 195
  modifications, 4, 6, 11, 12, 16, 24, 26, 27, 114, 130, 136, 137, 174
  signatures, 12
  therapies, 92
  variation, 222
Epigenome, 79
Epigenome-wide association study (EWAS), 147
Epigenomic alterations, 200
EpiMix, 157
Epimutations, 208–210, 212
Euchromatin, 7
EZH2 inhibitor, 93

F
Fabry disease, 200, 201, 212
Familial cold autoinflammatory syndrome (FCAS), 167
Familial Mediterranean fever, 164
Fertility, 31, 35, 36, 39, 40, 42
Fetal programming, 52
Finnish Allergy Programme, 158
5-methylcytosine (5mC), 6
Food allergy (FA), 146

G
Gaucher disease, 202
Gb3, 201
Gene expression, 25, 222, 258, 261, 264, 265
Genetics, 2
  architecture, 221
  historical overview, 1–3
  predisposition, 174
Genome-wide association studies (GWAS), 146–151
Gestational diabetes mellitus, 38, 42, 233
Giant cell arteritis (GCA), 113, 126, 137
Gliphosate, 41
Global DNA hypomethylation, 81
Glucokinase gene (GCK), 185
Glycosaminoglycans, 203
Granulomatosis with polyangiitis (GPA), 113
Gut microbiome, 252
Gut microbiota, 252, 253
  epigenetic mechanisms, 254–257
  metabolites and intestinal homeostasis, 253

H
Hematopoietic stem cell transplantation (HSCT), 213
Henoch-Schönlein purpura, *see* IgA vasculitis
Histone acetylation, 199, 213
Histone acetyltransferases (HATs), 21, 184
Histone deacetylase 5 (HDAC5) gene, 205
Histone deacetylases (HDAC), 93, 183
Histone methylation, 197, 199, 213
Histone methyltransferases (HMTs), 256
Histone modifications, 4, 8–10, 20–23, 31, 62, 63, 102, 195, 197, 203, 208, 213, 254, 256, 258–260, 262, 264
  in BD, 226
  in SCZ, 229
Histone proteins, 20, 62
Homocycteine, 207
Homocystinuria-megaloblastic anemia cblG type, 207, 208
Homoeo box, 49
Host-microbiota interactions, 258, 261
Human chorionic gonadotropin (hCG), 60
Human microbiome, 252
Hyperactivation of proteins, 173
Hypermethylation of CpG islands, 80

I
IgA vasculitis (IgAV), 136
  epigenetic mechanisms
    DNA methylation, 119
    histone modifications, 119
    miRNAs in, 118
  pathophysiology of, 117
  symptoms, 117
Imitation switch (ISWI), 22
Immune dysregulation, 170, 174
Immune-mediated diseases, 181
  neuro, 185–187
  rheumatoid arthritis, 183, 184
  systemic lupus erythematosus, 181–183
  type 1 diabetes, 184, 185
Immune-modulating pathways, 136
Inborn errors of metabolism (IEMs)
  cblG, 207, 208
  clear-cut epigenetic components, 195–199
  definition, 194
  epigenetics role in, 194, 195
  epigenomic alternations of, 200
  epimutation causative of, 208, 210–212
  Fabry disease, 200, 201
  Gaucher disease, 202
  homocystinuria-megaloblastic anemia, 207, 208
  ICIMD database, 194
  Krabbe disease, 202
  lysosomal storage disorders, 200
  mucopolysaccharidosi I, 202, 203
  mucopolysaccharidosi IIIB and IVA, 203
  multiple sulfatase deficiency, 204
  Niemann-Pick C, 204
  phenylketonuria, 206, 207
  Pompe disease, 204
  transmission, 194
  treatment for, 212–214
  Wilson disease, 205, 206
Inflammasome complex proteins, 172
Inflammation modulation, 166
Innate immune cells, 164
Innate immune system, 120
INO80 remodeler, 22
Insecticides, 35, 40
Interferon (IFN) signaling, 260

International Agency for Research on Cancer (IARC), 36, 38, 41
International Classification of Inherited Metabolic Disorders (ICIMD), 194
International Human Epigenome Consortium (IHEC), 10
Intrauterine growth restriction (IUGR), 59, 69
Intravenous immunoglobulins (IVIG), 120
Isocitrate dehydrogenases (IDH) inhibitors, 93

J
JAK/STAT signaling, 171

K
Kawasaki disease (KD), 113
  aetiology of, 120
  characteristics, 120
  DNA methylation in, 120–122
  genetic susceptibility, 120
  immunomodulation in, 123
  lncRNAs in, 125, 126
  miRNAs in, 122–125
  pathophysiology, 120
  symptoms, 120
Kinesins, 51
Krabbe disease, 202

L
Large-vessel vasculitis (LVV), 113
  characteristics, 126
  epigenetics of, 127–129
  GCA and TAK
    characteristics, 126
    genetic factors, 127
  pathophysiology of, 126
Learning Early About Peanut Allergy (LEAP), 156
Lifestyle, 101
Liver, 32, 34, 39, 42
Long non-coding RNAs (lncRNAs), 9, 25, 63, 64, 103, 180, 183
Lysosomal storage disorder, 204

M
Machine learning approach, 157
Maternal nutrition, 65
Maternal psychosocial stress, 60
MeDALL, 152
Mediterranean Fever (MEFV), 165
Messenger-RNA (mRNA), 112
Metabolomics, 234
Methylation, 21, 31, 32, 35–37, 165
Methylation profile in SCZ, 228, 229
Methyl-CpG binding proteins (MeCPs), 7
Methylome-wide association studies, 228
Microbial metabolites, 254, 258, 260, 263, 265
Microbiome
  in epigenetic pathogen-mediated repogram, 263
  gut microbiota
    epigenetic mechanism, 254, 255, 257–260
    metabolites and intestinal homeostasis, 253, 254
  human, 252, 253
  immune activity, 261–263
  lncRNAs, circRNAs, and miRNAs, 261
Microplastics, 42–43, 67, 69
MicroRNAs (miRNAs), 9, 23–25, 84, 94, 103, 112, 166, 180, 183, 185, 196–198, 201, 202, 213, 225, 261
  expression, 225
  miR-95, 204
Microscopic polyangiitis (MPA), 113
miRNA-induced silencing complex (miRISC), 23
Mono-omic approach, 88
Monoubiquitination, 22
Muckle–Wells, 167
Mucopolysaccharidosis IVA, 203
Mucopolysaccharidosis type I (MPS I), 202
Mucopolysaccharidosis type IIIB, 203
Multi-omic approach, 88
Multiple sclerosis (MS), 104, 186
Multiple sulfatase deficiency, 204

N
N-alpha-acetylglucosaminidase (NAGLU), 203
Nanoplastics, 67
Neonatal-onset multisystem inflammatory disease (NOMID), 167
Neural stem/progenitor cells, 48
Neurexins, 53
Neurodevelopmental disorders, 52
Neurogenesis, 48
Neurogenic niche, 49
Neuro-immuno-mediated diseases, 185, 187
Neuromyelitis optica (NMO), 187
Neuropsychiatric disorders, 61
Neurospora crassa, 2
Newborn screening (NBS) programmes, 194
Niemann-Pick, type C (NPC) disease, 204

Non-coding RNAs (ncRNAs), 16, 23, 63, 84, 103, 112, 225
in MDD, 227–228
in SCZ, 229
Nutritional epigenetics, 44

O
Obesogen effects, 32
Omics, 147–149
Oncogenes (OCG), 77
Oncohistones, 83
Orange agent, 37
Organochlorine, 41
Organophosphates, 41

P
Packaging, 34, 44
Parabens, 32, 38
Pax6, 51
Perfluoroalkyl substances (PFASs), 32, 38, 39
Peroxisome proliferator-activated receptors (PPARs), 206
Persistent organic pollutants (POPs), 36
Personalized medicine, 106, 234
Pesticides, 32, 35, 36, 40, 44
Phenylalanine, 206
Phenylalanine hydroxylase, 206
Phenylketonuria, 198, 206, 207
Phosphorylation, 22
Phthalates, 32, 40
PIWI-interacting RNAs (piRNAs), 9, 25
Placenta, 35, 39
development of, 58
epigenetics and, 62
aberrations, 61
chronic diseases, 61
cytotrophoblasts, 62
DNA methylation, 62
EDCs and environmental toxins, 67
environmental factors, 61
features, 59
histone modification, 63
implications for public health, 67
maternal BMI, 64
maternal diabetes, 64
maternal nutrition, 65
maternal smoking, 65, 66
mediating intergenerational effects, 67
methylation, 63
non-coding RNA, 64
regulation, 63, 70
stress and maternal health, 66
exposure to environmental factors, 60
hormone production, 60
insulin and fetal growth alterations, 60
metaphor, 71
mirco and nanoplastics in human, 67–70
nutrient perception and metabolism, 59
regulation of nutrient transport, 59
role in prenatal programming, 59, 61
self-redistribute maternal blood flow, 58
weight, 58
Placental microbiota, 60
Plastic burning, 36
Plasticizers, 40, 44
Plastics, 33, 34, 36, 37, 40, 42–44
Polycarbonates, 33, 34
Polychlorinated biphenyls (PCBs), 36, 41
Polychlorinated dibenzo-p-dioxins (PCDDs), 36
Polycomb (PcG), 84
Polymyalgia rheumatica, 126
Polyphenols, 264
Polyvinyl alcohol, 69
Pompe disease, 204
Post-translational modifications, 7, 8, 82, 225
PRDX1, 208, 209, 211, 212
Preeclampsia, 59
Pregnancy, 39, 42, 65, 71
adverse outcomes, 59
complications, 59, 61
conditions, 63
development of placenta, 58
environmental factors, 61
hormone production by placenta, 60
malnutrition, 61, 62
maternal psychosocial stress, 60
maternal smoking
cardiovascular risk factors, 66
changes in DNA epigenetics, 65
inflammatory bowel diseases and psychiatric morbidity, 66
reduced lung function, 66
Prenatal programming, 59, 60, 70
Preterm birth, 59, 66
Prevention, 31
Proinflammatory macrophages, 171
Psychiatric diseases, 220
Psychiatric nosology, 220
Public health agencies, 34
Pyrin, 165

R
Radial glial cells, 50
Radial glial progenitor, 48

Raman microspectroscopy techniques, 69
Rheumatoid arthritis (RA), 105, 183, 184
RNA interference (RNAi), 9, 24, 260
RNA methylation, 26
RNA modifications, 182
RNA-sequencing techniques, 8
RNA silencing, 23–25

S
S-adenosylmethionine (SAM), 257
Schizophrenia (SCZ), 229
Severe-therapy-resistant asthma (STRA), 153
Seveso (Italy), 37
Short-chain fatty acids (SCFAs), 156, 252–259, 262–265
Silent Spring, 35
Silk Route disease, *see* Behçet's disease
Single nucleotide polymorphisms (SNPs), 146
Site-specific DNA hypomethylation, 173
Small interfering RNAs (siRNAs), 24
Sonic hedgehog, 51
Sperm motility, 40
Steroid hormones, 60
Sulfatase-modifying factor 1 (SUMF1), 204
Switch/sucrose non-fermentable (SWI/SNF) remodeler, 22
Syncytiotrophoblasts, 58
Systemic juvenile idiopathic arthritis (sJIA), 170
Systemic lupus erythematosus (SLE), 104, 181–183
Systemic vasculitis
  AAV (*see* ANCA-Associated Vasculitis (AAV))
  Behçet's disease (*see* Behçet's disease)
  characteristics, 112
  clinical manifestations of, 112
  epigenetics mechanisms, 112
  etiology, 112
  future research, 137
  IgA vasculitis, 117–119
  Kawasaki disease (*see* Kawasaki disease)
  symptoms, 112

T
Takayasu arteritis (TAK), 113, 126, 137
Target therapies, 147
T-cell receptor (TCR) signaling pathway, 133
TET proteins, 20
2,3,7,8-Tetrachlorodibenzodioxin (TCDD), 36, 37
Therapeutic targeting, 106
Thioredoxin (TXN) system, 205
Thyroid, 32, 35, 38, 39, 41, 42
TLR/IL-1 signaling pathway, 132
TNFRSF1A gene, 169
Toll-like receptor (TLR), 117, 131, 133
Transcriptional interference, 208
Transcription factor FOXP3, 180
Transgenerational effects, 31
Transgenerational transmission, 224
Treatment response, 166
Tricarboxylic acid (TCA), 257
Triclosan, 42
Trithorax (TrxG), 84
Tumor hallmarks, 79
Tumor necrosis factor receptor-associated periodic syndrome (TRAPS), 169, 170
Tumorigenesis, 78
Tumour heterogeneity, 86
Tumour suppressor genes (TSG), 77
Type 1 diabetes (T1D), 105, 184, 185

U
UPF3B, 54

V
Vietnam, 37
Vitamin B2, 257
Vitamin D, 65

W
Waste management, 44
Water pollution, 44
WDFY3, 53
Wilson disease, 205, 206, 213

X
X inactivation, 9

GPSR Compliance

The European Union's (EU) General Product Safety Regulation (GPSR) is a set of rules that requires consumer products to be safe and our obligations to ensure this.

If you have any concerns about our products, you can contact us on ProductSafety@springernature.com

In case Publisher is established outside the EU, the EU authorized representative is:

Springer Nature Customer Service Center GmbH
Europaplatz 3
69115 Heidelberg, Germany

Batch number: 09107086

Printed by Printforce, the Netherlands